D0770453

San Diego Christian College
2100 Greenfield Drive
El Cajon, CA 92019

ABOUT ISLAND PRESS

Island Press is the only nonprofit organization in the United States whose principal purpose is the publication of books on environmental issues and natural resource management. We provide solutions-oriented information to professionals, public officials, business and community leaders, and concerned citizens who are shaping responses to environmental problems.

Since 1984, Island Press has been the leading provider of timely and practical books that take a multidisciplinary approach to critical environmental concerns. Our growing list of titles reflects our commitment to bringing the best of an expanding body of literature to the environmental community throughout North America and the world.

Support for Island Press is provided by the Agua Fund, The Geraldine R. Dodge Foundation, Doris Duke Charitable Foundation, The Ford Foundation, The William and Flora Hewlett Foundation, The Joyce Foundation, Kendeda Sustainability Fund of the Tides Foundation, The Forrest & Frances Lattner Foundation, The Henry Luce Foundation, The John D. and Catherine T. MacArthur Foundation, The Marisla Foundation, The Andrew W. Mellon Foundation, Gordon and Betty Moore Foundation, The Curtis and Edith Munson Foundation, Oak Foundation, The Overbrook Foundation, The David and Lucile Packard Foundation, Wallace Global Fund, The Winslow Foundation, and other generous donors.

The opinions expressed in this book are those of the author(s) and do not necessarily reflect the views of these foundations.

ABOUT THE SOCIETY FOR ECOLOGICAL RESTORATION INTERNATIONAL

The Society for Ecological Restoration International (SER) is an international nonprofit organization comprising members who are actively engaged in ecologically sensitive repair and management of ecosystems through an unusually broad array of experience, knowledge sets, and cultural perspectives.

The mission of SER is to promote ecological restoration as a means of sustaining the diversity of life on Earth and reestablishing an ecologically healthy relationship between nature and culture.

The opinions expressed in this book are those of the author(s) and are not necessarily the same as those of SER International. SER, 285 W. 18th Street, #1, Tucson, AZ 85701. Tel. (520)622-5485, Fax (207) 626-5485, e-mail, info@ser.org, www.ser.org.

OLD FIELDS

SOCIETY FOR ECOLOGICAL RESTORATION INTERNATIONAL

The Science and Practice of Ecological Restoration
James Aronson, EDITOR
Donald A. Falk, ASSOCIATE EDITOR

577.55
O44f

36.⁰⁰
FP

Old Fields

Dynamics and Restoration of Abandoned Farmland

Edited by

Viki A. Cramer and Richard J. Hobbs

Society for Ecological Restoration International

Washington · Covelo · London

Copyright © 2007 Island Press

All rights reserved under International and Pan-American Copyright Conventions.
No part of this book may be reproduced in any form or by any means without permission
in writing from the publisher: Island Press, 1718 Connecticut Avenue NW, Suite 300,
Washington, DC 20009, USA.

Island Press is a trademark of The Center for Resource Economics.

Library of Congress Cataloging-in-Publication Data

Old fields : dynamics and restoration of abandoned farmland / edited by Viki A. Cramer and
Richard J. Hobbs.
    p. cm. – (The science and practice of ecological restoration)
Includes bibliographical references.
ISBN-13: 978-1-59726-074-9 (cloth : alk. paper)
ISBN-10: 1-59726-074-6 (cloth : alk. paper)
ISBN-13: 978-1-59726-075-6 (pbk. : alk. paper)
ISBN-10: 1-59726-075-4 (pbk. : alk. paper)
1. Abandoned farms. 2. Reclamation of land. 3. Restoration ecology. I. Cramer, Viki A. II. Hobbs,
R. J. (Richard J.)
S606.O43 2007
577.5'5–dc22

                                                                                      2007025183

Printed on recycled, acid-free paper ✿

Manufactured in the United States of America

10 9 8 7 6 5 4 3 2 1

CONTENTS

ACKNOWLEDGMENTS

We wish to extend our warmest thanks to all of the people who have assisted us in bringing together this book. First, to the authors of the chapters, both for their contributions to the book and for their gracious responses to our requests; their cheerful assistance was much appreciated during the editing process. The encouragement and patient assistance of Barbara Youngblood, Barbara Dean, and Jessica Heise at Island Press guided us through making the book a whole from the sum of its parts. All chapters in the book were peer reviewed. We gratefully acknowledge those who reviewed chapters and provided further comment on the ideas presented within the book: Margarita Arianoutsou, Bob Bunce, Rodolfo Dirzo, Jennifer Fraterrigo, Patricia Holmes, Anke Jentsch, Jamie Kirkpatrick, Carlos Klink, John Morgan, Rafael Navarro Cerillo, Maria Ruiz-Jaen, Wolfgang Schmidt, and Vicky Temperton. We are grateful for financial support from the Australian Research Council during the preparation of this book. Finally, our heartfelt thanks to our families; their support and encouragement makes the load lighter and the path not so narrow.

Chapter 1

Why Old Fields? Socioeconomic and Ecological Causes and Consequences of Land Abandonment

Richard J. Hobbs and Viki A. Cramer

Land abandonment is a type of land use change that is increasing as human influence on the globe increases and ecological, social, and economic factors conspire to force the cessation of agriculture and other forms of land management. The "old fields" resulting from abandonment display a variety of dynamics and have been the subject of much study and consideration, particularly in ecology where they have played a pivotal role in the development of ideas and concepts of ecological succession (see chapter 2). Surprisingly, however, there have been no previous attempts to synthesize what is known about the dynamics of abandoned farmland in different parts of the world. Does abandonment result in the same set of general dynamics everywhere, regardless of the type of system or its location? Alternatively, are the post-abandonment dynamics inherently idiosyncratic and dependent on local context? What ecological theory do we have that can account for the observed dynamics of old fields? And what do old fields have to tell us that is of value to the growing field of restoration ecology? It is the purpose of this book to examine these questions and provide at least some answers.

Past and Current Trends in Land Abandonment

Land use change is a complex phenomenon, and various types of change can be identified, which can have markedly different outcomes, depending on the magnitude and abruptness of the change (Hobbs 2000). Houghton (1991) recognized seven broad types of land use change:

1. Conversion of natural ecosystems for permanent croplands
2. Conversion of natural ecosystems for shifting cultivation

3. Conversion of natural ecosystems to pasture
4. Abandonment of croplands
5. Abandonment of pastures
6. Harvest of timber
7. Establishment of tree plantations

Houghton discussed these in the context of impacts on carbon stocks and hence on atmospheric carbon dioxide, but they also provide a useful set of categories in a broader context. In addition to these categories, we can include urbanization as a factor that is growing in importance and likely to lead to the conversion of both natural and agricultural systems. In addition, restoration should be included as a factor that, although still relatively small scale compared to other types of conversion listed above, is growing in significance (e.g., Hobbs and Harris 2001; Cunningham 2002). In this book we cover primarily the changes encompassed by Houghton's (1991) categories 4 and 5.

Land abandonment has been a feature of humanity's relationship with the world's ecosystems for as long as history has been recorded. We know that past civilizations have developed agricultural systems that were subsequently abandoned for one reason or another (e.g., Lentz 2000; Diamond 2005). The abandonment of agriculture often coincided with the collapse of the civilization, and we are still today piecing together the evidence for the chain of events leading to both. Generally, the ancient civilizations had agricultural systems that were relatively localized and did not cover large areas of land. Here, we concentrate on more recent times, when humans have increasingly come to dominate the planet and its ecosystems (Turner et al. 1991; Vitousek et al. 1997). While the prevailing view of trends in land cover and land use over the last few decades is one of continuing deforestation and transformation of the earth's ecosystems, there is also a growing trend of abandonment of systems that were previously managed intensively.

It has proved very difficult to derive overview statistics on the rates and extent of land abandonment, either in particular parts of the world or globally, since relatively few attempts have been made to quantify these. Figure 1.1 presents data from Ramankutty and Foley (1999), which is the only published material we have found that includes a global analysis of abandonment of croplands. The key message from these data is the dramatic and ongoing increase in the amount of abandoned land worldwide over the twentieth century.

When considered on a regional basis (figure 1.1b), it can be seen that the first region to experience significant abandonment of cropland (in the

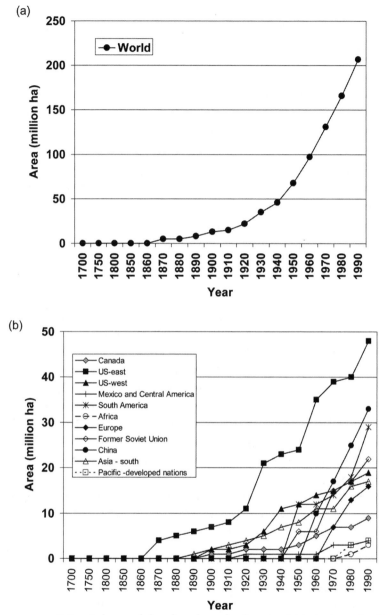

Figure 1.1. Estimated area of abandoned croplands, using historical cropland inventory data and remotely sensed land cover data, for (a) the world and (b) principal regions (as defined by the authors). Data from Ramankutty and Foley (1999); figures represent aggregate abandoned lands, including forest/woodland and savanna/grassland/steppe categories of the authors.

timeframe over which these data were collected) was eastern North America, with farmland being abandoned from 1870 onward and abandonment continuing at significant rates to the present time. Abandonment started later in all other regions, with the western United States, Canada, the former Soviet Union, and southern Asia showing initial abandonment during the late 1800s and early 1900s. Land abandonment has been a relatively recent feature in regions such as Europe, South America, and China, although rates of abandonment have increased significantly in these areas since the 1960s.

In eastern North America, land abandonment commenced in the 1870s in the northeastern part of the country as a result of improved transportation and the availability of agricultural goods from the more fertile Midwest and Great Plains (figure 1.2; Waisanen and Bliss 2002). The area of agricultural land declined in this area from a peak of 187,900 km^2 in 1880 to 48,800 km^2 in 1997, with individual states within the region showing dramatic reductions in area of agricultural land (figure 1.2b). The other regions showed smaller declines from peaks in 1940.

Considerable emphasis in the popular press and conservation literature is placed on recent levels of deforestation and conversion of tropical forest ecosystems (e.g., Laurance 1999; Oldfield 1989; Skole and Tucker 1993; Whitmore and Sayer 1992). However, it is also important to recognize that the agricultural system that replaces the natural ecosystem is often fairly short lived and is abandoned after only a few years, in some cases leading to the rapid development of secondary forest (Uhl et al. 1988; Nepstad et al. 1991; Houghton et al. 2000). Data from the Brazilian Amazon also suggest large differences in deforestation rates from year to year, as well as the presence of large areas of abandonment each year (Skole et al. 1994; Alves and Skole 1996). Recent estimates suggest that about 30% of the deforested area in the Amazon is now secondary forest resulting from land abandonment (Houghton et al. 2000). Limited temporal data indicate that the abandonment rate can be as high as 82% of the deforestation rate (1991–92, in Alves and Skole 1996).

It should also be noted that large areas of Southeast Asia have been subject to shifting cultivation over many years. Although not particularly evident in figure 1.1b, this process has contributed to vast areas of grassland across the region. One estimate suggests there are 9.7 million ha of grassland or shrubland in Vietnam alone (Gilmour et al. 2000), while Mittleman (1991) suggests that 100 million ha of forest have been significantly altered or removed (and then abandoned?) in the lower Mekong subregion over the last few decades.

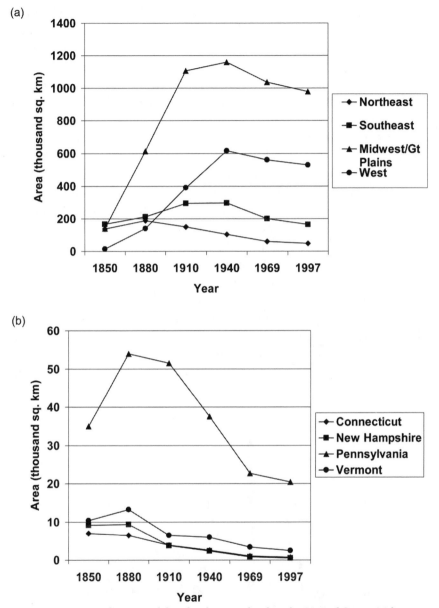

FIGURE 1.2. Area of improved farmland or cropland in the United States (a) by major region and (b) for selected states. Data from Waisanen and Bliss (2002).

In Europe, land abandonment has undoubtedly been occurring for many centuries, either locally and sporadically or possibly more generally such as in times of the great plague that swept many parts of the continent. In modern times, extensive land abandonment is a relatively recent phenomenon, but there is considerable regional variation similar to that illustrated for eastern North America. For instance, Bunce (1991) indicated that relatively little abandonment occurred in the United Kingdom. On the other hand, recent estimates suggest that 10%–20% of agricultural land in central and eastern European countries is now abandoned (Van Dijk et al. 2005) (table 1.1). Of particular interest in these countries is the abandonment of previously managed seminatural grasslands, with up to 60% of these grasslands now being abandoned in some countries (table 1.1). While statistics specifically relating to land abandonment are not available for Mediterranean countries, figure 1.3 presents data on reforestation, afforestation, and deforestation that show significant amounts of reforestation and afforestation in most countries. Most of this activity is associated with lands previously used for agriculture or grazing and gives an indication of the extent of abandonment in these areas.

There is a problem with interpretation of the different types of data available, and estimates based on indirect measures may or may not reflect a true rate of abandonment. For instance, figures on the cessation of one type of agricultural enterprise may not take into account some of the land being

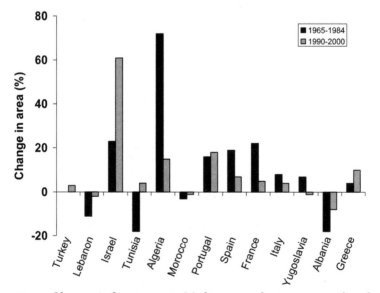

FIGURE 1.3. Changes in forest cover in Mediterranean basin countries, based on FAO statistics. Modified from Mazzoleni et al. (2004).

TABLE 1.1

Estimated extent of abandonment in central and eastern European countries (CEECs)

a. Total agricultural land

Country	Total area (million ha)	Area abandoned (million ha)	% of total abandoned
Estonia	1.5	.15	10.1
Latvia	2.5	.52	21.1
Lithuania	3.1	.32	10.3
Poland	18.7	3.29	17.6
Hungary	6.2	ca. 0.6	ca. 10
Total for all CEECs	57	5–11	10–20

b. Seminatural grasslands

Country	Total area (ha)	Area abandoned (ha)	% of total abandoned
Estonia	90,000	54,000	60
Latvia	17,323	10,394	60
Lithuania	167,933	100,760	60
Poland	1,955,000	1,000,000	51
Czech Rep.	550,000	82,500	15
Slovakia	294,900	38,337	13
Hungary	850,000	85,000	10
Romania	2,332,739	349,911	15
Bulgaria	444,436	66,665	15
Slovenia	268,402	40,260	15

Source: Data from Van Dijk et al. (2005).

transferred to another type of production rather than simply abandoned. There may also be definitional differences among regions. However, even given these caveats, the overall picture that emerges from both the global data and the regional analyses that are available is one of increasing levels of abandonment in most parts of the world.

Proximate and Ultimate Causes of Land Abandonment

What causes the abandonment of land previously under agricultural management of some kind? Causal factors are likely to vary regionally and are often a complex mix of social, economic, and ecological factors. Indeed, increasing recognition that agricultural systems can be viewed as complex social-ecological systems suggests that simple cause and effect relationships are unlikely to be apparent (Berkes and Folke 1998; Berkes et al. 2003; Allison and Hobbs 2006).

Ecological Factors

Ecological factors leading to the abandonment of land will often include declining soil fertility and productive capacity due to land degradation. Such degradation may result from the previous mismanagement of the land through overgrazing or inappropriate cultivation practices, fertilizer regimes, or fire treatments. It may also result from less proximate modifications resulting, for instance, from regional pollution problems, altered surface or groundwater hydrology regimes, or changing regional climatic conditions. Such factors are particularly important in areas where the agricultural enterprise was marginal in the first place.

Land abandonment may also occur, not because of ecosystem degradation, but rather because of ecosystem change, where changes in the management regime or biophysical settings result, for instance, in the loss of key pasture species or the invasion of aggressive weed species. Such processes are often set in motion by cessation of traditional management practices, such as mowing for hay in European meadows. This, in turn, is often the result of changed socioeconomic settings.

Changing Local and Global Economics

Current trends of globalization of markets and increased production capacity through technological advancements are leading to a loss of relevance and declining economic viability for traditionally produced local agricultural outputs, particularly on marginal agricultural lands. Declining terms of trade are likely to be a major driver of local enterprises going out of business and agriculture becoming locally nonviable. An opposing trend in some areas is one of increased land prices for nonagricultural purposes, such as development, hobby farms, or hunting, leading to small farms going out of production.

Changing Social and Political Settings

Depopulation of rural areas is an ongoing trend worldwide as the world's population becomes increasingly urbanized (United Nations Centre for Human Settlements 1996; McKinney 2002; Miller and Hobbs 2002). This results in the loss of infrastructure (schools, shops, banks, processing facilities, etc.), which in turn further accelerates the loss of people from rural areas. A flow-on effect of rural depopulation is the loss of traditional farming knowledge, making it more difficult for those wishing to return to rural areas, or those wanting to begin a career in farming. Additional underlying causes that can lead to forest loss and subsequent abandonment include rural poverty, a shortage of arable land, lack of land tenure, and limited or inappropriate institutional capacity.

In addition to the socioeconomic issues facing rural populations, national and supranational policies on farming and natural resource management can have large impacts on local and regional landscape management. In heavily regulated situations, such as the European Union, agricultural policies frequently dictate the course of land use change and to some extent determine the rate and extent of land abandonment or change from agriculture to other types of land use. In some cases, policies can lead to perverse incentives that result in poor environmental outcomes, and policy responses to environmental issues are often piecemeal and "Band-Aid" (Allison and Hobbs 2004; Allison and Hobbs 2006). There is increasing recognition of policy failure in dealing with such issues.

On top of the policies directed at agriculture at a national or supranational level, trade policies can have huge impacts on local economies. The current push for free trade and the removal of trade barriers has important implications for agriculture everywhere and will undoubtedly lead to the marginalization of agriculture in certain regions and, hence, increased land abandonment.

What Happens When Land Is Abandoned?

There are two predominant ways of viewing land abandonment. The first views abandonment as an opportunity for the redevelopment of an ecosystem similar to what was there prior to agricultural development, or at least as an opportunity for restoration activity of some description. The second views abandonment as more of a threat, resulting in the loss of specific ecosystem types that depended on ongoing agricultural management. Examples of the former include abandoned pastures in the tropics and old fields in eastern North America, while examples of the latter include hay meadows, seminatural grasslands, and dehesas in various parts of Europe. Each of these instances is explored in detail in later chapters.

The difference in perspective relates to the history and duration of agriculture in the region and the presence or absence of systems that are dependent on human management for their persistence. Over much of Europe, there are many systems that have developed almost entirely in the presence of human influence, and traditional management of these systems has shaped the European landscape for centuries. Economic, political, and social change is resulting in the intensification of agriculture in some areas and extensification or abandonment in others. The resultant removal of traditional management practices almost inevitably leads to ecosystem change, which in turn potentially leads to the loss of important biological or cultural values (see van Andel and Aronson 2006 for examples). Thus, for instance, ecosystems such as managed grasslands or hay meadows are often very rich in species, but these species depend on continued management to persist. Once management is removed, other species are able to become dominant, and the ecosystem may lose species or be transformed into another ecosystem type, particularly if woody species are able to establish. In this case, therefore, ecosystem development after abandonment is viewed as degradation because the system is developing away from something that is valued and becoming something less valued.

In New World ecosystems, on the other hand, agriculture has not necessarily been such a dominant force on the landscape for such lengthy periods. Abandonment in such situations may result in the rapid return of a native ecosystem, if conditions are suitable. Alternatively, the abandoned land may change more slowly back to some preexisting ecosystem or change toward a completely different system. Depending on the management objective, these alternatives may be appropriate and provide something of greater value than the previous agricultural system. Another possibility is that the abandoned system remains "stuck" in a similar state to when it was abandoned. This may lead to further degradation or may be viewed as an opportunity for restoration, depending on the context.

Both perspectives on land abandonment result from the fact that abandonment invariably leads to change of some description. In some cases the change is welcomed, and in others it is considered a threat. The systems that arise from the abandonment may resemble the systems present prior to agricultural development or they may represent entirely new systems that arise because of changed environmental conditions and the presence of species not previously present in the area. These novel or "emerging" ecosystems are likely to have novel and unpredictable dynamics that may profoundly impact ecosystem structure and function, and the ecosystem services they provide (Milton 2003; Aronson and van Andel 2006; Hobbs et al. 2006).

Relevance to Ecological Theory and Restoration Practice

We explore in detail the relevance of all this to both theory in practice via the individual case studies presented in later chapters, and in the final chapter, which aims to draw lessons from an aggregate of the case studies. However, it is important to consider up front what the likely relevance of studies of abandoned lands is to both ecological theory and restoration practice.

From the perspective of ecological theory, there is no doubt that old fields, particularly in eastern North America, have played an important role in the development of concepts and theories surrounding ecosystem dynamics (see chapter 2). The observation that abandoned farmland frequently displayed at least partially repeatable patterns of vegetation development led to a wide range of studies providing important insights for the theories of succession and ecosystem dynamics central to much of ecology. This process is ongoing, with more recent studies emphasizing the variations in postabandonment dynamics and considering the reasons for this variety of responses (chapter 3). In short, abandoned farmland provides an important outdoor laboratory within which many observational and experimental studies can be conducted that both feed into the conceptual development of ecology and provide useful test beds for existing theory.

From the perspective of restoration practice, abandoned farmland also provides many opportunities for increased understanding of the processes of ecosystem recovery and for improving practices relating to directing ecological succession or ecosystem assembly (Luken 1990; Temperton et al. 2004). In addition, as discussed earlier, abandonment of farmland can be viewed as either an opportunity or a threat, depending on the context. For instance, an interesting benefit of the abandonment of lands that subsequently develop or redevelop a forest cover is the increased sequestration of carbon, which has a net positive effect on the carbon balance (Watson et al. 2000). Deliberate or autogenic afforestation or reforestation can contribute significantly to the carbon balance, although the amounts currently remain low compared to the losses due to deforestation. On the other hand, if this afforestation takes place at the expense of ecologically important plant communities, such as hay meadows, then the benefits may not be so clear cut.

Hence, considerations of the dynamics of abandoned farmland and the management options available to direct these dynamics feed directly into current discussions about what appropriate goals for restoration are. For instance, with increased recognition of potentially irreversible changes in regional climates and the presence of a suite of introduced species likely to alter the course of postabandonment ecosystem development, it may no longer be appropriate to consider the preexisting vegetation as an appropriate restoration goal. Instead there may be alternative or novel ecosystems that may be more resilient and provide more value in terms of ecosystem services or conservation (see, for example, Aronson and van Andel 2006; Hobbs et al. 2006).

We revisit this and other key questions relating to ecological theory and restoration practice in later chapters.

Outline of the Book

In this book, we aim to synthesize past and current work on old fields to provide an up-to-date perspective on the ecological dynamics of abandoned land in different parts of the world and the relevance of this to ecology and restoration. The first two chapters provide a theoretical underpinning for considering old field dynamics. Chapter 2 provides a historical account of the development of ideas on ecological succession and the role studies of old fields played in this. Chapter 3 then considers more recent theoretical developments in the areas of ecosystem assembly and alternative stable states.

The following chapters present a series of case studies where old fields have been studied in detail in different parts of the world. Chapters 4–7 focus on different areas in the tropics, while chapters 8 and 9 examine recent studies in eastern North America. Chapter 10 provides an update of a classic, but almost inaccessible, study of old field dynamics in central Europe by Osbornová et al. (1990), while chapters 11–13 examine different parts of the Mediterranean basin. The last two case studies look at areas not traditionally known for old field studies. Chapter 14 examines old fields in South African renosterveld, and chapter 15 considers abandoned farmland in southwestern Australia.

Chapter 16 then considers the similarities and differences among the case studies and aims to draw out some generalities, particularly in relation to the theories discussed in chapters 2 and 3, and in relation to restoration practice. This final chapter considers how studies of old fields increase our understanding of ecosystem dynamics and improve restoration practice.

REFERENCES

Allison, H. E., and R. J. Hobbs. 2004. Resilience, adaptive capacity and the "lock-in" trap of the Western Australian agricultural region. *Ecology and Society* 9:3. http://www.ecologyandsociety.org/vol9/iss1/art3.

———. 2006. *Science and policy in natural resource management: Understanding system complexity.* Cambridge University Press, Cambridge, UK.

Alves, D. S., and D. L. Skole. 1996. Characterizing land cover dynamics using multitemporal imagery. *International Journal of Remote Sensing* 17:835–39.

Aronson, J., and J. van Andel. 2006. Challenges for restoration theory. In *Restoration ecology: The new frontier*, ed. J. van Andel and J. Aronson, 223–33. Blackwell, Oxford, UK.

Berkes, F., J. Colding, and C. Folke, eds. 2003. *Navigating social-ecological systems: Building resilience for complexity and change.* Cambridge University Press, Cambridge, UK.

Berkes, F., and C. Folke, eds. 1998. *Linking social and ecological systems: Management practices and social mechanisms for building resilience.* Cambridge University Press, Cambridge, UK.

Bunce, R. G. H. 1991. Ecological implications of land abandonment in Britain: Some comparison with Europe. *Options Méditerranéennes — Série Séminaires* 15:53–59.

Cunningham, S. 2002. *The restoration economy: The greatest new growth frontier — Immediate and emerging opportunities for businesses, communities and investors.* Berrett-Koehler, San Francisco, CA.

Diamond, J. M. 2005. *Collapse: How societies choose to fail or succeed.* Viking Penguin, New York.

Gilmour, D. A., V. S. Nguyen, and X. Tschalicha. 2000. *Rehabilitation of degraded forest ecosystems in Cambodia, Lao PDR, Cambodia and Thailand. Conservation issues in Asia.* IUCN, World Conservation Union, Gland, Switzerland.

Hobbs, R. J. 2000. Land use changes and invasions. In *Invasive species in a changing world*, ed. H. A. Mooney and R. J. Hobbs, 55–64. Island Press, Washington, DC.

Hobbs, R. J., S. Arico, J. Aronson, J. S. Baron, P. Bridgewater, V. A. Cramer, P. R. Epstein et al. 2006. Novel ecosystems: Theoretical and management aspects of the new ecological world order. *Global Ecology and Biogeography* 15:1–7.

Hobbs, R. J., and J. A. Harris. 2001. Restoration ecology: Repairing the earth's ecosystems in the new millennium. *Restoration Ecology* 9:239–46.

Houghton, R. A. 1991. Tropical deforestation and atmospheric carbon dioxide. *Climate Change* 19:99–118.

Houghton, R. A., D. L. Skole, C. A. Nobre, J. L. Hackler, K. T. Lawrence, and W. H. Chomentowski. 2000. Annual fluxes of carbon from deforestation and regrowth in the Brazilian Amazon. *Nature* 403:301–4.

Laurance, W. F. 1999. Reflections on the tropical deforestation crisis. *Biological Conservation* 91:109–17.

Lentz, D. L., ed. 2000. *Imperfect balance: Landscape transformations in the pre-Columbian Americas.* Columbia University Press, New York.

Luken, J. O. 1990. *Directing ecological succession.* Chapman and Hall, New York.

Mazzoleni, S., G. di Pasquale, and M. Mulligan. 2004. Conclusion: Reversing the consensus on Mediterranean desertification. In *Recent dynamics of the Mediterranean vegetation and landscape*, ed. S. Mazzoleni, G. di Pasquale, M. Mulligan, P. di Martino, and F. Rego, 281–5. Wiley, Chichester, UK.

McKinney, M. L. 2002. Urbanization, biodiversity and conservation. *BioScience* 52:883–90.

Miller, J. R., and R. J. Hobbs. 2002. Conservation where people live and work. *Conservation Biology* 16:330–7.

Milton, S. J. 2003. "Emerging ecosystems": A washing-stone for ecologists, economists and sociologists? *South African Journal of Science* 99:404–6.

Mittelman, A. 1991. Secondary forests in the lower Mekong subregion: An overview of their extent, roles and importance. *Journal of Tropical Forest Science* 13:671–90.

Nepstad, D. C., C. Uhl, and A. S. Serrão. 1991. Recuperation of a degraded Amazonian landscape: Forest recovery and agricultural restoration. *Ambio* 248–55.

Oldfield, S. 1989. The tropical chainsaw massacre. *New Scientist* 36–39.

Osbornová, J., M. Kovárová, J. Lepš, and K. Prach, eds. 1990. *Succession in abandoned fields. Studies in Central Bohemia, Czechoslovakia.* Kluwer Academic Publishers, Dordrecht, The Netherlands.

Ramankutty, N., and J. A. Foley. 1999. Estimating historical changes in global land cover: Croplands from 1700 to 1992. *Global Biogeochemical Cycles* 13:997–1027.

Skole, D., and C. Tucker. 1993. Tropical deforestation and habitat fragmentation in the Amazon: Satellite data from 1978 to 1988. *Science* 260:1905–10.

Skole, D. L., W. H. Chomentowski, W. A. Salas, and A. D. Nobre. 1994. Physical and human dimensions of deforestation in Amazonia. *BioScience* 44:314–22.

Temperton, V. M., R. J. Hobbs, T. J. Nuttle, and S. Halle, eds. 2004. *Assembly rules and restoration ecology: Bridging the gap between theory and practice.* Island Press, Washington, DC.

Turner, B. L. I., W. C. Clark, R. W. Kates, J. T. Mathews, J. R. Richards, and W. Mayer, eds. 1991. *The earth as transformed by human action.* Cambridge University Press, New York.

Uhl, C., R. Buschbacher, and E. A. S. Serrão. 1988. Abandoned pastures in eastern Amazonia. 1. Patterns of plant succession. *Journal of Ecology* 76:663–81.

United Nations Centre for Human Settlements. 1996. *An urbanising world: Global report on human settlements.* Oxford University Press, Oxford, UK.

Van Andel, J., and J. Aronson, eds. 2006. *Restoration ecology: The new frontier.* Blackwell, Oxford, UK.

Van Dijk, G., A. Zdanowicz, and R. Blokzijl, eds. 2005. *Land abandonment, biodiversity and the CAP. Land abandonment and biodiversity, in relation to the 1st and 2nd pillars of the EU's Common Agricultural Policy: Outcome of an international seminar in Sigulda, Latvia, 7–8 October 2004.* DLG, Government Service for Land and Water Management, Utrecht, The Netherlands.

Vitousek, P. M., H. A. Mooney, J. Lubchenco, and J. Melillo. 1997. Human domination of Earth's ecosystems. *Science* 277:494–99.

Waisanen, P. J., and N. B. Bliss. 2002. Changes in population and agricultural land in conterminous United States counties, 1790–1997. *Global Biogeochemical Cycles* 16:1–19.

Watson, R. T., I. R. Noble, B. Bolin, N. H. Ravindranath, D. J. Verardo, and D. J. Dokken, eds. 2000. *Land use, land-use change, and forestry.* Cambridge University Press, Cambridge, UK.

Whitmore, T. C., and J. A. Sayer, eds. 1992. *Tropical deforestation and species extinction.* Chapman and Hall, New York.

PART 1

Old Fields and the Development of Ecological Concepts

The study of old fields and the development of ecological concepts have been closely linked since the pioneering work of a number of ecologists—among them Clements, Gleason, and Egler—in the early twentieth century. Old fields have been pivotal to the development of succession theory, with old fields providing the classic example of secondary succession. These links continue today, with succession theory providing the underpinning for understanding the dynamics of vegetation in old fields in many of the case studies presented later in the book. In chapter 2, Hobbs and Walker provide a brief review of the development of successional theory, which has been extensively reviewed elsewhere, and highlight the main concepts that are relevant to succession in old fields.

As our understanding of ecological systems improved, the focus and scope of old field succession expanded to consider a wider variety of factors than were considered in the classic model. Ecologists began to recognize the limited utility of the classic model for understanding vegetation development in circumstances where the outcomes of succession were not deterministic, and since the 1970s succession theory has expanded to consider a wider range of factors that influence vegetation development. Hobbs and Walker describe how the study of old fields have informed this expansion of succession theory and describe the usefulness of the modern succession theory to the restoration of old fields.

As ecological theory makes the transition to using nonequilibrium dynamics to explain ecological systems, new concepts for describing and predicting the development of ecosystems after disturbance have emerged. These include, among others, discussions of the development of alternative ecosystem states, ecosystem assembly, and ecosystem resilience. These theories largely reject the deterministic nature of ecosystem development described by classical succession theory but have common conceptual links with modern succession theory. Many of these concepts have yet to be widely applied to vegetation development in old fields (but see chapters 10, 14, and 15). In chapter 3, Cramer considers the usefulness of using the ecological concepts derived from the study of complex, nonlinear systems to understanding old field dynamics and restoring old fields. The next step in our understanding of old field dynamics is likely to incorporate these new concepts, without discarding the successional theory that has been the foundation of understanding old field development.

Chapter 2

Old Field Succession: Development of Concepts

RICHARD J. HOBBS AND LAWRENCE R. WALKER

It has become a well-recognized phenomenon that when an agricultural field is abandoned, a series of processes are set in motion that result in changes in the composition and structure of the plant community within the field. Sometimes this change appears to be quite predictable, with most fields abandoned in any given area following a roughly similar pattern of development after abandonment. In other cases, the change is less predictable and more idiosyncratic, or there appears to be little discernable change over time at all. The various chapters in this book provide examples of each of these situations. However, the key consideration in this chapter is the observation of change in abandoned agricultural land and how these observations contributed to the development of the concept of ecological succession. This chapter examines the development of ideas on succession, the role of old field studies in this development, and the relevance of succession concepts to restoration.

Early Development of Ideas on Succession

First of all, we need to ask the question, "What is succession?" Examination of the literature on the topic reveals many different interpretations of the word *succession*, ranging from the relatively simple and inclusive "process of vegetation change" (e.g., Cooper 1926) through to more constrained definitions, such as "a hypothetically orderly sequence of changes in plant communities leading to a stable climax community" (Dodson et al. 1998). Such divergence is typical of the degree of debate and difference of opinion that has surrounded the concept of succession for over a century. Not wishing to become entangled in the definitional nightmare, here we adopt the more inclusive idea that succession is the process of vegetation change, which may include instances where apparently orderly sequences of changes in plant communities can occur.

Numerous detailed reviews of the development of successional ideas are available (e.g., Burrows 1990; McIntosh 1985; Glenn-Lewin et al. 1992; Miles and Walton 1993; Walker and del Moral 2003). We will not aim to provide a comprehensive account of that development here, but rather we will highlight the main points of discussion. The concept of vegetation change has been in evidence since antiquity, with recorded observations from as far back as 300 BC (Clements 1928). Hunters utilized fires to extend grasslands and improve conditions for their game. Gatherers of nuts and berries were dependent on an intuitive understanding of shifting vegetation patterns. However, it was the advent of agriculture that permanently linked successional dynamics to the human lifestyle. Conscious and persistent manipulation of the landscape was essential to agriculture. Where natural grasslands were not available, forests were cleared and swamps were drained to create fields for crops. Increasingly sophisticated agricultural practices reflected better knowledge of nutrient cycling, plant demography, plant physiology, and species interactions, all key elements of successional change. Formal development of succession was synchronous with the development of ecology in the late 1800s and early 1900s. Warming (1895) and Cowles (1899) linked species change to stabilization on dunes, and the concepts of succession received increasing attention in the literature thereafter. How communities coalesce is still considered one of the most important concepts in ecology (Cherrett 1989; Thompson 2001).

The work of Clements (1916, 1928, 1936) is perhaps the most well-known, early treatment of succession, and it laid the foundations for much future work in the area. Clements identified six basic processes that drive succession (table 2.1). These processes, with some modifications of course, still provide a solid, conceptual structure for succession (Walker and del Moral 2003). Disturbance (or abandonment, in the case of old fields) initiates changes in the available resources (e.g., nutrients and light) that then impact the preexisting and invading species. The differential life histories (dispersal, establishment, growth, and longevity) of each species provide the fundamental background for species change. Species interactions (e.g., via facilitation, competition, herbivory) and their impacts on the site further modify successional change.

Clements's work has received considerable criticism, mainly because of its preoccupation with the determinism of the successional process, its treatment of plant communities as holistic or "organismal" entities, and its strict

TABLE 2.1

Processes recognized by Clements as driving succession, with modern analogues and interpretations.

Clementsian process	Modern analogues
Nudation	Allogenic disturbances, stochastic events
Migration	Life history characteristics: dispersal
Ecesis	Life history characteristics: establishment, growth, longevity
Competition	Competition, allelopathy, herbivory, disease
Reaction	Site modification by organisms, facilitation
Stabilization	Development of climax

Source: Walker and del Moral 2003.

adherence to a single endpoint or "climax" community for any given bio-physical setting. Clements's (1936) work also resulted in a plethora of complex terminology aimed at dealing with the multiple possibilities for exceptions to the single endpoint in any given location. However, as McIntosh (1981, p.14) suggests, "careful rereading of his major accounts of succession is often required to distinguish his actual ideas from what they are said to be." Clements's work is admirable in its scope and formulation of a dynamic view of natural systems. Its influence is apparent throughout much of ecology during the 1900s, with reformulations of the main ideas being provided by Odum (1969) and Patten et al. (1995). Much of conservation and ecosystem management still focuses on deterministic outcomes and static endpoints (Lubchenco 1998; Scoones 1999; Wallington et al. 2005). Indeed, one of the pioneering books in the field of restoration ecology (Jordan et al. 1987) contains a chapter that considers succession almost entirely based on the work of Clements (MacMahon 1987).

Gleason (1926, 1927, 1939) and Cooper (1926) provided some of the earliest criticism of Clements's approach, shifting the focus from determinism and holism to the influence of individual plants and the role of processes, such as dispersal and competition, in the successional process. McIntosh (1981, p.15) suggests that the trend has continued of developing a "much less tidy concept of succession than vintage Clementsianism," starting with Whittaker (1953, 1975) and continuing to the present day. Van der Valk (1998) suggests that Gleason's early work was too vague to be of much practical use for predicting the changes in the composition of the vegetation in a specific area, and it was not until the late 1970s when the first mechanistic Gleasonian models of succession were developed. These models used species characteristics or attributes (such as seed dispersal, seed longevity, germination requirements, expected life spans, competitive abilities) to predict vegetation change resulting from disturbances (Van der Valk 1981), especially following fire (Noble and Slatyer 1980).

The underlying causes of succession have also been the subject of much debate, with a wide variety of views on which processes are important. Clements's original work incorporated the idea of *relay floristics*, or facilitation, in which one species or set of species modifies the environment and "paves the way" for another species or set of species. An alternative interpretation was later promoted, most prominently by Egler (1954, 1977). Egler suggested that in abandoned agricultural fields, the observed sequences of species had less to do with facilitation of one species by another and more to do with the life histories of the species involved. His idea of *initial floristic composition* suggests that most species that take part in a successional sequence are, in fact, present immediately following disturbance, but differential growth rates result in sequential conspicuousness. This type of succession emphasizes minimal interactions between species. A third scenario occurs where species inhibit their neighbors, the opposite of facilitation. Connell and Slatyer (1977) summarized the positive, neutral, and negative effects of an early successional species on a later one into three models (facilitation, tolerance, and inhibition; figure 2.1). Their models stimulated much research that equated evidence for one of these types of interactions with support for a particular model. It soon became clear, however, that the three possible interactions between species could occur at any given stage in succession and that the models were not mutually exclusive (McCook 1994) and should not apply to whole successional sequences (Connell et al. 1987; Walker and Chapin 1987). Instead, succession is likely to result from the consequences of a continuum of species impacts on each other, from positive to negative.

The recognition of this myriad of possible patterns and causes has resulted in an understanding of succession that is less deterministic and predictable than envisaged by Clements. While the overall processes discussed by Clements are still relevant today, their interpretation is vastly different. Glenn-Lewin et al. (1992) concluded that "contemporary ecologists view vegetation change as the outcome of populations interacting within fluctuating environmental conditions." Taken to its extreme, this has resulted in the development of individual-based models of succession, where the outcome is determined by the interactions between individuals of different species mediated by their growth rates and known environmental tolerances (e.g., Botkin 1993). Other ecologists focus on community dynamics or ecosystem changes as drivers of species change. For example, colonization at any given stage of succession may be driven by biotic filters that limit biodiversity (Diaz et al. 1999). Alternatively, the interplay of species adapted to high light versus high

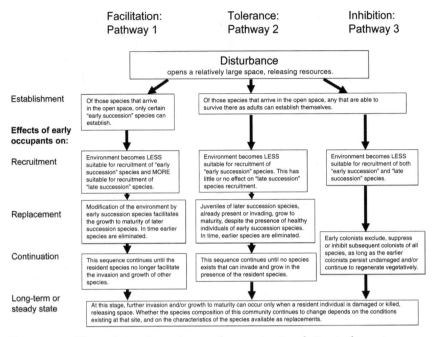

FIGURE 2.1. Three alternative pathways of succession in relation to the main successional processes. Diagram modified from Connell and Slatyer (1997) and McCook (1994).

soil fertility may drive species change (Tilman 1985). In fact, succession is driven by a combination of factors at individual, population, community, and ecosystem levels (Walker and del Moral 2003). The level of determinism or stochasticity perceived in any given situation may to some extent relate to the scale or grain of observation. For instance, exact patterns of change in species composition may vary greatly, but the overall pattern in terms of life-forms may be relatively predictable.

Other concepts such as *colonization* or *invasion windows* (Johnstone 1986) have been added, and ideas of vegetation change now consider the possibility of multiple pathways and endpoints. This includes consideration of state and transition models of vegetation dynamics and the idea of alternative stable states, where different community compositions are possible in any given area, and transitions between states are often driven by different combinations of factors, such as climatic events, fire, and herbivory (Westoby et al. 1989; Dublin et al. 1990; Hobbs 1994; Whalley 1994; Walker 1997; Beisner et al. 2003). This highlights the importance of disturbance events and regimes as drivers of vegetation dynamics.

Role of Old Fields in the Development of Succession Theory

Old fields provide the classical example of secondary succession—that is, succession on areas that once were vegetated but have been disturbed in some way so that some to all of the preexisting vegetation has been removed but where the soil is still more or less intact. Other examples include forest clear-cuts or burned grasslands. The alternative, primary succession, occurs on areas that have been completely denuded and where the soil has also been removed. Examples include new land exposed by retreating glaciers, or created by volcanoes, sand dunes, mudflows, river floodplains, or mine tailings (Miles and Walton 1993; Walker and del Moral 2003).

A bibliography on old field succession recently made available by Rejmánek and Van Katwyk (2005) listed 1,511 references over the period 1901–91. It is not our intention to provide an exhaustive review of this extensive literature here, but rather to summarize the early studies and highlight the main issues considered to be important in old field successions in eastern North America. This region is highlighted here because the pioneering studies in old field succession occurred there and many of the classical ideas relating to old field succession arose from this work. As indicated in chapter 1, this region also experienced the earliest and most extensive land abandonment in the modern age. We do, however, recognize the importance of land abandonment in other parts of the world and important research on the topic from European countries (as presented in later chapters).

Although succession has its ancient roots in agriculture, formal studies of succession in the early 1900s focused on dunes, glacial moraines, and other examples of primary succession. Clements and others worked largely in the central and western United States on grasslands and montane habitats. In Europe, a taxonomic approach to vegetation dynamics held sway through the early and mid-1900s (Braun-Blanquet 1932; Ellenberg 1956). Early old field studies had their most thorough examination in the middle 1900s in northeastern North America. Here, there were large numbers of abandoned fields within easy reach of major research centers. This proximity provided a convenient object of study, especially where the land use history was already also well known. Many of the classical ideas relating to old field succession arose from this area. Further discussion of research in old fields in northeastern North America is provided in chapters 8 and 9.

The classical textbook model of old field succession, from northeastern North America, is shown in figure 2.2. In this model, the initial colonization following abandonment is by annual species, which are replaced the following year by grasses and herbaceous perennials. These in turn are replaced by

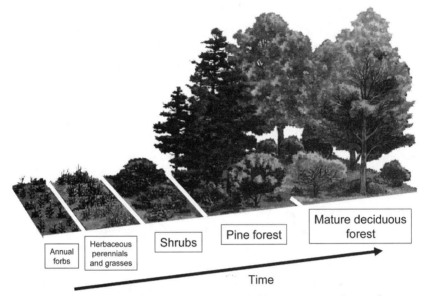

FIGURE 2.2. Textbook depiction of classical old field succession in the northeastern United States, from annual forbs, immediately after cessation of agriculture, to mature deciduous forest. Modified from Miller 1994.

shrub species, which persist until overtopped by pine species. Finally, a mixed deciduous hardwood community eventually replaces the pines. While the details of this succession and the exact species involved vary regionally, this is the generally accepted progression from abandonment to mature forest. Ironically, that acceptance does not mean global applicability. Secondary succession in old fields takes many forms, as later chapters in this book will explain, depending on the regional and local climate, disturbance regime, land use history, and species availability.

Eastern North America has a particular history. The northern part was scoured by glaciers and is now rebounding. These forces impacted patterns of sedimentation and soil composition. This in turn determined where fields were cleared by settlers (initially mostly on hilltops for easier plowing, then in clay-rich valley bottoms when better plows were invented). The species available for colonizing old fields were impacted by the use of fire by Indians and European colonists. White pine (*Pinus strobus*) dominated much of the terrain due in part to these past fire manipulations and therefore plays a predominant role in old field succession. In the central regions of eastern North America, the lack of glaciation led to a species-rich, mixed mesophytic forest, with different consequences for secondary succession in old fields.

Much of the early work in old fields used chronosequences of fields abandoned at different times to infer successional patterns and examined in detail the possible causal factors leading to the progressive changes observed. The ability to piece together successional sequences using a range of fields of different known ages since abandonment proved to be a potent tool in inferring dynamics. While this chronosequence method has its pitfalls (Pickett 1989), it is undoubtedly useful, especially in the face of the preponderance of short-term studies in ecology (Tilman 1989). Later studies thus supplemented the chronosequence approach with long-term observations of actual changes through time. When the chronosequence method is combined with experimentation and long-term observations, it becomes possible to consider pattern, process, and causation in considerable detail (e.g., Foster and Tilman 2000). The earliest experimental work on old field succession in the United States was by Keever (1950) and a college student (McCormick 1968). Following various pleas (Drury and Nisbet 1973; Connell and Slatyer 1977) for an experimental approach to understanding successional mechanisms, research on old fields accelerated. Recognition of the limited utility of the classic model of old field succession (figure 2.2) has helped direct further analysis of the factors leading to the diversity of successional patterns observed.

A wide range of factors has been found to influence the pattern of post-abandonment succession (table 2.2). Most of these factors affect the succession through their impacts either on the ability of individual species to disperse into the abandoned field or on the relative abilities of species to survive after arrival and following establishment. Factors affecting dispersal into the field include the timing of abandonment in relation to the production and dispersal of seeds, the size of the field, and its proximity to seed sources. Factors affecting subsequent survival include seed predation, soil conditions, competition from species already established in the field, or which arrive later, and herbivory. Subsequent disturbances and climatic variations may also provide additional "windows" for accelerated invasion of particular species.

Old fields have hence acted as the source of inspiration for ideas on succession and also as the test bed for alternative hypotheses. Early work on succession was predominantly on primary successions such as sand dunes, and only later did old fields become the object of study. The study of old field succession moved rapidly from the simple description of successional patterns through to trying to understand the mechanisms of species replacement and the reasons for the variations seen in the standard model. Subsequent chapters continue this process and report on recent studies in old fields both in eastern North America and in other parts of the world.

TABLE 2.2

Factors considered important in determining the pattern or rate of succession in abandoned fields in northeastern United States, with representative studies.

Causal factor	Representative studies
Season of abandonment	Keever 1950, 1979
Last crop sown prior to abandonment	Beckwith 1954; Myster and Pickett 1988, 1990, 1994
Plowed vs. left fallow after abandonment	Golley 1965; Collins and Adams 1983; Myster and Pickett 1990
Size of field	Golley et al. 1994
Distance from forest margin	Lawson et al. 1999; Myster and Pickett 1993; Dovciak et al. 2005
Soil type and condition (erosion, etc.)	Keever 1950; Beckwith 1954
Soil nutrient status	Inouye et al. 1987; Inouye and Tilman 1988, 1995; Gleeson and Tilman 1990
Herbivory	Myster and McCarthy 1989; De Steven 1991; Gill and Marks 1991; Myster 1993; Lawson et al.1999; Cadenasso et al. 2001
Seed predation	Gill and Marks 1991; Myster 1993; Myster and Pickett 1993
Invasive plant species	Fike and Niering 1999; Meiners et al. 2001, 2002
Postabandonment disturbance	Beckwith 1954; Inouye et al.1987; Collins et al. 2001; Bartha et al. 2003
Climate variation	Dovciak et al. 2005

Conclusion: Succession and Restoration

Davis et al. (2005) have recently called for a reunification of concepts and approaches in plant ecology that deal with various aspects of vegetation dynamics, because fragmentation of effort currently limits potential synergies and shared understanding. Certainly, the concepts of succession are, or should be, central to the science of restoration ecology (Walker et al. 2007), and one of the aims of restoration has been described as "directing ecological succession" (Luken 1990). Interestingly, however, the linkage between ecological succession and restoration concepts has been relatively little explored and rarely exploited in practice (Van der Valk 1998).

Bradshaw (1987, p.28) suggested that "the basic principles of land and ecosystem restoration are the same as the basic principles of succession," and that "the essential quality of restoration is an attempt to overcome artificially the factors that we consider will restrict ecosystem development." Van der Valk (1998) concluded simply that "restoration is accelerated succession." Certainly, restoration from a successional approach focuses on the manipulation and sometimes the acceleration of the order of species change to a desired endpoint (Luken 1990), although sometimes it may also be desirable to slow down or reverse successional processes (Prach et al. 2007). Restoration often occurs over shorter time scales than demanded by successional considerations, but it is, nevertheless, dependent on the broader successional dynamics for its success (Palmer et al. 1997; Walker et al. 2007). Patterns of successional development can also offer reference systems for the assessment of restoration actions and critical insights into species dispersal, species interactions, plant–soil interactions, and soil development (Hobbs and Norton 1996). In turn, restoration offers a practical test of successional theories (Bradshaw 1987).

The connections between restoration and succession are highlighted in the various case studies presented in subsequent chapters in this book and are explored further in Walker et al. (2007). In addition to concepts on succession, however, we also need to consider the alternative, complementary viewpoint of ecosystem assembly as another way of examining the dynamics of abandoned farmland: this is the subject of chapter 3.

References

Bartha, S., S. J. Meiners, S. T. A. Pickett, and M. L. Cadenasso. 2003. Plant colonization windows in a mesic old field succession. *Applied Vegetation Science* 6:205–12.

Beckwith, S. L. 1954. Ecological succession on abandoned farm lands and its relationship to wildlife management. *Ecological Monographs* 24:349–76.

Beisner, B. E., D. T. Haydon, and K. Cuddington. 2003. Alternative stable states in ecology. *Frontiers in Ecology and Environment* 1:376–82.

Botkin, D. B. 1993. *Forest dynamics: An ecological model.* Oxford University Press, Oxford, UK.

Bradshaw, A. D. 1987. Restoration: An acid test for ecology. In *Restoration ecology: A synthetic approach to ecological research*, ed. W. R. Jordan, M. E. Gilpin, and J. D. Aber, 23–30. Cambridge University Press, Cambridge, UK.

Braun-Blanquet, J. 1932. *Plant sociology: The study of plant communities* (English trans.). Hafner, New York.

Burrows, C. J. 1990. *Processes of vegetation change.* Unwin Hyman, London.

Cadenasso, M. L., S. T. A. Pickett, and P. J. Morin. 2001. Experimental test of the role of mammalian herbivores on old field succession: Community structure and seedling survival. *Journal of the Torrey Botanical Society* 129:228–37.

Cherrett, J. M. 1989. Key concepts: The results of a survey of our members' opinions. In *Ecological concepts: The contributions of ecology to an understanding of the natural world*, ed. J. M. Cherrett, A. D. Bradshaw, F. B. Goldsmith, P. J. Grubb, and J. R. Krebs, 1–16. Blackwell Scientific Publications, Oxford, UK.

Clements, F. E. 1916. *Plant succession: An analysis of the development of vegetation.* Publication No. 242. Carnegie Institution of Washington, Washington, DC.

———. 1928. *Plant succession and indicators.* Wilson, New York.

———. 1936. Nature and structure of the climax. *Journal of Ecology* 24:252–84.

Collins, B., G. Wein, and T. Philippi. 2001. Effects of disturbance intensity and frequency on early old-field succession. *Journal of Vegetation Science* 12:721–28.

Collins, S. L., and D. E. Adams. 1983. Succession in grasslands: Thirty-two years of change in a central Oklahoma tallgrass prairie. *Vegetatio* 51:181–190.

Connell, J. H., I. R. Noble, and R. O. Slatyer. 1987. On the mechanisms producing successional change. *Oikos* 50:136–37.

Connell, J. H., and R. O. Slatyer. 1977. Mechanisms of succession in natural communities and their role in community stability and organization. *American Naturalist* 111:1119–44.

Cooper, W. S. 1926. The fundamentals of vegetational change. *Ecology* 7:391–413.

Cowles, H. C. 1899. The ecological relations of the vegetation on the sand dunes of Lake Michigan. *Botanical Gazette* 27:95–117, 167–202, 281–308, 361–391.

Davis, M. A., P. Pergl, A.-M. Truscott, J. Kollmann, J. P. Bakker, R. Domenech, K. Prach et al. 2005. Vegetation change—A reunifying concept in plant ecology. *Perspectives in Plant Ecology, Evolution, and Systematics* 7:69–76.

De Steven, D. 1991. Experiments on mechanisms of tree establishment in old-field succession: Seedling emergence. *Ecology* 72:1066–75.

Diaz, S., M. Cabido, and F. Casanoves. 1999. Functional implications of trait–environment linkages in plant communities. In *Ecological assembly rules: Perspectives, advances, retreats*, ed. E. Weiher and P. Keddy, 338–62. Cambridge University Press, Cambridge, UK.

Dodson, S. I., T. F. H. Allen, S. R. Carpenter, A. R. Ives, R. L. Jeanne, J. F. Kitchell, N. E. Langston, and M. G. Turner. 1998. *Ecology*. Oxford University Press, New York.

Dovciak, M., L. E. Freelich, and P. B. Reich. 2005. Pathways in old-field succession to white pine: Seed, rain, shade, and climate effects. *Ecological Monographs* 75:363–78.

Drury, W. H., and I. C. T. Nisbet. 1973. Succession. *Journal of the Arnold Arboretum* 54:331–68.

Dublin, H. T., A. R. E. Sinclair, and J. McGlade. 1990. Elephants and fire as causes of multiple stable states in the Serengeti-Mara woodlands. *Journal of Animal Ecology* 59:1147–64.

Egler, F. 1977. *The nature of vegetation: Its management and mismanagement.* Aton Forest, Norfolk, CT.

Egler, F. E. 1954. Vegetation science concepts. 1. Initial floristic composition, a factor in old-field vegetation development. *Vegetatio* 4:412–17.

Ellenberg, H. 1956. *Aufgaben und Methoden der Vegetationskunde.* Eugen-Ulmer, Stuttgart, Germany.

Fike, J., and W. A. Niering. 1999. Four decades of old field vegetation development and the role of *Celastrus orbiculatus* in the northeastern United States. *Journal of Vegetation Science* 10:483–92.

Foster, B. L., and D. Tilman. 2000. Dynamic and static views of succession: Testing the descriptive power of the chronosequence approach. *Plant Ecology* 146:1–10.

Gill, D. S., and P. L. Marks. 1991. Tree and shrub seedling colonization of old fields in central New York. *Ecological Monographs* 6:183–205.

Gleason, H. A. 1926. The individualistic concept of the plant association. *Bulletin of the Torrey Botanical Club* 53:7–26.

———. 1927. Further views on the succession concept. *Ecology* 8:299–326.

———. 1939. The individualistic concept of the plant association. *American Midland Naturalist* 21:92–110.

Gleeson, S. K., and D. Tilman. 1990. Allocation and the transient dynamics of succession on poor soils. *Ecology* 71:1144–55.

Glenn-Lewin, D. C., R. K. Peet, and T. T. Veblen, eds. 1992. *Plant succession: Theory and prediction*. Chapman and Hall, London.

Golley, F. B. 1965. Structure and function of an old-field broomsedge community. *Ecological Monographs* 35:113–37.

Golley, F. B., J. E. Pinder III, P. J. Smallidge, and N. J. Lambert. 1994. Limited invasion and reproduction of loblolly pines in a large South Carolina old field. *Oikos* 69:21–27.

Hobbs, R. J. 1994. Dynamics of vegetation mosaics: Can we predict responses to global change? *Écoscience* 1:346–56.

Hobbs, R. J., and D. A. Norton. 1996. Towards a conceptual framework for restoration ecology. *Restoration Ecology* 4:93–110.

Inouye, R. S., N. J. Huntly, D. Tilman, and J. R. Tester. 1987. Pocket gophers (*Geomys bursarius*), vegetation, and soil nitrogen along a successional sere in east central Minnesota. *Oecologia* (Berlin) 72:178–84.

Inouye, R. S., N. J. Huntley, D. Tilman, J. R. Tester, M. Stillwell, and K. C. Zinnel. 1987. Old-field succession on a Minnesota sand plain. *Ecology* 68:12–26.

Inouye, R. S., and D. Tilman. 1988. Convergence and divergence of old-field plant communities along experimental nitrogen gradients. *Ecology* 69:995–1004.

———. 1995. Convergence and divergence of old-field vegetation after 11 yrs of nitrogen addition. *Ecology* 76:1872–87.

Johnstone, I. M. 1986. Plant invasion windows: A time-based classification of invasion potential. *Biological Review* 61:369–94.

Jordan, W. R. I., M. E. Gilpin, and J. D. Aber, eds. 1987. *Restoration ecology: A synthetic approach to ecological research*. Cambridge University Press, Cambridge, UK.

Keever, C. 1950. Causes of succession on old fields of the Piedmont, North Carolina. *Ecological Monographs* 20:229–50.

———. 1979. Mechanisms of plant succession on old fields of Lancaster County Pennsylvania. *Bulletin of the Torrey Botanical Club* 106:299–308.

Lawson, D., R. S. Inouye, N. Huntly, and W. P. Carson. 1999. Patterns of woody plant abundance, recruitment, mortality, and growth in a 65 year chronosequence of old-fields. *Plant Ecology* 145:267–79.

Lubchenco, J. 1998. Entering the century of the environment: A new social contract for science. *Science* 279:494–97.

Luken, J. O. 1990. *Directing ecological succession*. Chapman and Hall, New York.

MacMahon, J. A. 1987. Disturbed lands and ecological theory: An essay about mutualistic association. In *Restoration ecology: A synthetic approach to ecological research*, ed. W. R. Jordan, M. E. Gilpin, and J. D. Aber, 221–37. Cambridge University Press, Cambridge, UK.

McCook, L. J. 1994. Understanding ecological community succession: Causal models and theories, a review. *Vegetatio* 110:115–47.

McCormick, J. 1968. Succession. *Via* 1:22–35, 131–32.

McIntosh, R. P. 1981. Succession and ecological theory. In *Forest succession: Concepts and application*, ed. D. C. West, H. H. Shugart, and D. B. Botkin, 10–23. Springer-Verlag, New York.

———. 1985. *The background of ecology: Concept and theory*. Cambridge University Press, Cambridge, UK.

Meiners, S. J., S. T. A. Pickett, and M. L. Cadenasso. 2001. Effects of plant invasions on the species richness of abandoned agricultural land. *Ecography* 24:633–44.

———. 2002. Exotic plant invasions over 40 years of old field successions: Community patterns and associations. *Ecography* 25:215–23.

Miles, J., and D. W. H. Walton, eds. 1993. *Primary succession on land*. Blackwell Scientific Publications, Oxford, UK.

Miller, G. T. J. 1994. *Living in the environment. Principles, connections and solutions*. Wadsworth, Belmont, CA.

Myster, R. W. 1993. Tree invasion and establishment in old fields at Hutcheson Memorial Forest. *Botanical Review* 59:251–72.

Myster, R. W., and B. C. McCarthy. 1989. Effects of herbivory and competition on survival of *Carya tomentosa* (Juglandaceae) seedlings. *Oikos* 56:145–48.

Myster, R. W., and S. T. A. Pickett. 1988. Individualistic patterns of annuals and biennials in early successional old fields. *Vegetatio* 78:53–60.

———. 1990. Initial conditions, history and successional pathways in ten contrasting old fields. *American Midland Naturalist* 124:231–38.

———. 1993. Effects of litter, distance, density and vegetation patch type on post-dispersal tree predation in old fields. *Oikos* 66:381–88.

———. 1994. A comparison of rate of succession over 18 yrs in 10 contrasting old fields. *Ecology* 75:387–92.

Noble, I. R., and R. O. Slatyer. 1980. The use of vital attributes to predict successional changes in plant communities subject to recurrent disturbance. *Vegetatio* 43:5–21.

Odum, E. P. 1969. The strategy of ecosystem development. *Science* 164:262–70.

Palmer, M. A., R. F. Ambrose, and N. L. Poff. 1997. Ecological theory and community restoration ecology. *Restoration Ecology* 5:291–300.

Patten, B. C., S. E. Jørgensen, and S. I. Auerbach, eds. 1995. *Complex ecology. The part-whole relation in ecosystems*. Prentice-Hall, Englewood Cliffs, NJ.

Pickett, S. T. A. 1989. Space-for-time substitution as an alternative to long-term studies. In *Long-term studies in ecology. Approaches and alternatives*, ed. G. E. Likens, 110–35. Springer-Verlag, New York.

Prach, K., R. H. Marrs, P. Pysek, and R. van Diggelen. 2007. In *Linking restoration and succession in theory and in practice*, ed. L. R. Walker, J. Walker, and R. J. Hobbs. 121–49. Springer, New York.

Rejmánek, M., and K. P. Van Katwyk. 2005. *Old-field succession: A bibliographic review (1901–1991)*. http://botanika.bf.jcu.cz/suspa/pdf/BiblioOF.pdf.

Scoones, I. 1999. New ecology and the social sciences: What prospects for a fruitful engagement? *Annual Review of Anthropology* 28:279–507.

Thompson, J. N., O. J. Reichman, P. J. Morin, G. A. Polis, M. E. Power, R. W. Sterner, C. A. Couch et al. 2001. Frontiers of ecology. *BioScience* 51:15–24.

Tilman, D. 1985. The resource-ratio hypothesis of plant succession. *American Naturalist* 125:827–52.

———. 1989. Ecological experimentation: Strengths and conceptual problems. In *Long-term studies in ecology. Approaches and alternatives*, ed. G. E. Likens, 136–57. Springer-Verlag, New York.

Van der Valk, A. G. 1981. Succession in wetlands: A Gleasonian approach. *Ecology* 62:688–98.

———. 1998. Succession theory and restoration of wetland vegetation. In *Wetlands for the future*, ed. A. J. McComb and J. A. Davis, 657–67. Gleneagles Press, Adelaide, Australia.

Walker, L. R., and F. S. I. Chapin. 1987. Interactions among processes controlling successional change. *Oikos* 50:131–35.

Walker, L. R., and R. del Moral, eds. 2003. *Primary succession and ecosystem rehabilitation.* Cambridge University Press, Cambridge, UK.

Walker, L. R., J. Walker, and R. J. Hobbs, eds. 2007. *Linking restoration and succession in theory and in practice.* Springer, New York.

Walker, S. 1997. Models of vegetation dynamics in semi-arid vegetation: Application to lowland Central Otago, New Zealand. *New Zealand Journal of Ecology* 21:129–40.

Wallington, T. J., R. J. Hobbs, and S. A. Moore. 2005. Implications of current ecological thinking for biodiversity conservation: A review of the salient issues. *Ecology and Society* 10:15. http://www.ecologyandsociety.org/vol10/iss1/art15/.

Warming, E. 1895. *Plantesamfund: Grunträk af den Ökologiska Plantegeografi.* Philipsen, Copenhagen, Denmark.

Westoby, M., B. Walker, and I. Noy-Meir. 1989. Opportunistic management for rangelands not at equilibrium. *Journal of Range Management* 42:266–74.

Whalley, R. D. B. 1994. State and transition models for rangelands. 1. Successional theory and vegetation change. *Tropical Grasslands* 28:195–205.

White, P., and A. Jentsch. 2001. The search for generality in studies of disturbance and ecosystem dynamics. *Progress in Botany* 62:399–450.

Whittaker, R. H. 1953. A consideration of climax theory: The climax as population and pattern. *Ecological Monographs* 23:41–78.

———. 1975. *Communities and ecosystems.* 2nd ed. Macmillan, New York.

Chapter 3

Old Fields as Complex Systems: New Concepts for Describing the Dynamics of Abandoned Farmland

Viki A. Cramer

The problem of community assembly in vegetation has been examined in depth by plant ecologists since the early twentieth century, and understanding the processes that determine the patterns and numbers of species within a community remains central to community ecology (Booth and Larson 1999; Chase 2003). While succession continues to be the dominant conceptual framework for vegetation development in old fields (Ganade and Brown 2002; Hooper et al. 2002; Meiners et al. 2002; Buisson and Dutoit 2004; Myster 2004; Riege and del Moral 2004; El-Sheikh 2005), it is unlikely that contemporary authors are using "succession" in the strict Clementsian sense of a unidirectional and deterministic development of plant communities that leads to a single climax community. Contemporary succession theory incorporates all of the once-competing theories of early last century that related to vegetation development over time (chapter 2).

While Clements, Tansley, and Egler may have disagreed upon the exact meaning of the term succession, or whether it was an appropriate term to use at all, they agreed that succession (or vegetation development) was directional (although not unidirectional) and deterministic (Clements 1916; Tansley 1935; Egler 1954). Yet ecological science is in transition, with a shift in emphasis from equilibrium to nonequilibrium dynamics in theoretical ecology (Holling 1998). Ecologists have accepted that vegetation development after disturbance may be neither directional nor deterministic, and new frameworks of community assembly have been proposed. These frameworks are part of a web of related theories, including systems theory, assembly theory, models of alternative states, and assembly rules (table 3.1).

Yet does consideration of old fields within the context of these new frameworks provide for greater understanding of vegetation development and

TABLE 3.1

Comparison of three broad frameworks of community assembly in the context of complex systems theory and site conditions within which each framework may be most applicable.

Framework	Theoretical conditions under complex systems theory	Site conditions under which framework may operate	Related conceptual models/frameworks
Deterministic	Short term Initial conditions known	Abiotic conditions unchanged or restored completely Nearby source of propagules	Traditional succession theory Assembly rules for trait-based functional groups or guild proportionality
Stochastic	Initial conditions unknown Random events Order of arrival	Frequent and relatively intense disturbance Small spatial scales	Carousel models Persistent nonequilibrium states
Alternative states	Initial conditions unknown Historical contingency Environmental conditions important Random events	Variation in history of sites Environmental conditions changed Variation in propagule access Disturbance influences community assembly	Alternative stable states Multiple equilibrium states Regime shifts Stability landscapes and basins of attraction Resilience theory

improved opportunities for its management? Ecology has been recently criticized as a science where there is a lack of appreciation of past literature: ecologists frequently pose seemingly novel ideas without recognizing that the same, or similar, ideas were developed decades earlier by famous ecologists (Belovsky et al. 2004). For example, Booth and Larson (1999) argue that if "constraint" was substituted for "rule," and "development" substituted for "assembly," then a discussion of assembly rules is reduced to a discussion of development constraints on community structure: ideas that were fully explored by Clements, Gleason, Egler, and many others early last century. Here I explore community assembly in old fields as community assembly in a complex system that may exhibit both determinism and stochasticity. I consider how the new frameworks of community assembly may offer an understanding of old field dynamics that both complements and extends beyond contemporary succession theory, and how these frameworks may assist in the restoration of old fields.

Community Assembly in Complex Systems:
Determinism, Stochasticity, and Alternative States

Theoretical ecology increasingly recognizes the complex and nonlinear nature of ecosystems (Levin 1998; Drake et al. 1999; Schulze 2000). Ecosystem processes occur episodically, cyclically, ephemerally, continually, or stochastically, and the rates of these processes may be rapid, or occur over days, months, or longer. They also operate over different spatial scales. The variation in the rates of ecosystem processes over a variety of spatial scales introduces a high degree of nonlinearity and frequently chaotic dynamics into an ecosystem (Schulze 2000; Salthe 2001). Processes within complex nonlinear systems may range from completely stochastic to completely deterministic, and historical conditions play an important role in the development of the system (Drake et al. 1999). The three broad frameworks currently used to describe community assembly models—deterministic, stochastic, and alternative states (Temperton and Hobbs 2004)—also reflect this range of possibilities and, for the latter, the role of land use history. Traditionally these may be viewed as competing frameworks, but if we consider ecosystems as complex nonlinear systems, then all frameworks are applicable to the development of vegetation in old fields. The applicability of each of these frameworks is dependent upon the time scale under which the process of community assembly is considered and our knowledge of abiotic and biotic conditions at the time of abandonment.

Deterministic models of community assembly are allied closely with the traditional view of succession, whereby a highly ordered sequence of species assemblages moves toward a sustained climax whose characteristics are determined by climate and edaphic conditions (Holling 1995; but see chapter 2 for current views on succession). Theoretically, complex nonlinear systems are deterministic over short time scales and may be predictable over longer time scales when initial conditions are known in detail (Hastings et al. 1993). If we translate this to an ecosystem, then deterministic frameworks are most appropriately applied to community assembly in old fields where the dynamics of populations are predictable over short time scales, or where environmental and biotic conditions, as well as land use history, are known in detail. For example, Lockwood and Pimm (1999) reviewed thirty-four restoration projects that aimed to restore original species composition, yet only two of these projects succeeded. The successful projects exactly replicated the original physical conditions (i.e., the initial conditions) of the site, and there were undisturbed sources of colonizing species nearby. Others argue that community assembly is deterministic in terms of the general composition of trait-based functional groups or in terms of guild proportion-ality, but that species composition within these communities is historically contingent, due to priority effects based on the order of species arrivals (Wil-son and Gitay 1995; Wilson 1999; Fukami et al. 2005). Within plant com-munities, such trait-based assembly rules may only operate over small spatial scales because of the localized nature of plant interactions (Wilson and Whittaker 1995).

In stochastic models, community composition and structure is the result of essentially random processes, depending only upon the availability of va-cant niches and the order of arrival of organisms. Yet even completely chaotic systems are predictable over short time scales because nonlinear systems are deterministic in the short term. Their predictability in the longer term is lim-ited due to a lack of complete information about the exact location of initial conditions (Hastings et al. 1993). Stochastic models, such as the carousel model, are most commonly associated with the small-scale turnover of short-lived species in an homogenous environment (Van der Maarel and Sykes 1993; Palmer and Rusch 2001).

In the models dealing with alternative states, the development of the community may occur along a number of different trajectories, depending on the environmental history of the area, current environmental conditions, the availability of propagules, and the influence of random events. Theoret-ically, the differences between the trajectories of two communities with dif-ferent initial conditions will grow until this difference is essentially as large as the variation in either trajectory (Hastings et al. 1993), hence, the devel-opment of alternative states. Such sensitivity to initial conditions means that anything less than a perfect understanding of the initial conditions greatly decreases our ability to predict the trajectory of the ecosystem (Drake et al. 1999). For example, past land use may have persistent, long-term effects on the spatial heterogeneity of soil nutrients (Fraterrigo et al. 2005; Standish et al. 2006), which may directly affect native plant establishment, or indirectly affect native plant establishment through competition with weed species (Riege and del Moral 2004). As with the stochastic framework, the dynam-ics of populations are likely to be predictable over short time scales. Our ability to predict the trajectory of the system in the longer term is reliant on our understanding of both current conditions (abiotic and biotic) and the land use history of the old field. At present, frameworks dealing with alter-native states are popular with ecologists, and the deterministic idea of a community "climax" in the Clementsian sense is frowned upon as unrealis-tic (Young et al. 2001).

Development of Alternative States on Abandoned Farmland in a Theoretical Landscape

In a highly fragmented agricultural landscape an abandoned old field is likely to be distant from remnant native vegetation. It is likely that the seed sources of weed species will be more abundant than the seed sources of natives and propagule dispersal into the old field dominated by weed species. The length of time the field has been cropped and the agronomic management of the field will affect edaphic factors, such as nutrient availability, soil carbon and organic matter, and soil structure (Compton and Boone 2000; Fraterrigo et al. 2005; Standish et al. 2006). Therefore, the landscape context in which old fields occur may be described by the abundance and type of native species, the abundance of weed species, and the condition of the soil.

The "state space" of a landscape is defined by the n variables that constitute that system, and is the n dimensional space of all possible combinations of the amounts of those variables (Walker et al. 2004). In our theoretical landscape, the state space is the three-dimensional space of all possible combinations of the abundance of native species, the abundance of weed species, and soil conditions (sensu Walker et al. 2004). The particular combination of these variables present upon abandonment constitutes the initial conditions for old field recovery. If the system tends toward equilibrium, then old fields with similar initial conditions will converge toward an equilibrium state (the "attractor") (Hastings et al. 1993; Walker et al. 2004). The "basin of attraction" of an equilibrium state constitutes all of the initial conditions that will tend toward that equilibrium state and is the region in the state space where the system will tend to remain (Walker et al. 2004) (figure 3.1). There may be more than one basin for a given system, thus leading to the development of alternative equilibrium states. Stochastic events (e.g., disturbances) tend to move the system around within a basin of attraction, while large and intense disturbances may shift the system to an alternative basin of attraction. The various basins that a system may occupy and the boundaries that separate them are known as a stability landscape (Walker et al. 2004). Although we may not be able to predict precisely the species composition and abundance of a community in the long term, given some knowledge of initial conditions we do have some idea of what state the old field is likely to move toward.

For old fields in a stability landscape with two basins of attraction, for example, the system may tend toward a state with few native species, many weed species, and poor soil condition (basin 1, figure 3.1), or tend toward a state with many native annual and perennial species, few weed species, and

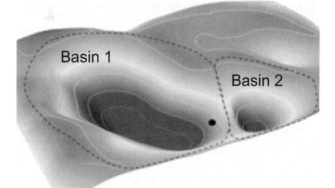

FIGURE 3.1. Three-dimensional stability landscape with two basins of attraction. For old fields within a highly fragmented landscape, the system (i.e., the old field, represented as a black dot) is likely to be drawn into the largest basin with the strongest attractor (basin 1), resulting in a system with few native species, many alien species, and poor soil condition. Basin 2, where the system would move toward a state with many native species, few alien species, and good soil conditions, will have a weaker attractor within this stability landscape. Adapted from the original figure from Walker et al. (2004), with the permission of Brian Walker.

good soil condition (basin 2, figure 3.1). The most desirable basin of attraction for community assembly in old fields will be basin 2, as this most closely reflects natural vegetation prior to clearing. In a highly fragmented landscape however, this basin is likely to be the smaller of the two and have the weakest attractor. Yet once the old field moves close to the attractor in this basin and a suite of native species have established in the old field, the resistance to perturbations, such as the invasion of weed species, may be quite high. Basin 1, where the old field becomes dominated by a community of invasive weed species, is likely to have the deepest and widest basin of attraction. Once the old field lies within this basin of attraction it will show great resistance to management inputs that aim to move the system to the other basin.

Ecological Filter and Systems Models: Exploring Causal Relationships in Community Assembly on Old Fields

The idea of basins of attraction within a stability landscape provides a conceptual model for understanding how alternative states develop in old fields. Yet how do we determine which state the old field may tend toward, given the initial conditions present upon abandonment? Ecological filter models and systems models allow us to investigate how environmental and biotic conditions will affect the trajectory of community assembly.

Ecological Filter Models

In an ecological filter model, the process of community assembly involves a series of filters that either allow or disallow species from the regional species pool to colonize and continue to inhabit a site (Diaz et al. 1999) (figure 3.2). The filters may be environmental (e.g., soil nutrient deficiency, soil compaction, soil salinity) or biological (e.g., seed dispersal limitation, competition between species, facilitation). Ecological filters can be related to other concepts describing community assembly, such as ecological thresholds and gradients (Hobbs and Norton 2004). Ecological filters may be considered to

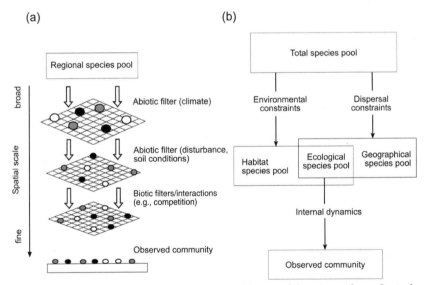

FIGURE 3.2. Schematic diagrams of ecological filter models: (a) simple ecological filter model whereby the traits of each species either allow them to pass, or prevent them from passing, through a series of abiotic and biotic filters. Species that pass through all filters are found in the observed community. Adapted from Diaz et al. (1999). And, (b) the scheme of Kelt et al. (1995) whereby different species pools are linked to particular filters. Environmental constraints will determine the habitat species pool, while dispersal constraints will determine the geographical species pool. The intersection of these two pools constitutes the ecological species pool (i.e., species that can overcome both environmental and dispersal constraints). Internal dynamics within the system (e.g., competition) will determine the observed community. Adapted from Belyea (2004).

be a form of assembly rule (Weiher and Keddy 1999; Temperton et al. 2004), because the combination of filters operating at a site will determine the composition of the community. For example, sets of plant attributes can be consistently associated with certain environmental conditions, allowing the effects of an ecological filter on community composition to be explicitly stated (Diaz et al. 1999). As a community develops at a site, it is unlikely that the filters will remain unchanged, and more recent frameworks of ecological filters consider that the filters are self-adjusting and will change over time because of feedback loops within the pool of established species (Fattorini and Halle 2004). From a theoretical viewpoint, the presence of a suite of slightly different ecological filters (i.e., different initial conditions) at similar sites will set the developing communities along different trajectories, culminating in the development of alternative states. From a practical viewpoint, identifying the ecological filters at a site allows their manipulation if some of the ecological filters will not allow the establishment of the desired suite of species. For example, soil can be ripped to overcome soil compaction, weeds can be removed to reduce competition, or seeds can be introduced where the soil seed bank is deficient and barriers to seed dispersal exist. The effects of many ecological filters, in the short term, will be deterministic and predictable, and thus amenable to management.

Yet it is unlikely that all ecological filters will be identifiable, and all of the interactions between filters are unlikely to be understood, thus introducing elements of unpredictability into the trajectory of community development. Further, disturbance may be considered to be an ecological filter (White and Jentsch 2004), and although the effects of particular disturbances may be known, it is unlikely that the timing, frequency, and intensity of disturbance are predictable. Thus the relative strengths of various filters may need to be reassessed after disturbances that may reset the assembly trajectory in an old field, and the ecological filter model for the system used adaptively.

Systems Models

A complementary approach to the ecological filter model is to use systems models to understand community assembly (Belyea 2004). One of the constraints of the ecological filter model as a conceptual framework for community assembly is the inherent hierarchical nature of the filter model, whereby climatic, disturbance, and interaction filters act at decreasing spatial scales (Diaz et al. 1999) (figure 3.2a). Hence, abiotic filters are often portrayed as operating at a level "above" biotic filters (figure 3.2). This "community perspective" of community assembly regards the environment as fixed in some sense (Beisner et al. 2003) and does not consider that biotic variables may influence environmental constraints. Yet plant establishment and growth is affected by a suite of interactions between the abiotic and biotic environment. For example, facilitative interactions may overcome or reduce environmental constraints to plant growth: low nutrient availability may not be an environmental constraint to a species when the appropriate mycorrhiza are present; the effects of low soil water availability and a harsh microclimate may be reduced when "nurse" plants are present (Bruno et al. 2003; Gómez-Aparicio et al. 2004). Whether the environment is considered as external or internal depends largely on the rate of feedback between biotic and abiotic processes (Beisner et al. 2003). Where the rate of feedback between certain biotic and abiotic processes is fast, then these environmental constraints may be considered as variables within the system, rather than independent, slowly changing parameters. Conversely, some important biotic filters are external to the system in question. One of the most important determinants of community assembly is the dispersal of propagules into the old field. The timing, amount, and type of propagules arriving at the site has important consequences for community development (Standish et al. 2007). A systems model approach allows us to represent environmental constraints or biotic processes as either internal or external to the system in question. For example, rainfall and seed dispersal are important determinants of community development, yet community development has no effect on rainfall and is likely to have limited influence on seed dispersal into the site. Therefore these constraints should be considered as external parameters independent of the system (figure 3.3).

Another advantage of a systems model approach is the identification of the relationships (feedback loops) between components of the ecosystem, whether they be related to process or structure, instead of focusing solely on species pools and environmental filters (Belyea 2004). In a restoration context, this allows the identification of interactions that either facilitate or deter the development of the desired community, some of which may be managed to reset the assembly trajectory. Constraints that produce strong feedback within the community can be identified as management priorities, and multiple constraints can be addressed at once (Suding et al. 2004). Incorporating stochastic events, such as disturbance, into systems models is problematic, as it is for ecological filter models. Systems models must also be used adaptively and reassessed after major changes to system dynamics that will alter the interactions between aspects of the system and the strength of various feedback loops.

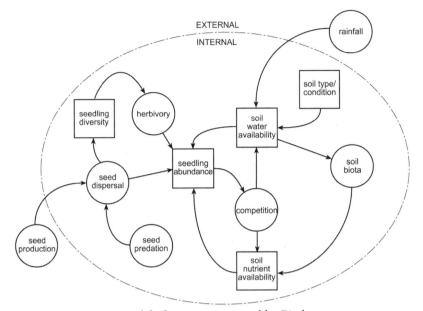

Figure 3.3. A systems model of community assembly. Circles represent ecosystem processes; squares represent aspects of ecosystem structure. Arrows indicate interactions between these variables. Parameters such as rainfall and seed production are external to the system (old field) under consideration and cannot be affected by the system. See text for further details.

Manipulating Community Assembly in Old Fields: Restoration of Degraded States Guided by a Systems Model

The degraded state of some abandoned farmland may represent an alternative state that is resistant to change (Suding et al. 2004). Understanding and manipulating feedbacks between biotic factors and the physical environment provides a management framework for "forcing" a community out of a resilient, degraded state and guiding community assembly in old fields along a desired trajectory. Rather than continue to assume deterministic change in the restoration trajectory of the old field, we should rely on monitoring programs to document actual trajectories of change (Suding and Gross 2006). Having documented the trajectory, we then need to be able to make management decisions to alter the trajectory if it appears unlikely that the desired outcome will be achieved.

The systems model presented in figure 3.3 describes community assembly in old fields in a highly fragmented landscape, such as the wheatbelt of Western Australia (chapter 15), where old fields are likely to be distant from remnant native vegetation and seed sources, the soil seed bank will be dominated by weed species after abandonment, and cultivation practices may have significantly altered soil structure and soil nutrient concentrations. In this model, aspects of both process (circles) and structure (squares) are included. Just as ecological processes will determine community structure, current structure may stifle the development of desired processes, and disturbance of this structure may be necessary to reset the assembly trajectory. In the model, the arrows indicate interactions between various processes and aspects of ecosystem structure. For example, an increase in the activity of soil biota (process) will lead to an increase in soil nutrient availability (structure), while an increase in herbivory (process) will lead to a decrease in seedling abundance (structure). Environmental constraints that lie outside of the dashed line are external to the system and cannot be influenced by the processes operating within the system. The systems model helps identify feedback loops in which the biotic community changes the environmental constraints within the system (Belyea 2004) and whether these feedback loops are reinforcing or balancing (Sterman 2000). For example, an increase in soil water availability will lead to greater seedling abundance. Increased seedling abundance will in turn lead to increased competition, while increased competition for soil water will lead to its depletion, ultimately leading to a decrease in seedling abundance. This constitutes a balancing feedback loop, whereby an increase in one or more variables in the loop is balanced by a decrease in other variables. In a reinforcing feedback loop, increases (decreases) in one variable will lead to increases (decreases) in the other variables, thus reinforcing the feedback within the loop. For example, increases in soil biota will improve soil condition in degraded soils, which in turn will increase soil water availability, and thus improve conditions for soil biota.

Having determined potential feedback loops within the system, suites of species' traits, such as resource acquisition and competitive ability, can be considered in the context of each of the feedback loops to determine the likelihood of that species establishing and surviving in the old field. Competition from introduced grasses or herbs is often the most important barrier to the establishment of native species in old fields (Humphrey and Schupp 2004; Riege and del Moral 2004; Standish et al. 2007). The establishment of dense stands of weed species will lead to high levels of competition with any native species that establish, and the superior competitive ability of weed species will lead to lower abundances of native species when soil resources such as water or nutrients are limiting. Thus although the feedback loop between seedling abundance, competition, and soil water is balancing, the resource acquisition and life history traits of certain weed species mean that they will be the most abundant species in the old field. In terms of managing community assembly in this old field, improving soil water availability by improving soil condition may not lead to better establishment of native species when the abundance of weed species remains high. Thus, improving soil water availability and reducing competition from weeds must be undertaken in tandem. The successful establishment of native species could be further enhanced by supplementing the seed supply of native species through direct seeding and limiting the input of seeds of weed species by controlling weeds before they set seed.

Do the New Frameworks of Community Assembly Improve Our Understanding and Management of Old Field Dynamics?

While current thinking in theoretical ecology is embracing the ideas from the theory of complex, nonlinear systems, it is important to recognize that many of these ideas, such as the importance of initial conditions and stochastic events on system development, are not new. In the 1920s, Gleason (1927) argued that we were unable to make predictions concerning the future of vegetation on the basis of observable causes alone, because "we are usually neglecting certain present conditions which we fail to recognize or which are impossible to discern at the present time" (302). He then argued that "physiographic or biotic processes may undergo unexpected changes of such magnitude that they seriously affect the vegetation and influence the trend of its succession" (303).

Undoubtedly, there is significant overlap between the newer and more traditional concepts (Young et al. 2001), and ideas developed in the early debates on succession have resurfaced. For example, argument over whether succession is due to biotic interactions only (autogenic succession) or involves abiotic factors (allogenic succession) (Tansley 1935) has been repeated in the context of assembly rules (Weiher and Keddy 1999). The idea of grazing pressure leading to the development of an alternative state in rangelands (Anderies et al. 2002) is not that different from Tansley's (1935) concept of a "biotic climax" caused by the incidence and maintenance of a decisive biotic factor, "such as the continuous grazing of animals" (292). While the new frameworks of community assembly may not be completely novel, the increasing prominence of these frameworks and the language that describes them has encouraged scientists and managers to take greater account of historical contingency, incomplete knowledge of initial conditions, and the role of stochastic events in vegetation development. Yet this does not mean that deterministic models of community assembly are no longer relevant and must be discarded in a complete paradigm shift. Rather, viewing old fields as complex, nonlinear systems allows for a paradigm expansion, where the interplay between determinism and stochasticity expressed over various time and spatial scales leads to the possibility of multiple assembly trajectories and ecosystem states. Such a paradigm expansion may also provide a better theoretical template for old field dynamics outside of the temperate ecosystems where traditional succession theory was developed. Although it remains difficult to incorporate stochastic events into either a dynamic filter model or a systems model of community assembly, if we remain cognizant of the complex, nonlinear nature of ecosystems then we will recognize the limitations of our predictive power. Acknowledging the likelihood of several possible outcomes, or that the assembly trajectory in old fields will not be completely predictable, means that adaptive management can be planned for at the outset, not when restoration of old fields fails to meet our predetermined goals.

Acknowledgments

I wish to extend my warm thanks to Rachel Standish, Mitch Aide, Vicky Temperton, and Richard Hobbs for their helpful and insightful comments on the manuscript, and to Helen Allison for useful discussions on systems models.

REFERENCES

Anderies, J. M., M. A. Janssen, and B. H. Walker. 2002. Grazing management, resilience, and the dynamics of a fire-driven rangeland system. *Ecosystems* 5:23–44.
Beisner, B. E., D. T. Haydon, and K. Cuddington. 2003. Alternative stable states in ecology. *Frontiers in Ecology and the Environment* 1:376–82.
Belovsky, G. E., D. B. Botkin, T. A. Crowl, K. W. Cummins, J. F. Franklin, M. L. Hunter Jr., A. Joern et al. 2004. Ten suggestions to strengthen the science of ecology. *BioScience* 54:345–51.
Belyea, L. 2004. Beyond ecological filters: Feedback networks in the assembly and restoration of community structure. In *Assembly rules and restoration ecology: Bridging the gap between theory and practice*, ed. V. M. Temperton, R. J. Hobbs, T. Nuttle, and S. Halle, 115–31. Island Press, Washington, DC.
Booth, B. D., and D. W. Larson. 1999. Impact of language, history and choice of system on the study of assembly rules. In *Ecological assembly rules: Perspectives, advances, retreats*, ed. E. Weiher and P. Keddy, 206–329. Cambridge University Press, Cambridge, UK.
Bruno, J. F., J. J. Stachowicz, and M. D. Bertness. 2003. Inclusion of facilitation into ecological theory. *Trends in Ecology and Evolution* 18:119–25.
Buisson, E., and T. Dutoit. 2004. Colonisation by native species of abandoned farmland adjacent to a remnant patch of Mediterranean steppe. *Plant Ecology* 174:371–84.
Chase, J. 2003. Community assembly: When should history matter? *Oecologia* 136:489–98.

Clements, F. E. 1916. *Plant succession: An analysis of the development of vegetation.* Publication N. 242. Carnegie Institution of Washington, Washington, DC.

Compton, J. E., and R. D. Boone. 2000. Long-term impacts of agriculture on soil carbon and nitrogen in New England forests. *Ecology* 81:2314–30.

Diaz, S., M. Cabido, and F. Casanoves. 1999. Functional implications of trait-environment linkages in plant communities. In *Ecological assembly rules: Perspectives, advances, retreats,* ed. E. Weiher and P. Keddy, 338–62. Cambridge University Press, Cambridge, UK.

Drake, J. A., C. R. Zimmerman, T. Purucker, and C. Rojo. 1999. On the nature of the assembly trajectory. In *Ecological assembly rules: Perspectives, advances, retreats,* ed. E. Weiher and P. Keddy, 233–50. Cambridge University Press, Cambridge, UK.

Egler, F. E. 1954. Vegetation science concepts. 1. Initial floristic composition, a factor in old-field vegetation development. *Vegetatio* 10:412–17.

El-Sheikh, M. A. 2005. Plant succession on abandoned fields after 25 years of shifting cultivation in Assuit, Egypt. *Journal of Arid Environments* 61:461–81.

Fattorini, M., and S. Halle. 2004. The dynamic environmental filter model: How do filtering effects change in assembling communities? In *Assembly rules and restoration ecology: Bridging the gap between theory and practice,* ed. V. M. Temperton, R. J. Hobbs, T. Nuttle, and S. Halle, 96–114. Island Press, Washington, DC.

Fraterrigo, J. M., M. G. Turner, S. M. Pearson, and P. Dixon. 2005. Effects of past land use on spatial heterogeneity of soil nutrients in southern Appalachian forests. *Ecological Monographs* 75:215–30.

Fukami, T., M. Bezemer, S. R. Mortimer, and W. H. van der Putten. 2005. Species divergence and trait convergence in experimental plant community assembly. *Ecology Letters* 8:1283–90.

Ganade, G., and V. K. Brown. 2002. Succession in old pastures of central Amazonia: Role of soil fertility and plant litter. *Ecology* 83:743–54.

Gleason, H. A. 1927. Further views on the succession-concept. *Ecology* 8:299–326.

Gómez-Aparicio, L., R. Zamora, J. M. Gómez, J. A. Hódar, J. Castro, and E. Baraza. 2004. Applying plant facilitation to forest restoration: A meta-analysis of the use of shrubs as nurse plants. *Ecological Applications* 14:1128–38.

Hastings, A., C. L. Hom, S. Ellner, P. Turchin, and H. C. J. Godfray. 1993. Chaos in ecology: Is Mother Nature a strange attractor? *Annual Review of Ecology and Systematics* 24:1–33.

Hobbs, R. J., and D. A. Norton. 2004. Ecological filters, thresholds and gradients in resistance to ecosystem assembly. In *Assembly rules and restoration ecology: Bridging the gap between theory and practice,* ed. V. M. Temperton, R. J. Hobbs, T. Nuttle, and S. Halle, 72–95. Island Press, Washington, DC.

Holling, C. S. 1995. What barriers? What bridges? In *Barriers and bridges to the renewal of ecosystems and institutions,* ed. L. H. Gunderson, C. S. Holling, and S. S. Light, 3–34. Columbia University Press, New York.

———. 1998. Two cultures of ecology. *Conservation Ecology* [online] 2(2): 4. http://www.consecol.org/vol2/iss2/art4/.

Hooper, E., R. Condit, and P. Legendre. 2002. Responses of 20 native tree species to reforestation strategies for abandoned farmland in Panama. *Ecological Applications* 12:1626–41.

Humphrey, L. D., and E. Schupp. 2004. Competition as a barrier to establishment of a native perennial grass (*Elymus elymoides*) in alien annual grass (*Bromus tectorum*) communities. *Journal of Arid Environments* 58:405–22.

Kelt, D. A., M. L. Tapper, and P. L. Meserve. 1995 Assessing the impact of competition on community assembly: A case study using small mammals. *Ecology* 76:1283–96.

Levin, S. A. 1998. Ecosystems and the biosphere as complex adaptive systems. *Ecosystems* 1:431–36.

Lockwood, J. L., and S. L. Pimm. 1999. When does restoration succeed? In *Ecological assembly rules: Perspectives, advances, retreats,* ed. E. Weiher and P. Keddy, 363–92. Cambridge University Press, Cambridge, UK.

Meiners, S. J., S. T. A. Pickett, and M. L. Cadenasso. 2002. Exotic plant invasions over 40 years of old field successions: Community patterns and associations. *Ecography* 25:215–23.

Myster, R. W. 2004. Regeneration filters in post-agricultural fields of Puerto Rico and Ecuador. *Plant Ecology* 172:199–209.

Palmer, M. W., and G. M. Rusch. 2001. How fast is the carousel? Direct indices of species mobility with examples from an Oklahoma grassland. *Journal of Vegetation Science* 12:305–18.

Riege, D. A., and R. del Moral. 2004. Differential tree colonization of old fields in a temperate rain forest. *American Midland Naturalist* 151:251–64.

Salthe, S. N. 2001. The natural philosophy of ecology: Developmental systems ecology. Accessed 8 August 2006. http://www.nbi.dk/~natphil/salthe/.

Schulze, R. 2000. Transcending scales of space and time in impact studies of climate and climate change on agrohydrological responses. *Agriculture, Ecosystems and Environment* 82:185–212.

Standish, R. J., V. A. Cramer, R. J. Hobbs, and H. T. Kobryn. 2006. Legacy of land-use evident in soils of Western Australia's wheatbelt. *Plant and Soil* 280:189–207.

Standish, R. J., V. A. Cramer, S. L. Wild, and R. J. Hobbs. 2007. Seed dispersal and recruitment limitation are barriers to native recolonisation of old-fields in Western Australia. *Journal of Applied Ecology* 44:435–45.

Sterman, J. D. 2000. *Business dynamics: Systems thinking and modelling for a complex world.* Irwin McGraw-Hill, Boston.

Suding, K. N., and K. L. Gross. 2006. The dynamic nature of ecological systems: Multiple states and restoration trajectories. In *Foundations of restoration ecology,* ed. D. Falk, M. Palmer, and J. Zedler, 190–209. Island Press, Washington, DC.

Suding, K. N., K. L. Gross, and G. R. Houseman. 2004. Alternative states and positive feedbacks in restoration ecology. *Trends in Ecology and Evolution* 19:46–53.

Tansley, A. G. 1935. The use and abuse of vegetational concepts and terms. *Ecology* 16:284–307.

Temperton, V. M., and R. J. Hobbs. 2004. The search for ecological assembly rules and its relevance to restoration ecology. In *Assembly rules and restoration ecology: Bridging the gap between theory and practice,* ed. V. M. Temperton, R. J. Hobbs, T. Nuttle, and S. Halle, 34–54. Island Press, Washington, DC.

Temperton, V. M., R. J. Hobbs, T. Nuttle, and S. Halle, eds. 2004. *Assembly rules and restoration ecology: Bridging the gap between theory and practice.* Island Press, Washington, DC.

Van der Maarel, E., and M. T. Sykes. 1993. Small-scale plant species turnover in a limestone grassland: The carousel model and some comments on the niche concept. *Journal of Vegetation Science* 4:179–88.

Walker, B., C. S. Holling, S. R. Carpenter, and A. Kinzig. 2004. Resilience, adaptability and transformability in social-ecological systems. *Ecology and Society* 9:5 [online]. http://www.ecologyandsociety.org/vol9/iss2/art5.

Weiher, E., and P. Keddy, eds. 1999. *Ecological assembly rules: Perspectives, advances, re-treats.* Cambridge University Press, Cambridge, UK.

White, P. S., and A. Jentsch. 2004. Disturbance, succession, and community assembly in terrestrial plant communities. In *Assembly rules and restoration ecology: Bridging the gap between theory and practice,* ed. V. M. Temperton, R. J. Hobbs, T. Nuttle, and S. Halle, 342–66. Island Press, Washington, DC.

Wilson, J. B. 1999. Assembly rules in plant communities. In *Ecological assembly rules: Perspectives, advances, retreats,* ed. E. Weiher and P. Keddy, 130–64. Cambridge University Press, Cambridge, UK.

Wilson, J. B., and H. Gitay. 1995. Community structure and assembly rules in a dune slack: Variance in richness, guild proportionality, biomass constancy and dominance/diversity relations. *Vegetatio* 116:93–106.

Wilson, J. B., and R. J. Whittaker. 1995. Assembly rules demonstrated in a saltmarsh community. *Journal of Ecology* 83:801–7.

Young, T. P., J. M. Chase, and R. T. Huddleston. 2001. Community succession and assembly: Comparing, contrasting and combining paradigms in the context of ecological restoration. *Ecological Restoration* 19:5–18.

PART 2

Case Studies from Around the World

Chapter 1 and part 1 provided an overview of land abandonment around the world (chapter 1) and a consideration of the theoretical aspects relating to old field dynamics, both from a classical succession perspective (chapter 2) and in light of recent developments in the areas of ecosystem assembly and alternative states (chapter 3). We also presented a brief overview of the types of factors likely to be important in influencing the course of old field development. In this section, we present a series of case studies that examines these factors in more detail in specific situations. As noted in chapter 1, land abandonment has been particularly prevalent in some parts of the world but is now increasing in extent and importance in many other places.

Given this increased incidence of land abandonment in differing locations and varying environmental and socioeconomic settings, how general are the patterns of ecosystem redevelopment following abandonment? How well do the models described in chapters 2 and 3 explain the various dynamics observed, and what factors are likely to be most important in driving the dynamics in different situations? Finally, what can an understanding of these factors tell us that will aid in a restoration context? In order to shed some light on these questions, we have assembled a series of case studies from different parts of the world. The aim of these case studies is to put some flesh on the ideas and theories already presented and provide concrete examples from which to develop a more coherent synthesis.

Elements Covered in the Case Studies

Connection Between Restoration Ecology and Ecological Restoration

We aim to make the book relevant to both scientists and restoration practitioners. Each chapter therefore addresses not only the dynamics of community development in old fields but also how this knowledge can better help practitioners in the active restoration of old fields.

Current Theories of Old Field Dynamics

We did not stipulate a "preferred" paradigm for authors to work within (e.g., succession, alternative stable states, assembly theory). However, we did ask that authors briefly describe what they mean if they use terms such as *old field succession* or *assembly theory*. In the literature, the same term appears to mean different things to different people; for example, succession is often used as a synonym for any type of vegetation development, rather than in its original sense of a directional and deterministic sequence of vegetation development. By having some explanation, we hope to avoid confusion about whether the reader's understanding of a term is the same as that of the authors.

Site History, Previous Land Use, and Landscape Context

Authors were asked to consider, where possible, the influence of site history, previous land use, and landscape context on the outcome of abandonment. Does knowledge of historical contingency provide greater predictability of old field dynamics, or does another layer of variability confound predictions of vegetation development? The case studies include contributions from authors working in areas with very different agricultural histories and current land uses, and while some authors have detailed knowledge of such information, others have only vague oral histories of past land use. Is detailed knowledge of site history and land use necessary for understanding old field dynamics? How do site history and landscape context interact to determine old field development? We hope to be able to draw some conclusions as to the role of history and landscape in the development of old fields.

Ecological Domain

While we recognize that most of the authors are plant ecologists, we encouraged consideration of the role other taxa and the interactions among groups of organisms in old field development. This not only need be related to the development of plant communities (e.g., the role of pollinators and seed dispersers, the importance of soil microbial communities for nutrient acquisition), but may focus on the dynamics of insects, birds, mammals, soil flora and fauna itself.

Geographic Representation

These case studies present detailed analyses of the dynamics of abandoned farmland in a variety of settings. Chapters 4–7 examine tropical systems. Chapters 8 and 9 examine recent studies in eastern North America, and chapter 10 presents a study from central Europe. Chapters 11–13 examine different parts of the Mediterranean basin: ostensibly areas with similar environmental constraints and socioeconomic settings, but with notable differences that have important impacts on old field dynamics. Following on from this, chapters 14 and 15 then examine two other Mediterranean-climate areas, South Africa and southwestern Australia, which are very different biogeographically and socioeconomically from the Mediterranean basin. Therefore, the case studies fall into the three broad categories of tropical, temperate, and Mediterranean systems, which allow comparisons both between these broad categories and between different situations within each category.

The case studies represent a sample of different environments that is not meant to be exhaustive and undoubtedly does not capture all the available, potential examples of land abandonment and old field dynamics. Nevertheless, they provide sufficient coverage of different situations that they can collectively be used to draw some general conclusions, which is done in the final section of the book.

Chapter 4

Implications of Land Use History for Natural Forest Regeneration and Restoration Strategies in Puerto Rico

JESS K. ZIMMERMAN, T. MITCHELL AIDE,
AND ARIEL E. LUGO

The study of succession following the abandonment of agriculture has contributed greatly to both a theoretical and a practical understanding of vegetation change (Connell and Slayter 1977; Bradshaw 1987; Picket et al. 1987; chapter 2). Much of this understanding, however, comes from the study of temperate ecosystems, while a similar appreciation of tropical ecosystems has lagged (Brown and Lugo 1990; Meli 2003; chapters 5–7). This difference is due to the fact that farm abandonment is mainly driven by socioeconomic changes that have only recently begun to occur in some tropical countries (Hall 2000; Chang and Tsai 2002; Rudel et al. 2002; Grau et al. 2003; chapter 1). Specifically, one of the many effects of the globalization process is the increase in rural to urban migration in tropical countries (Population Division 2002), which often results in the abandonment of marginal agricultural lands and secondary forest regeneration (Aide and Grau 2004). Understanding this relatively new form of tropical vegetation dynamics is essential for maximizing management and conservation efforts in the tropics (Brown and Lugo 1994).

Puerto Rico, a tropical Caribbean island, has had a recent history of farm abandonment (Rudel et al. 2000; Grau et al. 2003) that has spawned numerous studies of land use change and vegetation dynamics (Aide et al. 1995, 1996; Zimmerman et al. 1995; Lugo et al. 1996; Thomlinson et al. 1996; Rivera and Aide 1998; Aide et al. 2000; Pascarella et al. 2000; Thomlinson and Rivera 2000; López et al. 2001; Chinea 2002; Marcano Vega et al. 2002; Chinea and Helmer 2003; Helmer 2004; Lugo and Helmer 2004; Molina Colón and Lugo 2006). Many of these studies have employed chronosequences to suggest patterns of succession, here defined simply as the vegetation dynamics that follow a disturbance, human-caused or otherwise.

When striving for a mechanistic understanding of vegetation change, a barriers approach has often been utilized to understand the establishment of woody vegetation (Nepstad et al. 1990, 1996; Aide and Cavalier 1994; Zimmerman et al. 2000) rather than a direct test of ecological theory (Drury and Nisbet 1973; Connell and Slayter 1977; Pickett et al. 1987). A barriers approach leads directly to potential forest restoration methods (Aide 2000) and often touches on such concepts as facilitation and inhibition.

In this summary of research on forest recovery and restoration in Puerto Rico, we first place the island in the context of its climate, geology, and flora. We then summarize the history of land use change on the island. Specifically, we focus on secondary succession following the abandonment of three major land uses: pasture, shade coffee, and sugarcane. We describe key aspects of vegetation change and touch on the mechanisms of vegetation change where it is available. We conclude with a section outlining the major findings of the research on forest recovery in Puerto Rico and discuss the implications of our findings for tropical forest restoration.

The Puerto Rican Context

Despite Puerto Rico's small size (9,104 km^2), the island includes six Holdridge life zones (Ewel and Whitmore 1973). This is due to a strong rainfall gradient from the Luquillo Mountains (figure 4.1) in the northeast (~5,000 mm of rainfall per year) to the southwest portion of the island (<1,000 mm per year). The research that we review is conducted mostly in the subtropical moist and wet forest life zones within an annual range of rainfall of between 1,000 and 3,500 mm. A varied geological substrate, which in-

FIGURE 4.1. The physiography of Puerto Rico, showing regions and locations mentioned in the text. Locations of sites are also indicated: (a) Ciales, (b) Carite, (c) Sabana Seca, (d) Utuado.

cludes volcanoclastic, plutonic, and alluvial zones, as well as an extensive karst zone in the north central and northwest portion of the island (figure 4.1), ensures that all major soil orders are represented in Puerto Rico (Roberts 1942). Due to its insular nature and subtropical latitudinal position, the woody vegetation is less diverse in comparison with mainland, lowland tropical floras, but there is a strong floristic affinity with much of the neotropics (Little et al. 1974). There is little information on the original composition of the forest in Puerto Rico (Wadsworth 1950) because most forests had been converted to agriculture or were highly degraded before there were many scientific studies. The vegetation is frequently disturbed by hurricanes, which may predispose it to recover quickly from human disturbance (Lugo 1988). The faunal relationships of the woody vegetation appear generalized (Reagan and Waide 1996), not exhibiting the tendency for specialized pollinator and disperser relationships that are the hallmark of some lowland tropical forests (Primack and Corlett 2005).

The secondary forests that dominate Puerto Rico today bear the signature of human land use, in addition to gradients of rainfall and geological substrate (Chinea and Helmer 2003; Grau et al. 2003) (table 4.1; figure 4.2). Alien species are common (Lugo and Helmer 2004; Lugo 2004); approximately 200 of the 750 woody species on the island (Little et al. 1974) have been introduced since European colonization. The most common tree, *Spathodea campanulata*, is an alien, as is the seventh most common species, *Syzygium jambos* (table 4.1; Franco et al. 1997). Despite appearances, this does not represent an alien horror story, as has taken place in more isolated and younger tropical islands (e.g., MacDonald et al. 1988; Vitousek and Walker 1989; Meyer et al. 2003). *Spathodea campanulata* does not regenerate beneath its canopy (Aide et al. 2000; Lugo 2004) and, while *Syzygium jambos* persists in older forests (Aide et al. 2000; Heartsill Scalley and Aide 2003), from there it does not spread aggressively (Brown et al. 2006). The majority of the remaining alien species are rare and infrequent (Lugo 2004).

Summary of Land Use History of Puerto Rico

Deforestation of coastal forests in Puerto Rico started with the first inhabitants of the island, the Taíno Indians, who used these areas for settlements and agriculture (e.g., manioc and corn) (Gómez and Ballesteros 1980). Large-scale impact on coastal forested wetlands began with the arrival of the Spaniards in the early 1500s, when large areas were converted into pastures for cattle and sugarcane plantations (Picó 1986). Coffee was introduced in 1736 and became a dominant crop in montane areas throughout the island

TABLE 4.1

Species composition of secondary forests in Puerto Rico

Puerto Rico	Shade Coffee	Pasture

Carite

Puerto Rico	Shade Coffee	Pasture
1. Spathodea companulata	1. Guarea guidonia	1. Ocotea leucoxylon
2. Guarea guidonia	2. Cecropia schreberiana	2. Prestoea montana
3. Inga vera	3. Inga vera	3. Tabebuia heterophylla
4. Cecropia schreberiana	4. Andira inermis	4. Cecropia schreberiana
5. Andira inermis	5. Dendropanax arboreus	5. Casearia arborea
6. Tabebuia heterophylla		

Ciales

	Shade Coffee	Pasture
7. Syzygium jambos	1. Guarea guidonia	1. Spathodea companulata
8. Inga laurina	2. Coffea arabica	2. Guarea guidonia
9. Erythrina poeppigiana	3. Dendropanax arboreus	3. Andira inermis
10. Schefflera morototoni	4. Montezuma speciossima	4. Cupania americana
	5. Ocotea leucoxylon	5. Casearia sylvestris

Luquillo

Shade Coffee	Pasture
1. Guarea guidonia	1. Andira inermis
2. Prestoea montana	2. Schefflera morototoni
3. Cecropia schreberiana	3. Tabebuia heterophylla
4. Cordia borenquensis	4. Casearia guianensis
5. Casearia sylvestris	5. Miconia prasina

Utuado

Shade Coffee	Pasture
1. Guarea guidonia	1. Guarea guidonia
2. Cecropia schreberiana	2. Spathodea companulata
3. Spathodea companulata	3. Tabebuia heterophylla
4. Ocotea leucoxylon	4. Inga vera
5. Inga vera	5. Casearia sylvestris

Source: Grau et al. (2003).
Note: Most common species, ranked by basal area per hectare, are shown for all forested lands in Puerto Rico (left column; Franco et al. 1997). Similar data from four regions (see figure 4.1) show species composition of two major land uses: abandoned shade coffee and abandoned pastures. Ages of abandonment for coffee range from thirty to sixty years; pastures range from ten to sixty years. See Aide et al. (2000, 1996), Pascarella et al. (2000), Rivera and Aide (1998), and Zimmerman et al. (1995), for additional details.

(Dietz 1987). Forest cover islandwide exceeded 80% until the beginning of the nineteenth century (Birdsey and Weaver 1982; Grau et al. 2003), when the population of Puerto Rico was less than 200,000 individuals. As the population grew, to approximately one million individuals by the early twentieth century and to over two million by 1950, forest cover declined steadily to less than 6% of the island. A typical Puerto Rican landscape was one of nearly complete deforestation, save for small patches of riparian forest and kitchen gardens grown under a canopy of leguminous trees (figure 4.2). About one-half of the remaining forest cover was in shade coffee (Birdsey and Weaver 1982), and primary forest was limited to the Luquillo Mountains in northeastern Puerto Rico and in the highest portions of the Cordillera Central (figure 4.1).

FIGURE 4.2. Agricultural landscape in central Puerto Rico. Photograph taken by Frank Wadsworth in the 1960s.

Through the first half of the twentieth century, the Puerto Rican landscape was dominated by agriculture. Sugarcane covered the coastal plains and nearby hillsides. R. G. Tugwell, the governor of Puerto Rico from 1941 to 1946, described the island as "mountains surrounded by sugar cane" (Tugwell 1946). Shade coffee plantations, along with tobacco and subsistence agriculture, dominated the highlands, and cattle pastures extended from the lowlands to highlands.

The end of the Second World War heralded socioeconomic improvements in Puerto Rico driven by government incentives to bring industry to the island and the booming U.S. economy (Dietz 1987). This, in turn, led to a recovery of forest cover in Puerto Rico, as the rural population moved to urban areas to take advantage of the newly created employment opportunities (Rudel et al. 2000; Grau et al. 2003). Today, forest cover is approximately 44% of the island (T. Kennaway and E. H. Helmer, pers. comm.). Postwar farm abandonment was most rapid in montane areas of marginal agricultural value (Rudel et al. 2000), but this was preceded in the Luquillo Mountains by abandonment, which followed severe hurricanes that devastated the eastern portion of the island in 1928 and 1932 (Scatena 1989). Peak rates of farm abandonment islandwide were recorded in the 1960s through the 1980s (Grau et al. 2003). Sugarcane was often converted to pastureland before being abandoned (Thomlinson et al. 1996). Urban areas began to expand as the rural population moved to the urban centers. Urbanization occurred particularly in the coastal plains (e.g., Thomlinson and Rivera 2000; Grau et al. 2003) and especially around the capital, San Juan. As a result, between 1977 and 1994 the island lost 6% of its prime agricultural land to expanding urban areas (López et al. 2001). Efforts to increase tourism have led to the development of many resorts along the coast. By the middle of the twenty-first century, it is likely that Tugwell's Puerto Rico will be mountains surrounded by urban development and golf courses.

Land Use, Succession, and Mechanisms

We summarize below our understanding of succession following the abandonment of pastures, coffee plantations, and sugarcane in Puerto Rico. For pastures and coffee plantations, we describe possible mechanisms of succession, drawn from detailed observational and experimental studies. In the case of sugarcane, we describe current conditions that apparently inhibit succession.

Pastures

Pasture succession in Puerto Rico was studied by assembling chronosequences of abandoned pastures using aerial photographs dating to 1936. While the chronosequence approach had certain limitations, it provided a basis for more detailed study of some of the mechanisms controlling the early patterns of succession.

SUCCESSIONAL TRENDS

Studies of pasture chronosequences began along the flanks of the Luquillo Mountains (Aide et al. 1995, 1996; Zimmerman et al. 1995) and then were extended to include other parts of the island (Rivera and Aide 1998; Pascarella et al. 2000; Chinea 2002; Marcano Vega et al. 2002). The results indicated that forest recovery, in terms of tree density and biomass, occurred rapidly following a lag of approximately ten years (Aide et al. 2000). Tree density and biomass often returned to levels observed in the oldest identifiable forest stands (those apparent in 1936 aerial photographs) within thirty to forty years of abandonment. Species richness also recovered to levels observed in the oldest forest stands. Although there was rapid recovery, the comparison with reference sites (i.e., old forest stands) may be biased because these sites were probably left undisturbed due to their poor agricultural suitability (Chinea

2002), and these sites may have lower plant diversity than the original forests that were converted to pastures. However, given the relative constancy of species per area in Puerto Rico (approximately 60 species per ha; Lugo 2005), this bias may be minimal.

Chronosequences from the northern flank of the Luquillo Mountains illustrated possible successional series (Aide et al. 1996) (figure 4.3). At the highest elevations there are forests without any history of severe land use, which allowed comparison with the oldest abandoned pastures (Zimmerman

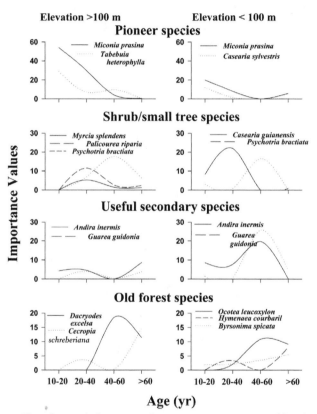

FIGURE 4.3. Hypothesized changes in forest vegetation suggested by chronosequences assembled on the northern flank of the Luquillo Mountains in northeastern Puerto Rico (Aide et al. 1996). Changes are divided into low and high elevations, corresponding closely to humid and wet subtropical forest (Ewel and Whitmore 1973), respectively. Note that the manner in which the data are presented suggests that abandoned pastures eventually recover the same species composition of the oldest, least-disturbed forests. Our research now indicates that species assemblages in abandoned agricultural lands may never return to their original species composition, or they may be very slow to return.

et al. 1995). *Miconia prasina* and *Tabebuia heterophylla* are important pioneers following pasture abandonment, as they are in the Cayey Mountains and the Cordillera Central (Pascarella et al. 2000; Marcano Vega et al. 2002). Once established, stands of *M. prasina* and *T. heterophylla* appear to accumulate species of shrubs and small trees in the understory, such as *Myrcia splendens, Palicourea riparia, Pychotria bractiata,* and *Casearia guinensis.* Elsewhere in Puerto Rico, abandoned pastures of intermediate age are often dominated by the alien *Spathodea companulata* (table 4.1), but as already noted, this species does not regenerate beneath itself (Aide et al. 2000; Lugo 2004). The forest vegetation of the oldest abandoned pastures, sixty to eighty years postabandonment, is often distinct from that of nearby areas of relatively undisturbed forest at similar elevations (Zimmerman et al. 1995; Aide et al. 1996; Pascarella et al. 2000). This suggests that the composition of abandoned pastures is slow to return to that of undisturbed, mature forest, if at all. Thus, in the example of the Luquillo Mountains, the reality may be very different from that implied by the chronosequences shown in figure 4.3—that old forest species will replace the secondary species found in forty- to sixty-year-old abandoned forests.

Mechanisms

The chronosequence studies suggested that there was often a lag of approximately ten years between the time a pasture was abandoned and the establishment of woody species (Aide et al. 1995). This was presumed to be because pastures first became overgrown with herbs, vines, and ferns, which inhibited the establishment of woody species. Zimmerman et al. (2000) attempted to confirm this in a recently abandoned pasture by studying the fate of seeds of eleven woody species important in pasture succession. Experimental removal of herbaceous vegetation sometimes retarded seed germination (in four species), enhanced the survivorship of only two species, and enhanced the growth of only one species (Zimmerman et al. 2000). This study did not, however, control for belowground competition, shown by Holl (1998) and others to be an important factor in abandoned pastures. A subsequent study by Netherton (2003) combined clipping and trenching treatments in a factorial experiment and looked at the growth of planted seedlings of three species of woody plants (including two studied by Zimmerman et al. 2000). Netherton (2003) showed that reducing both aboveground and belowground competition by herbaceous vegetation in pastures significantly affected the growth rate of all three species. Aboveground competition had a much stronger effect on seedling growth than belowground competition in all three species. Thus, there was some evidence that pasture vegetation inhibited the growth, but not the establishment, of tree seedlings.

Dispersal limitation is undoubtedly a key factor determining the rate of arrival and species composition of establishing woody vegetation in abandoned pastures (e.g., Aide and Cavelier 1994; Wunderle 1997; Holl 1998, 1999; Wijdeven and Kuzee 2000; Slocum 2001). In Puerto Rico, seed rain and abundance of seeds in the seed bank decline sharply from the edges of pastures (Zimmerman et al. 2000; Cubiña and Aide 2002) and seed additions to pasture plots increased the abundance of some woody species (Zimmerman et al. 2000). At a local level, one would expect that the composition of nearby forest stands would strongly impact the composition of the establishing woody vegetation (Gill and Marks 1991), but no one has conducted a similar study in Puerto Rico or elsewhere in the neotropics. At the scale of the municipality, no effect of distance to nearest forest patch (<50 vs. >50 m distance at time of abandonment) was found on either the species composition or the species richness of Luquillo pastures (Aide et al. 1996; cf. Chinea 2002). Chinea and Helmer (2003), analyzing islandwide forest inventory data, were able to detect an effect of distance to the largest forest patches on both species diversity and composition but noted that the effect was not large (i.e., a loss of ~0.25 species for each 10 km increase in distance). Given how rapidly seed dispersal declines with distance (Holl 1999; Zimmerman et al. 2000; Cubiña and Aide 2001), it is not surprising that landscape-level effects are undetectable or, when detected, small in magnitude.

Another factor that helps explain the lag in vegetation recovery after pasture abandonment is site degradation. It is possible that soil erosion and compaction and resultant low soil fertility, coupled with higher temperatures due to the absence of a forest canopy change conditions for germination and growth sufficiently to preclude the establishment of mature forest species. For example, Silander (1979) found reduced germination of the pioneer *Cecropia schreberiana* in open pastures where air temperature was higher and more variable and relative humidity lower than in canopy gaps. Gleason (unpublished data) found that the growth of *C. schreberiana* was improved greatly by fertilization with supplemental nutrients. Under extreme conditions of degradation, succession to forest vegetation can be arrested (Parrotta 1995).

Shade Coffee

The abandonment of shade coffee plantations usually occurs at the same time within a region, which prevents the development of chronosequences.

For example, many coffee plantations in the Luquillo Mountains were abandoned in 1928, following a severe hurricane (Scatena 1989; Zimmerman et al. 1995). Shade coffee was abandoned in the karst region in Ciales (figure 4.1) in the 1950s (Rivera and Aide 1998), likely spurred by the socioeconomic changes taking place throughout Puerto Rico at that time (Grau et al. 2003). Only in the Cordillera Central, where cultivation of shade coffee continues, were abandoned coffee stands of differing age identified for study (Marcano Vega et al. 2002). At the scale of the island, forest inventory studies conducted by the USDA Forest Service have been invaluable in showing general trends of forest succession following the abandonment of shade coffee (Weaver and Birdsey 1986). Marcano Vega et al. (2002) also documented the recovery forest vegetation in abandoned sun coffee, albeit in stands at a fairly young age (eight to twelve years).

SUCCESSIONAL TRENDS

In Puerto Rico, shade coffee is grown under canopies of *Inga vera* or *Erythrina poeppigiana*, and these species, along with persistent coffee shrubs, dominate the woody vegetation following abandonment (Marcano Vega et al. 2002). Within forty to sixty years, diversity and basal area of abandoned coffee stands are similar to those of abandoned pastures of similar age (Rivera and Aide 1998; Marcano Vega et al. 2002), and, where comparisons are available, (Zimmerman et al. 1995; Pascarella 2000) to relatively undisturbed forest. Popper et al. (1999) found that shade coffee forests in the coffee region of Utuado had large canopy trees and high density of understory vegetation with higher species richness than the canopy. The stands, however, become dominated by *Guarea guidonia* (Weaver and Birdsey 1986; Zimmerman et al. 1995; Rivera and Aide 1998; Grau et al. 2003) (table 4.1) and other native secondary species as the fast-growing shade trees mature and die out (Weaver and Birdsey 1986; Zimmerman et al. 1995). The increasing dominance of *G. guidonia* may result from its unique ability (Fernández 1997) to take advantage of high soil nutrients found in abandoned shade coffee plantations (Popper et al. 1999). The important pioneer species, *Cecropia schreberiana*, absent from even the oldest identified pastures, is abundant in abandoned coffee plantations of all ages (table 4.1). Other species common in abandoned shade coffee in Puerto Rico include *Dendropanax arboreus, Ocotea leucoxylon*, and in the karst area near Ciales (figure 4.1), *Montezuma speciossisima* (table 4.1). The predominance of native tree species in the understory of coffee shade forest was observed by Wadsworth and Birdsey (1983), who found that many of these regenerating species also had a high timber value.

TABLE 4.2

Twenty-two-year change in species composition and dominance of shade coffee forests in Puerto Rico

Parameter	1980	1990	2002
Number of tree species	101	87	62
Alien species (% of total)	19.8	20.7	19.3
Importance value of alien species	25.2	28.0	27.8

Source: Lugo and Brandeis 2005.

However, in spite of the abundance of native species in the regeneration of coffee shade forests, Lugo and Brandeis (2005) found that over a period of twenty-two years, the dominance of alien species in these forests was sustained (table 4.2) while the total number of tree species declined after abandonment.

MECHANISMS

Kalwani et al. (unpublished) recently studied the factors that might maintain the unique species composition of abandoned shade coffee. Working in a seventy-year-old abandoned shade coffee plantation in the Luquillo Mountains, they were able to map the edge of the plantation using aerial photographs from 1936. To assess the role of dispersal limitation in explaining the different species composition, they placed transects perpendicular to the land use border and looked for evidence that species dominating one side of the border or the other were invading the adjacent land use type. Examples of the data obtained (figure 4.4) indicate, with exceptions, that species were not invading the adjacent land use type (as would be evidenced by seedlings and saplings appearing in the adjacent land use type). In particular, the dominant species of undisturbed forest of that region and elevation, *Dacryodes excelsa* and *Manilkara bidentata*, have not established in the abandoned coffee plantation, even though adult trees of the species have existed within tens of meters of the area for the last seventy years. Thus, in this case, dispersal limitation does not explain the inability of these forests to recover a species composition similar to the mature forest.

Kalwani et al. (unpublished) then considered soil factors. They began with the assumption that the history of growth of nitrogen-fixing trees had permanently altered the nitrogen economy of the soils, but soon discovered that liming (indicated by limestone gravel found in portions of the plantation) was an added factor. There was evidence of this in the lowered C:N ratios of the soils in the abandoned shade coffee and in changes in the relative

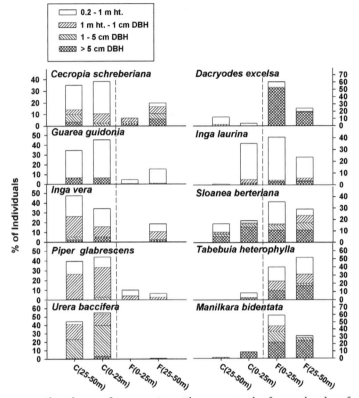

FIGURE 4.4. Abundance of tree species with respect to the former border of an abandoned coffee plantation in the Luquillo Mountains, Puerto Rico. Abundances are shown by size class in the abandoned coffee plantation (C) and the adjacent forest (F) at two distances from the border. From Kalwani et al. (unpublished data).

amounts of ammonium and nitrate in the surface soils, consistent with soils occurring under a nitrogen-fixing canopy (Erickson et al. 2001). Soils from the abandoned coffee had higher levels of both N mineralization and N nitrification. Abandoned coffee soils also had higher pH (5.4 vs. 4.9) and greater Mg and Ca saturation, indicative of liming. Leaves of plant species spanning the two land use types had consistently higher N and P contents, indicating that the higher fertility of the soils in the abandoned shade coffee translated into higher plant fertility. Kalwani et al. (unpublished) argued that the more fertile soils of the abandoned shade coffee promoted a community of species able to take advantage of higher nutrient availability and outcompete species common in undisturbed forests, like *Dacryodes excelsa* and *Manilkara bidentata*. The results suggest that abandoned shade coffee represents an alternative stable state (now increasingly called a "regime shift" to indicate that ecosystems continue to be dynamic but fluctuate among a completely different set of conditions) (Scheffer and Carpenter 2003) maintained by changes in soil fertility (Dupouey et al. 2002).

Sugarcane

Historically, the coastal plains of Puerto Rico were covered with forest (Wadsworth 1950). During the eighteenth and nineteenth centuries, these forests were cleared for agriculture, mainly sugarcane (Wadsworth 1950; Picó 1986). The sugarcane industry continued to expand during the beginning of the twentieth century, but during the 1940s and 1950s a change in government policy, which encouraged small industry, led to the decline of the sugar industry and the abandonment of many plantations (Dietz 1987). Different from the abandoned pastures or coffee plantations, forest recovery has been rare in abandoned sugarcane plantations. Today most of the abandoned cane fields have become urban areas, cattle pastures (Thomlinson et al 1996; López et al 2000), or are dominated by herbaceous vegetation. An exception to this pattern has been the rapid recovery of *Spathodea campanulata* in abandoned alluvial soils free of saltwater intrusion.

This variation in recovery patterns is due to the interaction between the soil and hydrological conditions and management practices. When wetland coastal forests were converted to sugarcane, canals were constructed and pumps were used to lower the water level. In addition to removing surface water, the canals dried the substantial peat layer produced by the original forest. When fire was used during the harvesting process, in addition to burning the sugarcane it also burned the peat and effectively lowered the soil level. Once the sugar plantation, canals, and pumps were abandoned, the reduction of the peat layer led to longer periods of flooded conditions, which favored the colonization of herbaceous vegetation, specifically *Typha domingensis*. These conditions of extended flooding and a high density of herbaceous vegetation are inhibiting forest recovery. In addition, limited seed sources exacerbate the problem.

In the Sabana Seca region of Puerto Rico, most of the coastal forest that was converted to sugarcane was dominated by *Pterocarpus officinalis*. Today, the remnant stands of *P. officinalis* occur at the extreme of their salinity tolerance near mangroves. At this extreme, the forest stands have lower productivity and recruitment, and higher mortality than areas of lower salinity (Eusse and Aide 1999; Rivera Ocasio et al., forthcoming). Furthermore, if salinity levels increase more with the threat of sea level rise, these remnant forests could be doomed. Although the adjacent sugarcane fields have been abandoned for more than fifty years, the herbaceous vegetation (e.g., *Typha domingensis*) appears to be resistant to invasion by forest trees, as attested to by the stability of the forest boundaries seen in the 1937 and 1995 aerial photographs (figure 4.5). In this case, natural regeneration has not been effective, and intervention will be necessary to restore a forest habitat.

A very different situation is found in abandoned sugarcane plantations occurring on alluvial soils that are not subject to salinity intrusion. In these sites,

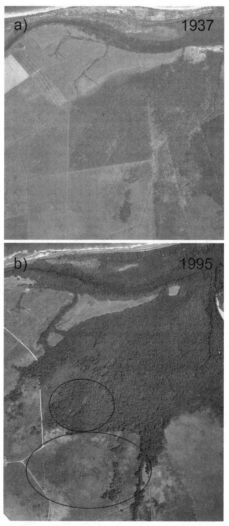

FIGURE 4.5. Aerial photographs of Sabana Seca from (a) 1937 and (b) 1995. The remnant forested wetland dominated by *Pterocarpus officinalis* and the restoration site are outlined in (b).

forest recovery has been rapid, and the secondary forests are often monodominant stands of the alien *Spathodea campanulata*. These forests rapidly develop high basal areas and produce nutrient-rich litter (Lugo 2004). The rapid structural and nutrient recovery facilitates the recovery of the fauna (e.g., healthy earthworm populations) and flora (e.g., diverse understory community). In these conditions, *S. campanulata* is very successful at outcompeting the herbaceous vegetation, and it creates an understory environment that is more appropriate for the colonization of other species.

Implications for Restoration

In order to understand the restoration implications of vegetation dynamics following farmland abandonment, one needs to first define why the restoration is needed. Narrowly defined (Brown and Lugo 1994), restoration signifies replacing vegetation and associated fauna as it appeared before human disturbance altered the landscape. Where the goal is to increase populations of rare or endangered plant species or habitat types, or to provide habitat for taxa that have specialized associations with particular plant species or habitats, this would be an obvious goal. However, in many cases the establishment of any type of vegetation, no matter its composition, may suffice to reestablish critical ecosystem services. Secondary forests of any composition provide habitat for most wildlife, and they provide complete requirements for watershed protection and other beneficial ecosystem characteristics that accrue to any forested land (Brown and Lugo 1990, 1994; Lugo 2002). Particularly where the land is severely degraded, as it sometimes is following agricultural abandonment, this may be the least costly and most practical and achievable goal.

As we have concluded elsewhere (Aide et al. 2000; Zimmerman et al. 2000; Lugo 2002; Lugo and Helmer 2004), the Puerto Rican experience with pasture succession strongly suggests that passive restoration is the most cost-effective strategy for reestablishing forest cover. In thirty to forty years these secondary forests have a structure (i.e., biomass and density) and plant richness similar to mature forest. If restoration activities are to be considered, we suggest, depending on initial conditions and goals, accelerating the initial phase of forest establishment, enrichment planting with mature forest species, or some combination of both.

Eliminating competition with herbaceous vegetation by establishing fast-growing tree species, which will shade out the herbaceous vegetation, will often accelerate the initial phase of forest establishment in abandoned pastures. When restoring forests to pastures where succession is arrested, in particular, it is necessary to plant species of trees capable of surviving in degraded lands

(Lugo 1997). For example, the quick establishment of forest cover in moist and wet forests could be done in Puerto Rico by planting wildings of the native species *Tabebuia heterophylla* (Wadsworth 1943) or *Calophyllum brasilense* (Holdridge 1940; Marrero 1947, 1950). Nitrogen-fixing species, like *Inga fagifolia*, can be used to improve nutrient capital. In the case of Puerto Rico, some of these choices for using native species to promote forest regeneration in abandoned pastures were the result of field trials with many species (Marrero 1947, 1950). Martínez-Garza et al. (2005), working in southern Mexico, have recently identified some simple leaf characteristics that may help identify late-successional species that rapidly establish in areas of abandoned agriculture.

Alien species may aide the initial establishment of forest cover in cases where the land is badly degraded or a rapid establishment of vegetation is desired. Ewel and Putz (2004) recently offered a general guide for the use of alien species in forest restoration. *Spathodea campanulata*, because it does not establish beneath its own canopy, is an obvious candidate for restoration of abandoned pastures in Puerto Rico. In pastures on infertile, degraded soils, which have remained treeless, establishment of an initial plantation of the alien, nitrogen-fixing *Albizia lebbek* can promote the establishment of many native species. This occurs once an overstory has been established to shade out the pasture grasses (Parrotta 1995). Like *S. campanulata*, *A. lebbek* does not establish in its understory. Silver et al. (2004) provide additional documentation of the success of this type of restoration in the Luquillo Experimental Forest.

Without intervention, abandoned pastures appear to maintain a distinct species composition long after abandonment. Where aesthetic reasons or specialized wildlife needs dictate a native species forest composition in a long-abandoned pasture, it will be necessary to outplant saplings in order to improve the mix of native species in these forest stands.

Like abandoned pastures, abandoned shade coffee quickly establishes a forest of similar structure and diversity to undisturbed forest. However, succession and carbon accumulation is faster in abandoned shade coffee than in pastures (Silver et al. 2000). Thus, passive restoration is an effective management tool in humid and wet environments where coffee is grown. However, like the abandoned pastures, the forest composition is often quite distinct from native forest and, where desired, supplemental planting of native species will be necessary. In view of the high soil fertility in abandoned coffee plantations in the Luquillo Mountains and in Utuado (Kalwani et al., unpublished; Popper et al. 1999), research is needed to show the generality of these results, as well as the implications for the success of any supplemental plantings of native species, which may not be able to compete in fertile soils with the established vegetation.

In the case of abandoned sugarcane in a former *Pterocarpus officinalis* swamp, a much greater level of intervention is usually necessary to restore forest cover. *Pterocarpus officinalis* stands exhibit hummock and swale topography with most trees clumped at the hummocks. Tree regeneration generally takes place on the hummocks. In the case of Sabana Seca, to restore this critical feature of the forest, a bulldozer was employed to remove the *Typha* and to build up a series of 3 × 4 m mounds in an 18-ha area. Mounds were planted with either *P. officinalis* or *Annona glabra* (figure 4.6). Each mound

FIGURE 4.6. Results of efforts to restore forested wetland at Sabana Seca, Puerto Rico, showing artificially created mounds covered with barrier plastic and planted with *Pterocarpus officinalis* and *Annona glabra*.

was covered with ground cover to inhibit the resprouting of *Typha*, and then planted with at least three seedlings (20 to 40 cm tall). Creating mounds and covering them with ground cover increased the time that roots experienced aerobic conditions and reduced the competition with *Typha*, which resulted in high survivorship and growth. Three years after the mounds were created, they remained, on average, approximately 35 cm above the soil surface and provided a site of aerobic soil during periods of high water. The ground cover was removed from all mounds by three years later; at this time the overall survivorship was 82%, and the average tree height was 382 cm (figure 4.6). Although the ground cover has been removed from the mounds, *Typha* has not recolonized because the shade of trees now inhibits its growth. Thus, the intervention was largely successful and, because the forest is composed of a very few dominant species, forest composition was restored at the same time.

Alluvial soils without the threat of saltwater intrusion present a more favorable situation for restoration of forest in abandoned sugarcane. Under this scenario, restoration activities could focus on reducing the density of naturally established *Spathodea campanulata* (i.e., stand thinning) as a mechanism to select for desired species. Alternatively, these stands can be allowed to mature and develop into new forest communities, as described by Lugo and Helmer (2004). Because historical deforestation in Puerto Rico cleared alluvial forests before scientists could study them, we lack information on the species composition and structural complexity of these forests. The fast biomass accumulation in *Spathodea* stands reflects the high productivity of alluvial soils and offers an opportunity to experiment with the restoration of native species in these locations.

Conclusion

The study of forest recovery in Puerto Rico offers some simple lessons for understanding the potential and needs of tropical forest restoration. The studies of abandoned pastures and coffee plantations demonstrate that under moist to wet conditions in Puerto Rico these land use practices did not lead to an irrevocable loss of the ability of tropical forests to recover biomass, diversity, and related ecosystem services (Gómez-Pompa et al. 1972; Brown and Lugo 1994). The rapid recovery of forests following the abandonment of these land uses suggests that a restoration approach of "no intervention" is appropriate, particularly given the low costs. A potential disadvantage of this approach is that these land use practices leave a strong legacy that is most clearly reflected in the species composition of the recovering forests (Zimmerman et al. 1995; Grau et al. 2003), which may represent alternative stable states/regime change. But, if the major objective of these forests is to offer ecosystem services (e.g., soil protection, nutrient retention), these recovering forests can be as effective as mature forests (Lugo and Helmer 2004). In other cases, the added expense of intervention is worth the costs. For example, planting fast-growing trees can greatly accelerate the recovery process (Parrotta 1995). If soil or hydrology conditions have been altered, alternative, unwanted communities may be dominant, and intervention will be required to restore native forest structure and function.

Acknowledgments

The authors thank Karen Holl and an anonymous reviewer for comments on a previous draft of the manuscript. Thanks to the U.S. Geological Survey for providing the physiography of Puerto Rico used in figure 4.1. Research described in this chapter was supported by an Institutional Research Award from the National Aeronautics and Space Administration, by funds provided to the University of Puerto Rico's Center for Excellence in Science and Technology; by the National Science Foundation (NSF, HRD-9353546); and by funds provided by NSF to the Luquillo Long-Term Ecological Research Program (DEB-9411973, 9705814, 00805238, and 0218039), which is also supported by the University of Puerto Rico and the USDA Forest Service's International Institute of Tropical Forestry. This chapter was developed while Jess Zimmerman was serving at the National Science Foundation. Any opinions, findings, and conclusions or recommendations expressed are those of the authors and do not necessarily reflect the views of the National Science Foundation.

REFERENCES

Aide, T. M. 2000. Clues for tropical forest restoration. *Restoration Ecology* 8:327.

Aide, T. M., and J. Cavelier. 1994. Barriers to lowland tropical forest restoration in the Sierra Nevada de Santa Marta, Colombia. *Restoration Ecology* 2:219–29.

Aide, T. M., and H. R. Grau. 2004. Globalization, migration, and Latin American ecosystems. *Science* 305:1915–16.

Aide, T. M., J. K. Zimmerman, L. Herrera, M. Rosario, and M. Serrano. 1995. Forest recovery in abandoned tropical pastures in Puerto Rico. *Forest Ecology and Management* 77:77–86.

Aide, T. M., J. K. Zimmerman, J. Pascarella, L. Rivera, and H. Marcano-Vega. 2000. Forest regeneration in a chronosequence of tropical abandoned pastures: Implications for restoration ecology. *Restoration Ecology* 8:328–38.

Aide, T. M., J. K. Zimmerman, M. Rosario, and H. Marcano. 1996. Forest recovery in abandoned cattle pastures along an elevational gradient in northeastern Puerto Rico. *Biotropica* 18:537–48.

Birdsey, R. A., and P. L. Weaver 1982. *The forest resources of Puerto Rico*. Resource Bulletin SO-85, USDA Forest Service, Southern Forest Experiment Station, New Orleans, LA.

Bradshaw, A. D. 1987. Restoration: An acid test for ecology. In *Restoration ecology*, ed. W. R. Jordan, M. Gilpin, and J. D. Aber, 23–29. Cambridge University Press, Cambridge, UK.

Brown, K. A., F. N. Scatena, and J. Gurevitch. 2004. Effects of an invasive tree on community structure and diversity in a tropical forest in Puerto Rico. *Forest Ecology and Management* 226:145–52.

Brown, S., and A. E. Lugo. 1990. Tropical secondary forests. *Journal of Tropical Ecology* 6:1–32.

———. 1994. Rehabilitation of tropical lands: A key to sustaining development. *Restoration Ecology* 2:97–111.

Chang, C. Y., and B. W. Tsai. 2002. *Land use/cover change in Taiwan*. In China (Taipei), IGBP Committee, Academia Sinica; National Science Council, Executive Yuan; National Taiwan University; National Central University; and National Sun Yat-sen University. Global Change Research in Taiwan: The China-Taipei IGBP National Report.

Chinea, J. D. 2002. Tropical forest succession on abandoned farms in the Humacao municipality of eastern Puerto Rico. *Forest Ecology and Management* 167:195–207.

Chinea, J. D., and E. H. Helmer. 2003. Diversity and composition of tropical secondary forests recovering from large-scale clearing: Results from the 1990 inventory in Puerto Rico. *Forest Ecology and Management* 180:227–40.

Connell, J. H., and R. O. Slatyer. 1977. Mechanisms of succession in natural communities and their role in community stability and organization. *American Naturalist* 111:1119–44.

Cubiña, A., and T. M. Aide. 2001. The effect of distance from forest edge on seed rain and soil seed bank in a tropical pasture. *Biotropica* 33:260–67.

Dietz, J. L. 1987. *Economic history of Puerto Rico*. Princeton University Press, Princeton, NJ.

Dupouey, J. L., E. Dambrine, J. D. Laffite, and C. Moares. 2002. Irreversible impact of past land use on forest soils and biodiversity. *Ecology* 83:2978–84.

Drury, W. H., and I. C. T. Nisbet. 1973. Succession. *Journal of the Arnold Arboretum* 54:331–68.

Erickson, H., M. Keller, and E. A. Davidson. 2001. Nitrogen oxide fluxes and nitrogen cycling during postagricultural succession and forest fertilization in the humid tropics. *Ecosystems* 4:67–84.

Eusse, A. M., and T. M. Aide. 1999. Patterns of litter production across a salinity gradient in a *Pterocarpus officinalis* tropical wetland. *Plant Ecology* 145:307–15.

Ewel, J. J., and F. E. Putz. 2004. A place for alien species in ecosystem restoration. *Frontiers in Ecology and Environment* 2:354–60.

Ewel, J. J., and J. L. Whitmore. 1973. *The ecological life zones of Puerto Rico and the U.S. Virgin Islands*. Research Papers ITF-18. USDA Forest Service, Southern Forest Experiment Station, New Orleans, LA.

Fernández del Viso, D. S. 1997. Contrasting light environments and response flexibility of trees in the Luquillo Mountains of Puerto Rico. PhD dissertation. University of Puerto Rico, Rio Piedras, PR.

Franco, P. A., P. L. Weaver, and S. Eggen McIntosh. 1997. *Forest resources of Puerto Rico, 1990*. Resource Bulletin SRS-22. USDA Forest Service, Southern Research Station, Asheville, NC.

Gill, D. S., and P. L. Marks. Tree and shrub seedling colonization of old fields in central New York. *Ecological Monographs* 61:183–205.

Gómez-Pompa, A., C. Vázquez-Yanes, and S. Guevara. 1972. The tropical rain forests: A nonrenewable resource. *Science* 177:762–65.

Gómez, L., and M. Ballesteros. 1980. *Vida y Cultura Precolombina de Puerto Rico.* Editorial Cultural, Río Piedras, PR.

Grau, H. R., T. M. Aide, J. K. Zimmerman, J. R. Thomlinson, E. Helmer, and X. Zou. 2003. The ecological consequences of socioeconomic and land-use changes in post-agriculture Puerto Rico. *Bioscience* 53:1159–68.

Hall, C. A. S. 2000. *Quantifying sustainable development: The future of tropical economies.* Academic Press, San Diego, CA.

Heartsill Scalley, T., and T. M. Aide. 2003. Riparian vegetation and stream condition in a tropical agriculture-secondary forest mosaic. *Ecological Applications* 13:225–34.

Helmer, E. 2004. Forest conservation and land development in Puerto Rico. *Landscape Ecology* 19:29–40.

Holdridge, L. R. 1940. *Calophyllum antillanum*, a desirable tree for difficult planting sites. *Caribbean Forester* 1:27–28.

Holl, K. D. 1998. Do bird perching structures elevate seed rain and seedling establishment in abandoned tropical pasture? *Restoration Ecology* 6:253–61.

———. 1999. Factors limiting tropical rain forest regeneration in abandoned pasture: Seed rain, seed germination, microclimate, and soil. *Biotropica* 31:229–42.

Little, E. L., R. O. Woodbury, and F. H. Wadsworth. 1974. *Trees of Puerto Rico and the Virgin Islands.* Vol. 2. USDA Forest Service Agricultural Handbook 449. Washington, DC.

López, T. M., T. M. Aide, and J. R. Thomlinson. 2001. Urban expansion and the loss of prime agricultural lands in Puerto Rico. *Ambio* 49–54.

Lugo, A. E. 1988. Estimating reductions in the diversity of tropical forest species. In *Biodiversity*, ed. E. O. Wilson and F. M. Peter, 58–70. National Academy Press, Washington, DC.

———. 1997. The apparent paradox of re-establishing species richness on degraded lands with tree monocultures. *Forest Ecology and Management* 99:9–19.

———. 2002. Can we manage tropical landscapes? An answer from the Caribbean perspective. *Landscape Ecology* 17:601–15.

———. 2004. The outcome of alien tree invasions in Puerto Rico. *Frontiers in Ecology and the Environment* 2:265–73.

———. 2005. Los bosques. In *Biodiversidad de Puerto Rico: Vertebrados terrestres y ecosistemas*, ed. R. L. Joglar, 395–548. Editorial del Instituto de Cultura Puertorriqueña, San Juan, PR.

Lugo, A. E., and T. J. Brandeis. 2005. A new mix of alien and native species coexist in Puerto Rico's landscapes. In *Biotic interactions in the tropics: Their role in the maintenance of species diversity*, ed. D. F. R. P. Burslem, M. A. Pinard, and S. E. Hartley, 484–509. Cambridge University Press, Cambridge, UK.

Lugo, A. E., and E. Helmer. 2004. Emerging forests on abandoned land: Puerto Rico's new forests. *Forest Ecology and Management* 190:145–61.

Lugo, A. E., O. Ramos, S. Molina, F. N. Scatena, and L. L. Vélez-Rodríquez. 1996. A *fifty-three year record of land-use change in the Guanica forest biosphere reserve and its vicinity.* International Institute of Tropical Forestry, USDA Forest Service, Rio Piedras, PR.

MacDonald, I. A., W. L. Ortiz, J. E. Lawesson, and J. B Nowak. 1988. The invasion of highlands in Galapagos by the red quinine tree *Cinchona-succirubra*. *Environmental Conservation* 15:215–20.

Marcano Vega, H., T. M. Aide, and D. Baez. 2002. Forest regeneration in abandoned coffee plantations and pastures in the Cordillera Central of Puerto Rico. *Plant Ecology* 161:75–87.

Marrero, J. 1947. A survey of the forest plantations in the Caribbean National Forest. MS thesis. University of Michigan, Ann Arbor.

———. 1950. Results of forest planting in the insular forests of Puerto Rico. *Caribbean Forester* 11:107–14.

Martínez-Garza, C., V. Peña, M. Ricker, A. Campos, and H. Howe. 2005. Restoring tropical biodiversity: Leaf traits predict growth and survival of late-succesional trees in early successional environments. *Forest Ecology and Management* 217:365–79.

Meli, P. 2003. Tropical forest restoration, twenty years of academic research. *Interciencia* 28:581–89.

Meyer J. Y., J. Florence, and V. Tchung. 2003. The endemic *Psychotria* (Rubiaceae) of Tahiti (French Polynesia) threatened by the invasive *Miconia calvescens* (Melastomataceae): Status, distribution, ecology, phenology, and conservation. *Revue D Ecologie-La Terre et la Vie* 58:161–85.

Molina Cólon, S., and A. E. Lugo. 2006. Recovery of a subtropical dry forest after abandonment of different land uses. *Biotropica* 38:354–64.

Nepstad, D. C., C. Uhl, and E. A. S. Serrao. 1990. Surmounting barriers to forest regeneration in abandoned, highly degraded pastures: A case study from Paragominas, Para, Brazil. In *Alternatives to deforestation: Steps toward sustainable use of the Amazon rain forest*, ed. A. B. Anderson, 215–29. Columbia University Press, New York.

Nepstad, D. C., C. Uhl, C. A. Pereira, and J. M. DaSilva. 1996. A comparative study of tree establishment in abandoned pasture and mature forest of Eastern Amazonia. *Oikos* 76:25–39.

Netherton, J. A. 2003. The effects of above- and belowground competition on the growth of *Guarea guidonia* (L.) Sleumer, *Inga laurina* (Sw.) Willd., and *Tabebuia heterophylla* (DC.) Britton seedlings in shrub and herb patches in an abandoned pasture in Puerto Rico. MS thesis. University of Kentucky, Lexington.

Parrotta, J. A. 1995. Influence of overstory composition on understory colonization by native species in plantations on a degraded tropical site, *Journal of Vegetation Science* 6:627–36.

Pascarella, J. B., T. M. Aide, M. I. Serrano, and J. K. Zimmerman. 2000. Land-use history and forest regeneration in the Cayey Mountains, Puerto Rico. *Ecosystems* 3:217–28.

Pickett, S. T. A., S. L. Collins, and J. J. Armesto. 1987. Models, mechanisms, and pathways of succession. *Botanical Review* 53:335–71.

Picó, F. 1986. *Historia General de Puerto Rico*. Ediciones Huracán, Río Piedras, PR.

Popper, N., C. Domínguez Cristóbal, A. Santos, N. Méndez Irizarry, E. Torres Morales, A. E. Lugo, Z. Z. Rivera Lugo et al. 1999. A comparison of two secondary forests in the coffee zone of central Puerto Rico. *Acta Científica* 13:27–41.

Population Division of the Department of Economic and Social Affairs of the United Nations Secretariat. 2002. *World urbanization prospects: The 2001 revision*. United Nations, New York.

Reagan, D. P., and R. B. Waide. 1996. *The food web of a tropical rain forest*. University of Chicago Press, Chicago.

Primack, R., and R. Corlett. 2005. *Tropical rainforests: An ecological and biogeographical comparison*. Blackwell Publishing, Oxford, UK.

Rivera, L. W., and T. M. Aide. 1998. Forest recovery in the karst region of Puerto Rico. *Forest Ecology and Management* 108:63–75.

Rivera-Ocasio, E. T., T. M. Aide, and N. Rios-López. Forthcoming. The effects of salinity on the dynamics of a *Pterocarpus officinalis* forest stand in Puerto Rico. *Journal of Tropical Ecology.*

Roberts, R. C. 1942. *Soil Survey of Puerto Rico.* U.S. Printing Office, Washington, DC.

Rudel, T. K., D. Bates, R. Machinguiashi. 2002. A tropical forest transition? Agricultural change, out-migration, and secondary forests in the Ecuadorian Amazon. *Annals of the Association of American Geographers* 92:87–102.

Rudel, T. K., M. Perez-Lugo, and H. Zichal. 2000. When fields revert to forest: Development and spontaneous reforestation in post-war Puerto Rico. *Professional Geographer* 52:86–397.

Scatena, F. N. 1989. *An introduction to the physiography and history of the Bisley experimental watersheds in the Luquillo Mountains of Puerto Rico.* General Technical Report SO-72. USDA Forest Service, Southern Forest Experiment Station, New Orleans, LA.

Scheffer, M., and S. R. Carpenter. 2003. Catastrophic regime shifts in ecosystems: Linking theory to observation. *Trends in Ecology and Evolution* 18:648–56.

Silander, S. R. 1979. A study of the ecological life history of *Cecropia peltata* L., an early secondary successional species in the rain forest of Puerto Rico. MS thesis. University of Tennessee, Knoxville.

Silver, W. L., L. M. Kueppers, A. E. Lugo, R. Ostertag, and V. Matzek. 2004. Carbon sequestration and plant community dynamics following reforestation of tropical pasture. *Ecological Applications* 14:1115–27.

Silver, W. L., R. Ostertag, and A. E. Lugo. 2000. The potential for carbon sequestration through reforestation of abandoned tropical agricultural and pasture lands. *Restoration Ecology* 8:394–407.

Slocum, M. G. 2001. How tree species differ as recruitment foci in a tropical pasture. *Ecology* 82:2547–59.

Thomlinson, J. R., and L. Y. Rivera. 2000. Suburban growth in Luquillo, Puerto Rico: Some consequences of development on natural and semi-natural systems. *Landscape and Urban Planning* 49:15–23.

Thomlinson, J. R., M. I. Serrano, T. López, T. M. Aide, and J. K. Zimmerman. 1996. Land-use dynamics in a post-agricultural Puerto Rican landscape (1936–1988). *Biotropica* 18:525–36.

Tugwell, G. R. 1946. *This stricken land: The story of Puerto Rico.* Greenland Press, New York. Repr. 1968.

Vitousek, P. M., and L. R. Walker. 1989. Biological invasion by *Myrica faya* in Hawaii: Plant demography, nitrogen fixation, ecosystem effects. *Ecological Monographs* 59:47–265.

Wadsworth, F. H. 1943. Roble, a valuable forest tree in Puerto Rico. *Caribbean Forester* 4:59–76.

———. 1950. Notes on the climax forests of Puerto Rico and their destruction and conservation prior to 1900. *Caribbean Forester* 11:38–37.

Wadsworth, F. H., and R. A. Birdsey. 1983. Un nuevo enfoque de los bosques de Puerto Rico. In *Puerto Rico Department of Natural Resources 9th symposium on natural resources,* 12–27. Puerto Rico Department of Natural Resources, San Juan, PR.

Weaver, P. L., and R. A. Birdsey. 1986. Tree succession and management opportunities in coffee shade stands. *Turrialba* 36:47–58.

Wijdeven, S. M. J., and M. E. Kuzee. 2000. Seed availability as a limiting factor in forest recovery processes in Costa Rica. *Restoration Ecology* 8:414–24.

Wunderle, J. M. 1997. The role of animal seed dispersal in accelerating native forest regeneration on degraded tropical lands. *Forest Ecology and Management* 99:223–35.

Zimmerman, J. K., T. M. Aide, M. Rosario, M. Serrano, and L. Herrera. 1995. Effects of land use management and a recent hurricane on forest structure and composition in the Luquillo Experimental Forest, Puerto Rico. *Forest Ecology and Management* 77:65–76.

Zimmerman, J. K., J. B. Pascarella, and T. M. Aide. 2000. Barriers to forest regeneration in an abandoned pasture in Puerto Rico. *Restoration Ecology* 8:350–60.

Chapter 5

Processes Affecting Succession in Old Fields of Brazilian Amazonia

GISLENE GANADE

The Brazilian Amazon forest covers an area of 5.8 million km^2 and hosts one of the highest diversity of plants and animals in the world. However, approximately 15% of the tropical rain forest has been cleared since the 1970s. Clearings occurred due to agriculture, road implementation, and urban expansion, but the most widespread type of commercial land use in the Brazilian Amazon is the conversion of tropical forest to pasture for cattle ranching (Fearnside 1990). The establishment of these pasturelands was the result of heavy fiscal incentives from the Brazilian government that promoted increases of land price after forest cover was removed. This governmental procedure enhanced profitability of cattle ranching in the Amazon (Arima et al. 2005). However, it also induced some landowners to convert tropical forest to pastures for the purposes of land speculation (Fearnside 1990).

Pastures in this region became continuously threatened by decreases in productivity due to soil compaction and erosion, declining levels of phosphorus in the soil, and invasion by plants that are inedible to cattle (Dias-Filho 2005). Therefore, the agricultural sustainability of these pastures could be maintained only by means of heavy and sometimes uneconomic input (Fearnside 1990). As a result, many of these pastures were abandoned without being used for more than ten years (Buschbacher 1988). One recent estimate of the degree of abandonment revealed that as much as 35.8% of cleared land in central Amazonia has been neglected and remains in different stages of old field succession (Lucas et al. 2002).

Barriers to forest restoration in degraded pastures have been extensively studied in the neotropics (Nepstad et al. 1991; Aide and Cavelier 1994; Miriti 1998; Holl et al. 2000). The most important barriers are low propagule availability, seed and seedling predation, and competition with pasture grasses (Nepstad et al. 1991; Holl 1999). In the Brazilian Amazon, however, the rate of recovery of these old fields to a forest habitat is inversely related to the intensity of pasture use prior to its abandonment. In eastern Amazonia, highly degraded pastures that were frequently slashed and burned have a very low chance of being naturally reconverted to forest (Uhl et al. 1988). In central Amazon, on the other hand, most pastures were younger and active for only a few years, after which they were completely or partially abandoned. These lightly used pastures, dominated by the exotic African grasses of the genus *Brachiaria*, can be totally invaded by pioneer tree species after six years of abandonment (Mesquita et al. 2001; Ganade 2001). This process has been creating the secondary forests that have become a common landscape feature from western to eastern Amazonia (Roberts et al. 2003).

Amazon old fields can accumulate woody species at a relatively fast rate and are recognized as important carbon sinks (Mesquita et al. 2001; Houghton 2000). In central Amazonia, old field areas can sustain up to 50% of the aboveground biomass of primary forest fourteen years after abandonment (Feldpausch et al. 2004). Similar patterns were recorded in western Amazonia, where eighteen-year-old secondary forests that had developed on lightly used pastures achieved up to 60% of the aboveground biomass found in the primary forest (Alves et al. 1997). However, areas in eastern Amazonia that were used as pastures for longer periods of time and were more intensively managed may take more than forty years to acquire such biomass levels (Vieira et al. 2003). Saldarriaga and Uhl (1991) estimate that it may take about 240 years for areas abandoned from slash-and-burn agriculture to gain the biomass standards of primary forests. Processes that influence tree species composition in secondary forest areas are, nevertheless, only beginning to be understood (Mesquita et al. 2001; Ganade 2001).

Although forest regeneration in old fields improves a number of ecosystem functions such us carbon uptake, soil fertility, and hydrological cycle, most of these areas are not economically viable and are frequently cleared by landowners (Fearnside and Guimarñes 1996; Vieira et al. 2003). As the pathways and processes that modulate the rates of old field conversion to primary rain forest are unveiled, restoration programs that promote fast rates of recovery and economical sustainability become possible (Ganade 2001). In this chapter I review the processes that may influence forest recovery in old fields of Brazilian Amazon and discuss possible techniques and policies to improve restoration practice and provide incentives for forest restoration programs.

Seed Dispersal Limiting Old Field Regeneration

Seed arrival is an essential process for restoring disturbed forests to structured primary forest. To colonize new areas, tropical trees depend on a variety of dispersal agents such as wind, water, and gravity. However, 92% of the neotropical tree species are dispersed by animals (Howe and Smallwood 1982). Therefore, the restoration of abandoned pasture to tropical forest may depend on the ability of animals to bring seeds to this new type of habitat. Silva and colleagues (1994) recorded low activity of frugivorous bird species in areas of abandoned pastures that are distant from a forest source. Additionally, only a small proportion of forest understory, fruit-eating birds were recorded visiting secondary forest areas in central Amazonia (Stouffer and Borges 2001). These studies indicate that when large areas of forest are clearcut, potential seed dispersers tend to avoid these disturbed areas. Thus seed dispersal may become one of the main factors restricting forest expansion into abandoned pastures (Janzen and Vásquez-Yanes 1991).

Distance from a primary forest can strongly influence the invasion of new plant species into old fields because forests are sources of propagules. However, a variety of works undertaken in tropical areas have shown that seed rain decreases substantially in the first 5 to 15 meters from the forest edge into the pasture (Aide and Cavelier 1994; Holl 1999; Cubiña and Aide 2001). Rates of seed arrival are, nevertheless, dependent on the dispersal strategies of each plant species. Pioneer tree species normally produce fruits with hundreds of seeds that are typically dispersed by bats and habitat generalist birds, and they can be spread rapidly over long distances within abandoned fields (Nepstad et al. 1991). Primary forest tree species, which normally produce a smaller number of larger seeds, depend on large vertebrates for dispersion and are expected to arrive in isolated sites at a much lower rate than in sites near forest sources.

To investigate how propagule invasion in secondary forest areas can be modified by proximity to a forest edge, I studied patterns of seed rain in secondary forest areas at various distances from a tropical rain forest source in central Amazon, Brazil. Seed rain was studied over a fourteen-month period by collecting seeds in 1 m^2 seed traps placed in a primary forest and an adjacent secondary forest at distances of 5, 25, 50, and 100 m from the edge within both vegetation types. Surveys of seeds were conducted at approximately two-week intervals for the first six months. Arriving seeds were then left to accumulate in the traps for eight months until March 1995, when all seeds, fruits, and germinated seedlings that were present inside the traps were collected. All seeds collected during the surveys were counted and

morphotyped with the help of local taxonomic specialists. In order to estimate the possible variation in the size of seeds that fall in pristine and secondary forest sites, seed dimensions were classified according to six size categories: 0–2, 2–4, 4–8, 8–16, 16–32, and 32–64 mm. Each seed was then
deemed to possess a size that is equal to the midpoint of its size category. The
total number of seeds, the total seed richness, and the mean seed size were
calculated for each trap using the data accumulated from the surveys.

I recorded ninety-six seed morphotypes of which 91% were found in the
pristine forest. The vast majority of seeds collected in the secondary forest
were the small and numerous ones belonging to the fruits of *Vismia* (Clusiaceae) and *Bellucia* (Melastomataceae) pioneer trees. On the other hand, a
wider variety of seed sizes and morphotypes were sampled in the forest inte-
rior at distances of 25, 50, and 100 m from the edge (figure 5.1). These results
show that even though seeds of late-successional trees were available in the
forest, they were not able to arrive in the secondary forest at distances as close
as 5 meters to the forest source. This narrow penetration of primary forest species into secondary forest sites is mostly due to restrictions in seed dispersal.
Therefore, seed arrival is confirmed to be an important constraint limiting
the invasion of late-successional trees in these old fields.

Seeds of late-successional tree species do not arrive in old field areas even
though widespread pioneer tree species can function as potential perches. It
is likely that the seed vectors that promote a quick regeneration from abandoned pasture to secondary forest are not the same ones that would bring the
seeds of late-successional species to these secondary forest sites. In central
Amazonia, seeds of the pioneer species of the genus *Vismia* can be dispersed
considerable distances from forest gaps to open pasture sites due to the long-
ranging distances of their bat and bird dispersers (Nepstad et al. 1991). As
soon as these pioneer species reach maturity, they produce large quantities of
fruits that help to maintain high populations of their seed dispersers. Indeed,
the population of bat species that feed on fruits of pioneer trees are much
higher in abandoned pastures than in adjacent forest (R. Gribel, pers.
comm.). Most frugivorous birds found in these secondary forests eat fruits of
pioneer species and seem to be more important in maintaining the seed pool
of pioneer species than in bringing new types of propagules to these secondary forest areas (Borges 1995).

Seed dispersal by means other than animals, for example, wind dispersal,
could also be an important trait that allows species invasion into secondary
forest sites. In Costa Rica, large clearings or pastures adjacent to forests are
initially colonized by wind-dispersed tree species (Janzen and Vásquez-Yanes
1991). However, wind dispersal is efficient over only relatively short dis-

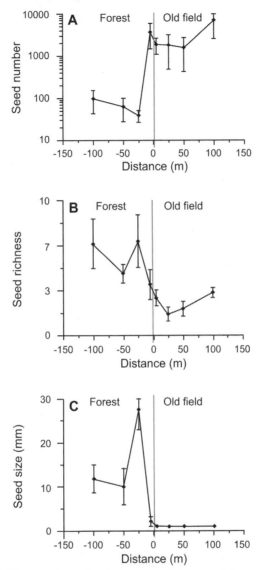

FIGURE 5.1. (a) Mean seed number in log scale, (b) mean richness of seed morphotypes, and (c) mean seed size, for seeds collected in seed traps placed at different distances from the edge between pristine Amazon forest and a six-year-old field dominated by species of the pioneer tree *Vismia* (Clusiaceae). Error bars represent ± 1 SE.

tances. In Panama, for example, *Platypodium elegans* seeds reach distances of no more than 100 m from their parent trees (Augspurger 1983), and in Malaysia, even high-canopy dipterocarp trees have their fruits dispersed by wind no farther than 60 m (Appanah and Nor 1991). Wind-dispersed seeds are important colonizers of Australian old fields. Nevertheless, their seed shadows decrease considerably with increasing distance from the forest source, while seeds that are dispersed by vertebrates continue to arrive in high densities in areas that are far from the forest source (Willson and Crome 1989). In this study, only six of the ninety-six seed morphotypes collected in seed traps were dispersed by wind. Furthermore, none of these wind-dispersed seeds were collected in traps located in secondary forest sites. I conclude that due to their lower availability and lower capability of dispersing long distances, wind-dispersed species have only a limited relevance to the early stages of natural forest restoration in central Amazonia.

There was a tendency for seeds of pioneer species from secondary forests to invade the forest edge (figure 5.1). Seeds collected in traps placed in the forest at 5 m from the edge were mostly from *Cecropia* (Cecropiaceae) and *Vismia* pioneer tree species. These results show that seed dispersal not only limits the invasion of late-successional species into secondary forest sites but also may limit the pool of tree species that could invade forest edges (Benitez-Malvido 1998). This process has important consequences for old field restoration, because when forest edges become degraded, seed penetration in adjacent old fields becomes increasingly shorter.

The distance from a forest source will determine the rates and routes along which secondary forests will become structured primary forests. This work has discussed that the narrow penetration of seeds of late-successional species into disturbed sites is most likely due to restrictions in seed dispersal. I then consider seed dispersal of primary forest species the main factor influencing the rapidity with which secondary forests can be converted to highly structured rain forests.

Interactions Between Pioneer Vegetation and Later Colonizers

Interactions between established pioneer vegetation and late-successional colonizers can affect primary forest restoration in old field areas and may influence community structure and floristic composition of restored forests. Connell and Slatyer (1977) suggested three mutually exclusive models in which early established species would have either positive (*Facilitation model*), negative (*Inhibition model*), or little effect (*Tolerance model*) on the establishment of later-invading species.

The Facilitation model states that early-successional species are the only ones that could become established in a given disturbed area. They then modify the environment in such a way that it becomes more suitable for the establishment of late-successional species and less suitable for the establishment of early-successional species. Over time, late-successional species grow to maturity and early colonists are eliminated by competition. According to the Inhibition model, any species could become established in a given disturbed area, and modification of the environment by early occupants makes the environment less suitable for the recruitment of subsequent invaders, whether these are early-successional or late-successional species. Further invasion can occur only when the resident individual is damaged or killed, thereby freeing the previously occupied space. The Tolerance model proposes that any species could invade the site and that the presence of early colonists neither reduces nor increases the rates of establishment and growth of later colonists. Thus, juveniles of later-successional species grow to maturity despite the presence of healthy individuals of early-successional species.

Experimental tests of these three models in old field areas can be extremely useful for improving our understanding of the processes that drive vegetation succession and direct forest restoration. Because a great deal of the influence of pioneer species is operating during the seedling establishment phase, tests can be performed by introducing seeds and/or seedlings of later species in old field areas with and without earlier species. When the early vegetation is removed, any performance of late invaders, such as germination, survival, and growth would (1) decrease in the case of the Facilitation model, (2) increase according to the Inhibition model, or (3) be unaffected under the Tolerance model.

Experimental works testing the nature of interactions between early- and late-successional old field species have been extensively performed. Wilson (1999) reviewed a series of experimental studies of plant interactions during old field succession and concluded that Facilitation occurred much less frequently than competitive interactions. It is argued that Facilitation is more likely to occur in environments that experience harsh conditions, such as deserts and extremely cold areas (Bertness and Callaway 1994; Callaway and Walker 1997). Nevertheless, recent works have shown that Facilitation is a rather frequent process in old fields of neotropical productive ecosystems (Ganade and Brown 2002; Zanini et al. 2006). Presence of pioneer vegetation can improve microclimatic conditions for germination and establishment (Ganade and Brown 2002; Zanini and Ganade 2005), and can even improve growth and reduce seed predation of late-successional species (Zanini et al. 2006). Pioneer species in Amazonia old pastures can also attract seed dispersers and improve soil nutrient conditions for later colonizers (Vieira et al. 1994). Processes of Inhibition due to resource competition between seedlings and pasture grasses are, however, widespread in tropical old field areas (Nepstad et al. 1991; Aide and Cavelier 1994; Miriti 1998; Holl et al. 2000).

I tested which successional model explains the interaction between established pioneer tree species and late-colonizing tree species in an old, central Amazonian pasture. This pasture was grazed for approximately five years before abandonment and was abandoned for six years. During this short time, a dense community of pioneer trees, mostly dominated by species of *Vismia* (Clusiaceae), developed in the site (Ganade 2001). Seeds of four tropical tree species were introduced in plots where old field vegetation had been experimentally removed or left intact. Eight replicate blocks were randomly placed within a 2 ha old field site. Each block was divided into two 10 × 10 m plots, 5 m apart from each other. The two plots in each block were randomly assigned to have the vegetation either intact or removed. The vegetation removal treatment comprised the removal of all trees, shrubs, herbs, emerging roots, and plant litter from the soil surface. Groups of thirty-two seeds from each species were randomly placed in each vegetation treatment. Seeds were sown 10 cm apart and marked with wooden sticks. Seedling establishment was recorded for ten months, after which seedlings were harvested, dried, and weighed to ascertain total growth in biomass.

Four species of tropical trees were used in the experiment: *Aspidosperma discolor* (Apocynaceae), *Inga edulis* (Mimosoideae), *Socratea exorrhiza* (Arecaceae), and *Oenocarpus bataua* (Arecaceae). Seeds of all species were collected at the beginning of the dry season when the experiment began. *Aspidosperma discolor* is a canopy tree of upland primary forest with wind-dispersed seeds (2.5 × 2.5 × 0.01 cm) that germinate within a month of arriving in the soil. *Inga edulis* is a fast-growing tree species that is widely cultivated in the Amazon region. Its large, soft seeds (3.5 × 1.2 × 0.8 cm) germinate so quickly that it is common to find seeds germinating inside the pods. *Socratea exorrhiza* and *O. bataua* are palm trees that occur frequently in the forest, around the freshwater streams. *Socratea exorrhiza* has smaller seeds (2.0 × 1.4 × 1.4 cm) than *O. bataua* (3.0 × 1.8 × 1.8 cm), and both species have hard seed coats with a nutritious nut tissue inside. Seeds of these palm species germinate within three to five months after landing on the soil.

Facilitation occurred during the early phase where the presence of old field vegetation enhanced seedling establishment of *A. discolor, S. exorrhiza,* and *O. bataua,* but *I. edulis* had a similar number of seedlings established in both vegetation treatments (figure 5.2a). Facilitation occurred probably because the better microclimatic conditions under old field vegetation im-

TABLE 5.1

Microclimatic and soil fertility variables measured under old field pioneer vegetation and in plots (10 × 10 m) where vegetation was removed.

Measurements	Vegetation intact		Vegetation removed		Statistical test	
	Mean	SE	Mean	SE	$F_{1,14}$	p-value
Light (PAR)	83.3	18.4	1,735.0	28.8	2,340.0	<0.001
Air humidity (%)	49.4	1.6	53.5	2.7	1.8	ns
Air temperature (°C)	38.6	.4	41.8	.5	24.0	<.001
Soil temperature (°C)	21.8	.1	28.5	.4	190.7	<.001
N (%)	.2	.01	.1	.01	1.4	ns
P ($\mu g \cdot g^{-1}$)	2.2	.2	2.0	.2	2.3	ns
K ($\mu g \cdot g^{-1}$)	23.3	3.3	18.0	1.0	2.2	ns
Ca (me %)	.5	.1	.7	.3	.3	ns
Mg (me %)	.3	.04	.2	.1	.03	ns
Al (me %)	1.5	.1	1.5	.1	.01	ns
Soil pH	4.7	.07	4.6	.1	0.5	ns

Note: Microclimatic measurements represent: photosynthetic active radiation (PAR), air humidity (%), and air temperature (°C). Soil fertility measurements were taken for macro- (NPK) and micronutrients (Ca, Mg, Al) and soil acidity (pH). All measurements were performed in eight plots randomly placed within a 2 ha site located in each vegetation treatment. Photosynthetic Active Radiation (PAR, $\mu mol.m^{-2}.s^{-1}$) was measured at soil surface using a Delta T sunflecks ceptometer and air humidity (%) using a whirling hygrometer. The light and air measurements were performed twice, in two consecutive days of clear blue sky, between 1100 and 1300 h. Each replicated soil sample was a compound of five random samples collected at 0–15 cm depth within a 100 m^2 area.

proved seed germination and seedling survival (table 5.1). Previous work at this old field site has shown that the moist microsites created by the presence of plant litter are one of the main causes of this improvement (Ganade and Brown 2002). However, all the species studied incorporated more biomass when pioneer species were removed, showing that Inhibition started to occur during the growth phase (figure 5.2b). These results support the hypothesis that Facilitation effects tend to decrease through vegetation succession, while competition tends to increase (Wilson 1999).

Resource Competition Influencing Successional Processes

Resource competition between pioneer and late-successional species may explain the process of growth Inhibition. The availability of soil nutrients and light is crucial for plant establishment and survival. In wide open areas, such as abandoned pastures, light availability is high, and the soil nutrient concentrations can also be high, depending on the intensity of pasture utilization (Fearnside 1990). Nevertheless, when the early-successional vegetation becomes established in abandoned pastures, it can promptly modify light and soil nutrient conditions (Lu et al. 2002). How these modifications alter the

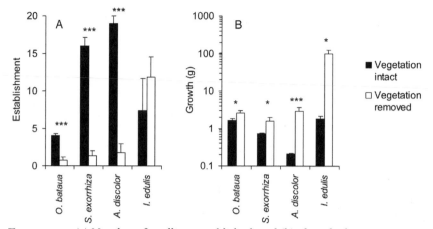

FIGURE 5.2. (a) Number of seedlings established, and (b) plant dry biomass produced, of late- and midsuccessional species in an old field of central Amazonia. Open bars represent plots where vegetation was removed; closed bars represent plots where vegetation was left intact. Bars represent means + 1 SE. * $p < .05$; *** $p < .001$.

patterns of plant establishment will depend on how rapidly pioneer trees reduce the light levels and limit the soil resources.

The availability of a limiting resource, such as a soil nutrient, can be an important factor controlling the organization and structure of plant communities (Newman 1973; Tilman 1986). Indeed, nutrient addition in plant communities commonly leads to an increase in plant biomass combined with a decrease in species diversity (Tilman 1986; Berendse and Elberse 1990; Goldberg and Miller 1990). This pattern is due to an enhancement of the aboveground biomass production of species that are the best competitors at that soil nutrient level (Tilman and Wedin 1991). This in turn results in a decline of light availability to subcanopy species and seedlings (Newman 1973; Goldberg and Miller 1990), leading to an overall decrease in the species diversity of the plant communities.

In the Amazon *terra firme* forest, soils are poor, phosphorus is in very low concentration, and it is generally considered to be the most limiting resource for plant growth (Vitousek 1984). However, factorial field experiments in this old field area have shown that processes of growth Inhibition are not mediated by competition for P and K between pioneer and late-successional species (Ganade and Brown 2002). Other experimental studies confirmed that P does not limit seedling growth in tropical forests (Lewis and Tanner 2000; Burslem et al. 1995). A study performed in gaps and in the understory of the tropical forest of Costa Rica revealed that light and not phosphorus or other nutrients was the limiting resource controlling the growth and reproduction of seven tropical species (Denslow et al. 1990). Additionally, light availability increased substantially when pioneer vegetation was removed, while nutrient availability did not (table 5.1). This evidence suggests that light may be the limiting resource that drives the Inhibition process in these old fields.

Seed Predators and Herbivores Influencing Successional Processes

The Facilitation, Inhibition, and Tolerance models were intended to explain the net effect that early-successional species could impose on later ones. Indeed, this net effect is the output of a long, continuous, and complex process involving interactions between many ecological factors (Walker and Chapin 1987; Pickett et al. 1987). For instance, there might be an increase in the population of seed predators and herbivores when the early-successional vegetation establishes itself, and so seed and seedling consumption by animals could be the main reason why Inhibition takes place in a particular successional system. Understanding how seed predation and herbivory affect the successive phases of plant establishment through experimental studies allows us to make more accurate predictions of how succession proceeds.

Herbivory by vertebrates and invertebrates can greatly influence the rates and directions of plant succession, leading to different species composition in plant communities (Gessaman and MacMahon 1984; Brown 1985; Huntly 1991). In most cases, herbivores act directly on seeds and seedlings, which are the most vulnerable stages of the plant life cycle (Fenner 1987). Postdispersal seed predation by mammals is a major factor limiting seedling emergence and recruitment in tropical forests (Schupp 1988; Sork 1987; De Steven and Putz 1984). Therefore, herbivores can be important factors influencing forest restoration in abandoned pastures. In eastern Amazonia, attack by seed predators and leaf-cutting ants are important barriers to forest regeneration in abandoned and degraded pastures (Nepstad et al. 1991). Moreover, the density of generalist herbivores such as leaf-cutting ants can dramatically increase in abandoned pastures, causing a reduction in the establishment of woody species (Nepstad et al. 1990; Vasconcelos and Cherrett 1995; Vasconcelos and Cherrett 1997; Vasconcelos 1997).

Tests of Connell and Slatyer's models specifically addressing the role of herbivores have been performed in old fields of temperate regions. These tests have shown the extreme importance of vertebrate herbivores in determining the patterns of succession (Gill and Marks 1991; De Steven 1991a,

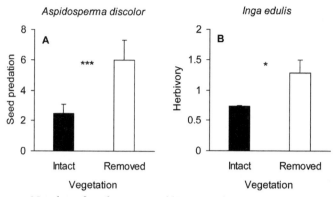

FIGURE 5.3. Number of seeds consumed by (a) predators and (b) leaf herbivory, in old field of central Amazonia. Open bars represent plots where vegetation was removed; closed bars represent plots where vegetation was left intact. Bars represent means + 1 SE. * $p < .05$; *** $p < .001$.

b). Measurements of seed predation and herbivory in seedlings of the four tree species studied in central Amazonia revealed that, at least for two of the species studied, seed predation and herbivory increased when pioneer vegetation was removed (figure 5.3a, 5.3b). This indicates that herbivores can contribute to the process of Facilitation in these old fields, and their influence must be taken into account during restoration programs.

Restoration Practices

In this section I indicate possible procedures that could assist restoration in old fields that have been previously colonized by pioneer tree species in the Brazilian Amazonia.

Improving Natural Seed Arrival

There is a scarcity of seed dispersers invading old field areas, thus if restoration is solely dependent on natural seed arrival, areas to be restored should be as close as possible to a forest source. Vegetation surveys in degraded Amazon areas have shown that distance from a forest source can be crucial to define the floristic structure of naturally colonized plant species (Mesquita et al. 2001; Rodrigues et al. 2004). In central Amazonia, plant species richness can be reduced by half in secondary forests located 100 m from forest areas (Mesquita et al. 2001). This work has shown that old field areas that are much farther than 100 m from a forest source may receive negligible input of seeds from late-successional trees. I suggest that landscape planning should be an important strategy for restoration procedures that rely on natural seed arrival. If possible, restoration sites should be located adjacent to, or no farther than 100 m away from, a primary forest.

Seed Sowing Versus Seedling Transplantation

Forest restoration campaigns are frequently implemented by transplanting seedlings of desirable trees in target areas. This procedure, however, can be labor intensive and expensive. Although direct seeding is rarely used in restoration programs, this is a low-cost procedure that can sometimes be very effective (Engel and Parrota 2001; Camargo et al. 2002). In central Amazon, plants of *Inga edulis* established from seed-sowing experiments achieved 6 m in height and started to produce fruits only ten months after seeds were sown in managed old field areas (Ganade and Brown 2002). Additionally, direct sowing has proved to be a more efficient technique for large- than for small-seeded tree species (Camargo et al. 2002). Given that large-seeded species are less likely to colonize disturbed areas due to dispersal limitations, direct sowing of large-seeded trees seems to be an efficient method to promote forest restoration in old fields of central Amazonia. An additional advantage of this procedure is that large and inaccessible areas could be enriched by primary forest species by simply throwing seeds from a plane.

Procedures for Enhancing Seedling Growth

When seeds are able to arrive in old fields, seedling establishment can readily occur, because seed predation rates in old fields are lower in comparison to the forest sites, and old field vegetation improves microclimatic conditions for establishment (Ganade 2001). However, management procedures should be taken into account in order to enhance seedling growth. After seedlings have established, the canopy of pioneer trees should be partially removed for light levels to be improved, while protection against desiccation, wind, and other climatic hazards would be maintained by the trees and plant litter that remain at the site (Ganade and Brown 2002). Mesquita (2000) has shown that selective canopy removal in an Amazonian old field increased sapling growth threefold after only two years from gap formation. Canopy removal should be applied by cutting down one or a few pioneer trees for each gap created. This procedure would maintain a moist microclimate and may prevent increases in seed predation and herbivory that could occur when larger gaps are formed (figure 5.3).

What Can We Learn from Agroforestry Experiences?

Complete restoration of old fields to Amazon forest may take hundreds of years to achieve. To date, published experiments on long-term Amazonian forest restoration are nonexistent. It seems that most of the scientific effort to comprehend restoration in this region is focused, first, on understanding factors that influence the early stages of old field succession and, second, on unveiling the rates of biomass accumulation in secondary forests. Agroforestry experiences can contribute a great deal to our understanding of how manipulations performed at the first stages of old field succession could influence forest structure and future tree survival. Agroforestry programs in Rondonia have shown that after ten years of tree planting, nine of the native species studied showed high rates of survival and growth, achieving sizes higher than 18 cm in diameter at breast height (Browder et al. 2005). Therefore, once a seedling is able to overcome its first stages of establishment it is very likely to survive to maturity.

Restoration practices could use commercially valuable native tree species as an incentive for landowners (Lugo 1995; Ganade 2001). Many desirable species occur in the forest at low densities, thus forest restoration can provide the ideal amount of desirable species for sustainable management. Additionally, restoration sites could be implemented near urban areas, making it easier for forest products to be processed, stored, and commercialized. Another important matter is the choice of species to be introduced. The Amazon forest has a large array of commercially valuable native species, and some of them, such us the Brazil nut (*Bertholletia excelsa*), açaí (*Euterpe oleraceae*), and cupuaçu (*Theobroma grandiflorum*), are good candidates for commercial use because they are already sold on the international market. The use of such commercially valuable species could make landowners commit themselves to large-scale restoration programs in Amazonian old fields.

Acknowledgments

I thank José Eremildes da Silva for field assistance, and Ingo Hubel, Guilherme Mazzochine, and Carlos Fonseca for improvements on the manuscript. This study was supported by the Smithsonian Institution and the Instituto Nacional de Pesquisas da Amazônia (INPA) and represents publication 485 in the Biological Dynamics of Forest Fragments Project technical series.

REFERENCES

Aide, T. M., and J. Cavelier. 1994. Barriers to lowland tropical forest restoration in the Sierra Nevada de Santa Marta, Colombia. *Restoration Ecology* 2:219–29.

Alves, D. S., J. V. Soares, S. Amaral, E. M. K. Mello, S. A. S. Almeida, O. F. da Silva, and A. M. Silveira. 1997. Biomass of primary and secondary vegetation in Rondônia, Western Brazilian Amazon. *Global Change Biology* 3:451–61.

Appanah, S., and S. Mohd. Nor. 1991. Natural regeneration and its implications for forest management in the dipterocarp forests of peninsular Malaysia. In *Rain forest regeneration and management*, ed. A. Gómez-Pompa, T. C. Whitmore, and M. Hadley, 361–69. Man and the Biosphere Series. Parthenon Publishing Group, Carnforth, Lancashire, UK

Arima, E., P. Barreto, and M. Brito. 2005. *Pecuária na Amazônia: Tendências e implicações para a conservação ambiental.* Imazon, Belém, Pará.

Augspurger, C. K. 1983. Seed dispersal of the tropical tree, *Platipodium elegans*, and the escape of its seedlings from fungal pathogens. *Journal of Ecology* 71:759–71.

Benitez-Malvido, J. 1998. Impact of forest fragmentation on seedling abundance in a tropical rain forest. *Conservation Biology* 12:380–89.

Berendse, F., and W. T. Elberse. 1990. Competition and nutrient availability in heathland and grassland ecosystems. In *Perspectives on plant competition*, ed. J. B. Grace and D. Tilman, 93–117. Academic Press, New York.

Bertness, M. D., and R. M. Callaway. 1994. Positive interactions in communities. *Trends in Ecology and Evolution* 9:191–93.

Borges, S. H. 1995. Comunidades de aves em dois tipos de vegetação secundária da Amazônia central. MS thesis, Instituto Nacional de Pesquisas da Amazônia, Manaus, Brazil.

Browder, J. O., R. H. Wynne, and M. A. Pedlowski. 2005. Agroforestry diffusion and secondary forest regeneration in the Brazilian Amazon: Further findings from the Rondônia agroforestry pilot project (1992–2002). *Agroforestry System* 65:99–111.

Brown, V. K. 1985. Insect herbivores and plant succession. *Oikos* 44:17–22.

Buschbacher, R., C. Uhl, and E. A. S. Serrão. 1988. Abandoned pastures in eastern Amazonia, 2. Nutrients stocks in the soil and vegetation. *Journal of Ecology* 76:682–99.

Burslem, D. F. R. P., P. J. Grubb, and I. M. Turner. 1995. Responses to nutrient addition among shade-tolerant tree seedlings of lowland tropical rain forest in Singapore. *Journal of Ecology* 83:113–22.

Callaway, R. M., and L. R. Walker. 1997. Competition and facilitation: A synthetic approach to interactions in plant communities. *Ecology* 78:1958–65.

Camargo, J. L. C., I. D. K. Ferraz, and A. M. Imakawa. 2002. Rehabilitation of degraded areas of central Amazonia using direct sowing of forest tree seeds. *Restoration Ecology* 10:636–44.

Connell, J. H., and R. O. Slatyer. 1977. Mechanisms of succession in natural communities and their role in community stability and organization. *American Naturalist* 111:1119–44.

Cubiña, A., and T. M. Aide. 2001. The effect of distance from forest edge on seed rain and soil seed bank in a tropical pasture. *Biotropica* 33:260–67.

De Steven, D. 1991a. Experiments on mechanisms of tree establishment in old-field succession: Seedling emergence. *Ecology* 72:1066–75.

———. 1991b. Experiments on mechanisms of tree establishment in old-field succession: Seedling survival and growth. *Ecology* 72:1076–88.

De Steven, D., and F. E. Putz. 1984. Impact of mammals on early recruitment of a tropical canopy tree, *Dipteryx panamensis*, in Panama. *Oikos* 43:207–16.

Denslow, J. S., J. C. Schultz, P. M. Vitousek, and B. R. Strain. 1990. Growth responses of tropical shrubs to treefall gap environments. *Ecology* 71:165–79.

Dias-Filho, M. B. 2005. Degradação de pastagens: Processos, causas e estratégias de recuperação. Embrapa, Belém, Pará.

Engel, V. L., and J. A. Parrota. 2001. An evaluation of direct seeding for reforestation of degraded lands in central São Paulo, Brazil. *Forest Ecology and Management* 162:169–81.

Fearnside, P. M. 1990. Predominant land uses in Brazilian Amazonia. In *Alternatives to deforestation: Steps toward sustainable use of the Amazon rain forest*, ed. A. B. Anderson, 233–51. Columbia University Press, New York.

Fearnside, P. M., and W. M. Guimarães. 1996. Carbon uptake by secondary forests in Brazilian Amazonia. *Forest Ecology and Management* 80:35–46.

Fenner, M. 1987. Seedlings. *New Phytologist* 106:35–47.

Feldpausch, T. R., M. A. Rondon, E. C. M. Fernandes, S. J. Riha, and E. Wandelli. 2004. Carbon and nutrient accumulation in secondary forests regenerating on pastures in Central Amazonia. *Ecological Applications* 14:S164–S176.

Ganade, G. 2001. Forest restoration in abandoned pastures of Central Amazonia. In *Lessons from Amazonia: The ecology and conservation of a fragmented forest*, ed. R. O. Bierregaard, C. Gascon, T. E. Lovejoy, and R. Mesquita, 313–24. Yale University Press, New Haven, CT.

Ganade, G., and V. K. Brown. 2002. Succession in old pastures of Central Amazonia: Role of soil fertility and plant litter. *Ecology* 83:743–54.

Gessaman, J. A., and J. A. MacMahon. 1984. Mammals in ecosystems: Their effect on the composition and production of vegetation. *Acta Zoologica Fennica* 172:11–18.

Gill, D. S., and P. L. Marks. 1991. Tree and shrub seedling colonization of old-fields in central New York. *Ecological Monographs* 61:183–205.

Goldberg, D. E., and T. E Miller. 1990. Effects of different resource additions on species diversity in an annual plant community. *Ecology* 71:213–25.

Holl, K. D. 1999. Tropical forest recovery and restoration. *Trends in Ecology and Evolution* 14:378–79.

Holl, K. D., M. E. Loik, E. H. V. Lin, and I. A. Samuels. 2000. Tropical montane forest restoration in Costa Rica: Overcoming barriers to dispersal and establishment. *Restoration Ecology* 8:339–49.

Houghton, R. A., D. L. Skole, C. A. Nobre, J. L. Hackler, K. T. Lawrence, and W. H. Chomentowski. 2000. Annual fluxes of carbon from deforestation and regrowth in the Brazilian Amazon. *Nature* 403:301–4.

Howe, H. F., and J. Smallwood. 1982. Ecology of seed dispersal. *Annual Review of Ecology and Systematics* 13:201–28.

Huntly, N. 1991. Herbivores and the dynamics of communities and ecosystems. *Annual Review of Ecology and Systematics* 22:477–503.

Janzen, D. H., and C. Vásquez-Yanes. 1991. Aspects of tropical seed ecology of relevance to management of tropical forest wildlands. In *Rain forest regeneration and management*, ed. A. Gómez-Pompa, T. C. Whitmore, and M. Hadley, 137–57. Man and the Biosphere Series. Parthenon Publishing Group, Carnforth, Lancashire, UK.

Lewis, S. L., and E. V. J. Tanner. 2000. Effects of above- and belowground competition on growth and survival of rain forest tree seedlings. *Ecology* 81:2525–38.

Lu, D., E. Moran, and P. Mausel. 2002. Linking Amazonian secondary succession forest growth to soil properties. *Land Degradation and Development* 13:331–43.

Lucas, R. M., M. Honzak, I. D. Amaral, P. J. Curran, and G. M. Foody. 2002. Forest regeneration on abandoned clearances in central Amazonia. *International Journal of Remote Sensing* 23:965–88.

Lugo, A. E. 1995. Management of tropical biodiversity. *Ecological Applications* 5:956–61.

Mesquita, R. C. G. 2000. Management of advanced regeneration in secondary forests of the Brazilian Amazon. *Forest Ecology and Management* 130:131–40.

Mesquita, R. C. G., K. Ickes, G. Ganade, and G. B. Williamson. 2001. Alternative successional pathways in the Amazon Basin. *Journal of Ecology* 89:528–37.

Miriti, M. N. 1998. Regeneração florestal em pastagens abandonadas na Amazônia central: Competição, predação e dispersão de sementes. In *Floresta Amazônica: Dinâmica, regeneração e manejo*, ed. C. Gascon and P. Montinho, 179–91. Instituto Nacional de Pesquisas da Amazônia, Manaus, Brazil.

———. 1991. Recuperation of a degraded Amazonian landscape: Forest recovery and agricultural restoration. *Ambio* 20:248–55.

Nepstad, D., C. Uhl, and E. A. Serrão. 1990. Surmonting barriers to forest regeneration in abandoned, highly degraded pastures: A case study from Paragominas, Para, Brazil. In *Alternatives to deforestation: Steps toward sustainable use of the Amazon rain forest*, ed. A. B. Anderson, 215–29. Columbia University Press, New York.

Newman, E. I. 1973. Competition and diversity in herbaceous vegetation. *Nature* 244:310.

Pickett, S. T. A., S. L. Collins, and J. J. Armesto. 1987. Models, mechanisms and pathways of succession. *Botanical Review* 53:335–71.

Roberts, D. A., M. Keller, and J. V. Soares. 2003. Studies of land-cover, land-use, and biophysical properties of vegetation in the Large Scale Biosphere Atmosphere experiment in Amazônia. *Remote Sensing of Environment* 87:377–88.

Rodrigues, R. R., S. V. Martins, and L. C. Barros. 2004. Tropical rain forest regeneration in an area degraded by mining in Mato Grosso state, Brazil. *Forest Ecology and Management* 190:323–33.

Saldarriaga, J. G., and C. Uhl. 1991. Recovery of forest vegetation following slash-and-burn agriculture in the upper Rio Negro. In *Rain forest regeneration and management*, ed. A. Gomez-Pompa, T. C. Whitmore, and M. Hadley, 303–12. Man and the Biosphere Series. Parthenon Publishing Group, Carnforth, Lancashire, UK.

Schupp, E. W. 1988. Seed and early seedling predation in the forest understory and in treefall gaps. *Oikos* 51:71–78.

Silva, J. M. C., C. Uhl, and G. Murray. 1994. Plant succession, landscape management, and the ecology of frugivorous birds in abandoned Amazonian pastures. *Conservation Biology* 10:491–503.

Sork, V. L. 1987. Effect of predation and light on seedling establishment in *Gustavia superba*. *Ecology* 68:1341–50.

Stouffer, P. C., and S. H. Borges. 2001. Conservation recommendations for understory birds in Amazonian forest fragments and second-growth areas. In *Lessons from Amazonia: The ecology and conservation of a fragmented forest*, ed. R. O. Bierregaard, C. Gascon, T. E. Lovejoy, and R. Mesquita, 248–61. Yale University Press, New Haven, CT.

Tilman, D. 1986. Resources, competition and dynamics of plant communities. In *Plant ecology*, ed. M. J. Crawley, 51–75. Blackwell Scientific Publications, Oxford, UK.

Tilman, D., and D. Wedin. 1991. Plant traits and resource reduction for five grasses growing on a nitrogen gradient. *Ecology* 72:685–700.

Uhl, C., R. Buschbacher, and E. A. S. Serrão. 1988. Abandoned pastures in eastern Amazonia. Patterns of plant succession. *Journal of Ecology* 76:663–81.

Vasconcelos, H. L. 1997. Foraging activity of an Amazonian leaf-cutting ant: Responses to changes in the availability of woody plants and to previous plant damage. *Oecologia* 112:370–78.

Vasconcelos, H. L., and J. M. Cherrett. 1995. Changes in leaf-cutting ant populations (Formicidae: Attini) after the clearing of mature forest in Brazilian Amazonia. *Studies on Neotropical Fauna and Environment* 30:107–13.

———. 1997. Leaf-cutting ants and early regeneration in central Amazonia: Effects of herbivory on tree seedling establishment. *Journal of Tropical Ecology* 13:357–70.

Vieira, I. C. G., A. S. de Almeida, E. A. Davidson, T. A. Stone, C. J. R. Carvalho, and J. B. Guerrero. 2003. Classifying successional forests using Landsat spectral properties and ecological characteristics in eastern Amazônia. *Remote Sensing of Environment* 87:470–81.

Vieira, I. C. G., C. Uhl, and D. Nepstad. 1994. The role of the shrub *Cordia multispicata* Cham. as a "succession facilitator" in an abandoned pasture, Paragominas, Amazônia. *Vegetation* 115:91–99.

Vitousek, P. M. 1984. Litterfall, nutrient cycling, and nutrient limitation in tropical forests. *Ecology* 65:285–98.

Walker, L. R., and F. S. Chapin III. 1987. Interactions among processes controlling successional change. *Oikos* 50:131–35.

Willson, M. F., and F. H. J. Crome. 1989. Patterns of seed rain at the edge of a tropical Queensland rain forest. *Journal of Tropical Ecology* 5:301–8.

Wilson, S. D. 1999. Plant interactions during secondary succession. In *Ecosystems of disturbed ground*, ed. L. R. Walker, 611–32. Elsevier, New York.

Zanini, L., and G. Ganade. 2005. Restoration of *Araucaria* forest: The role of perches, pioneer vegetation, and soil fertility. *Restoration Ecology* 13:507–14.

Zanini, L., G. Ganade, and I. Hübel. 2006. Facilitation and competition influence succession in a subtropical old field. *Plant Ecology* 185:179–90.

Chapter 6

Old Field Vegetation Succession in the Neotropics

Karen D. Holl

Tropical forests have been cleared extensively over the past fifty years, and much of that land has been converted to agriculture. While rates of tropical deforestation are difficult to calculate, they are unquestionably high. During the 1990s an average of 5.8 million ha of tropical forest were cleared annually, with another 2.3 million ha degraded (Mayaux et al. 2005). The causes of deforestation are complex and interrelated, and their relative importance varies by region (Geist and Lambin 2002). In Latin America, a primary cause of deforestation has been conversion of forests to pastures, and to a lesser degree for cash crops, such as coffee. Most of this land is still used for agriculture, and agricultural lands are increasing in many areas of the neotropics. But, in some regions, successional forest cover is increasing due to land being taken out of agriculture for a range of reasons, such as declines in commodity prices, changes in agricultural policy and subsidies, lack of productivity, industrialization, civil war, or land being set aside for conservation of species and/or ecosystem services (Campos 2001; Grau et al. 2003; Arroyo-Mora 2005; Hecht et al. 2006; chapter 5).

As land is abandoned, it provides an opportunity to learn more about tropical forest recovery in old agricultural fields. As a result, the number of studies on neotropical old field succession have increased substantially in the past decade, but most of those studies either focus on the first few years after abandonment (e.g., Uhl 1987; Aide and Cavelier 1994; Holl et al. 2000; Ferguson et al. 2003), or have relied on a chronosequence of abandoned sites (e.g., Hughes et al. 1999; Pascarella et al. 2000; Steininger 2000; Kennard 2002; but see Capers et al. 2005). Such time-for-space substitution studies are important to understanding succession but can be problematic because of difficulties in documenting conditions at the time of abandonment, variation in gradients other than just time since abandonment, and the relatively small number of old abandoned sites. Moreover, most of the past studies have focused only on the changes in tree dynamics, overlooking the nonwoody species (Chazdon 2007). One of the few locations in the neotropics where an extensive land area was abandoned more than ten years ago is Puerto Rico, which is discussed in detail in chapter 4.

Successional Theory

Although research on tropical old field succession has increased in the past decade, the vast majority of research on old field succession and nearly all the theoretical discussions on the topic have focused on the temperate zone, the eastern United States in particular. This narrow geographical focus largely reflects a different temporal history of land use and abandonment; abandonment of old fields in the United States was common beginning in the early 1900s. In the tropics, such abandonment is relatively recent. It is important to note that there is increasing evidence that agricultural use of tropical lands was quite extensive in certain regions prior to European colonization (e.g., Uhl et al. 1989; Clement and Horn 2001; Hecht et al. 2006), but because the forest regenerated in most of these areas prior to modern science, these lands are often considered to have been pristine prior to the most recent period of tropical forest clearing.

The history of ideas in the successional literature is reviewed in chapter 2. Therefore, here I only briefly review three ongoing debates in the successional literature, which I will evaluate in the context of tropical old fields.

One discussion in the successional literature questions whether forests are initially colonized by a group of pioneer species that are subsequently replaced by later-successional species (*relay floristics model*; Egler 1954) or whether the species that initially colonize the site remain later in the successional process (*initial floristic composition*; Gleason 1917). This question has important implications for restoration, because the model that best describes succession in a given system will affect when it is appropriate to introduce different species. As ecologists test this idea in tropical forests, they find that forests differ greatly in the degree to which each of these models applies. A number of authors (e.g., Ewel 1980; Uhl et al. 1989; Finegan 1996; Chazdon 2007) who work in tropical moist or wet forest describe a series of stages of forest succession in old fields. Soon after abandonment, fields are rapidly colonized by herbs, shrubs, and lianas. Within a few years, seedlings of fast-growing, short-lived pioneer species establish and live for ten to thirty years. Over time, long-lived pioneer species increasingly dominate the canopy, and some mature-forest, shade-tolerant species establish, although many are lacking even in older secondary forests (Finegan 1996; Chazdon et al. 2007). In contrast, some scientists studying succession in tropical dry forests report few successional stages (Ewel 1980; Kennard 2002) and rapid establishment of mature forest species due to resprouting, even in ecosystems subjected to extensive human disturbance (Sampaio et al., forthcoming; Vieira et al. 2006).

A second related question is whether ecosystem recovery is deterministic given the environmental conditions, essentially random as a result of a series of stochastic events, or a mixture of the two, which can result in alternative stable states (reviewed in Temperton and Hobbs 2004). This debate began with Clements's (1916) view of ecosystem progression toward a climax community compared with the focus of Gleason (1917), Tansley (1935), and Egler (1954) on how the initial suite of colonizing species might result in different successional trajectories at a site. This discussion has resurfaced among restoration ecologists in recent years as they seek to determine whether there are assembly rules that can be identified to help in directing the trajectory of ecological recovery (Lockwood 1997; Young et al. 2001; Temperton and Hobbs 2004). These ideas are also discussed among tropical ecologists (Finegan 1996; Hubbell et al. 1999; Aide et al. 2000; Chazdon 2007) who have generally concluded that tropical succession is not deterministic, which raises the question of what factors affect the numerous potential successional trajectories toward different mixtures of tree species in highly diverse tropical systems.

Therefore, a third primary question I address is what factors affect the rate at which neotropical old fields recover from disturbance and the direction of the successional trajectory. Past studies in the neotropics show that the rate of forest recovery is highly variable (figure 6.1). Over a period of fifteen to sixty years, a number of forest species establish and accumulate biomass levels that approach those of intact forests (Guarigata et al. 1997; Aide et al. 2000; Finegan and Delgado 2000; Silver et al. 2000). In contrast, many have argued that given the current extensive scale of tropical forest clearing, succession will proceed extremely slowly, if at all, on highly degraded sites (Buschbacher et al. 1988; Sarmiento 1997; Zahawi and Augspurger 1999). Many studies report numerous obstacles to old field recovery, such as low seed dispersal, high seed predation, soil degradation, herbivory, stressful microclimatic conditions, and aggressive exotic vegetation (reviewed in Holl 2002).

One of the reasons for support of a wide variety of models of succession is the immense diversity in types of tropical forests (Ewel 1980). The initial successional debates were largely focused on hardwood forests in the eastern United States. As these models have been applied to various temperate ecosystems, support for the different models has varied. It seems likely that much

(a)

(b)

FIGURE 6.1. Differences in rate of early tropical old field regeneration: (a) four-year-old abandoned pasture in montane moist forest in southern Costa Rica, showing dense grass cover and only scattered shrubs and trees present at the time of abandonment; (b) two-year-old abandoned swidden field in lowland dry forest in the Yucatan Peninsula, Mexico, showing dense woody regeneration throughout.

of the divergence in successional processes observed in the neotropics can be explained in terms of a few underlying abiotic gradients (rainfall, temperature, soil), as well as the severity of anthropogenic disturbance both surrounding and within old fields. The vast majority of past tropical forest research has focused on the lowland wet tropics, although they account for less than a third of tropical forest cover. In recent years, however, research on tropical forests along a range of rainfall and elevation gradients has emerged, enabling more rigorous comparison of tropical forest succession that can inform restoration efforts.

I aim to summarize the literature on tropical old field succession focusing on the three interrelated questions outlined earlier: Do tropical forests follow an initial floristics or relay floristics model? Is tropical old field succession deterministic or stochastic? And, what factors affect the rate and direction of succession in tropical old fields? In the following sections, I first briefly discuss mechanisms of tropical plant regeneration and then provide a detailed review of factors at a range of scales that affect the rate and direction of neotropical forest recovery. The final section draws on the literature review to summarize the answers to the three questions and address the implications of these trends for restoration of neotropical old fields. In this chapter I focus on vegetation succession and discuss fauna only in the context of how they affect vegetation regeneration.

Regeneration Mechanisms

Plants in tropical forests regenerate by one of four mechanisms. Seedlings can establish from propagules within the site, either from the seed bank, from seedlings established at the time of land abandonment (advance regeneration), or by resprouting from roots or stems. Alternately, seedlings can establish from seeds dispersed primarily from sources outside the site.

In tropical old fields, few later-successional species regenerate from seeds or seedlings upon abandonment. Seeds of later-successional trees in tropical wet forests often lack dormancy and do not develop a seed bank, although seed dormancy is more common in drier tropical forests (Vieira and Scariot 2006). Therefore, most seeds that establish from the seed bank at the time of old field abandonment are from shrubs and short-lived pioneer trees, although a few longer-lived pioneers have dormant seeds (Uhl 1987; Finegan 1996; Benítez-Malvido et al. 2001). In less-disturbed tropical forests, later-successional species often establish as seedlings in the understory and grow slowly until there is a disturbance, such as a treefall. Whereas this mode of regeneration is important in less-disturbed tropical forests, it is nearly nonexistent in lands that have been heavily used for agriculture where abiotic and biotic conditions are highly altered (Uhl et al. 1989; Finegan 1996).

The degree to which resprouting is an important regeneration mechanism in tropical old fields varies greatly, depending on the climate and intensity of land disturbance (discussed later). In some old fields, upwards of 75% of the forest species regenerate by resprouting (Sampaio et al., forthcoming; Vieira et al. 2006), whereas in other old fields no forest species regenerate by resprouting (Holl 1999). Certainly, in old fields with higher resprouting, recovery of predisturbance vegetation composition is likely to be much more rapid.

Limited and variable establishment of forest species from propagules within recently abandoned old fields means that recovery in many tropical old fields is dependent on colonization by seeds dispersed from plants outside the site. The majority of trees in tropical wet forests are dispersed by animals (Howe and Smallwood 1982). Numerous studies of seed rain in neotropical forest clearly demonstrate that seed rain of animal-dispersed forest species is low in old fields with minimal woody structure, and that most seeds are from species already present in the old fields or from a few small-seeded pioneer species (reviewed in Holl 2002; chapter 5). Wind-dispersed species are often less dispersal limited than animal-dispersed seeds, which may result in a predominance of wind-dispersed species early in old field succession (Janzen 2002; Finegan and Nasi 2004; Hooper et al. 2004). In the following sections I discuss how a number of factors affect these regeneration mechanisms and the implications for old field succession.

Factors Affecting Forest Succession in Old Fields

It is necessary to identify factors that affect the rate and direction of succession at a range of spatial scales in order to develop restoration strategies to accelerate or redirect successional trajectories. I categorize the factors affecting forest succession into three broad categories (figure 6.2): the underlying abiotic gradients (rainfall, temperature, and soil), the composition of surrounding land use mosaic, and the type and intensity of past land use. I discuss in detail how each of these general factors affects the rate and direction of tropical old field succession.

Underlying Abiotic Gradients

Much of the difference in the rate and direction of succession in neotropical old fields can be explained by anthropogenic factors, such as surrounding

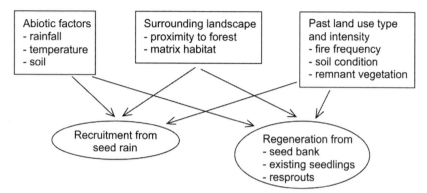

FIGURE 6.2. Factors affecting (boxes) different modes of regeneration (ellipses) of tropical old fields.

land use and land use history (discussed later), but these factors are overlain on a number of abiotic gradients, primarily rainfall, temperature, and soil type, which strongly influence forest recovery. Tropical forests fall along a rainfall gradient ranging from up to 7–8 m of rainfall evenly distributed throughout the year in the wettest sites to <1.5 m of rainfall with >6 months of dry season in the driest sites. The amount and seasonality of rainfall are primary evolutionary drivers for tropical forest plants, and they profoundly influence old field succession.

A number of tropical plant life-history traits that vary across the rainfall gradient strongly affect recovery. First, drier tropical forests have a higher percentage of wind-dispersed seeds (Ewel 1977; Janzen 2002; Vieira and Scariot 2006). For example, in a review of studies across a precipitation gradient, Viera and Scariot (2006) report 30%–63% wind-dispersed species in tropical dry forest and <16% wind-dispersed species in tropical moist or wet forest. Dry forest recovery may therefore be less dispersal limited than moist tropical forests (Janzen 2002). Second, although resprouting after disturbance is a common plant adaptation throughout the tropics, resprouting is more common in tropical dry forests, where water limitation favors an increased energy investment in roots (Ewel 1977; Vesk and Westoby 2004; Vieira and Scariot 2006). Third, seedling mortality due to desiccation is much higher in dry forests and varies a great deal interannually, depending on rainfall fluctuations (Vieira and Scariot 2006), which makes establishment from seed less predictable. Likewise, seedling growth may be slower due to lack of water. There is some evidence that dry forests may be more resilient because of lower structural complexity, greater wind dispersal, and more frequent resprouting; nonetheless, it is impossible to generalize about the rate of recovery of all forests across a rainfall gradient, given the many other factors that may influence the rate of recovery.

A second underlying, large-scale, climatic gradient is temperature. Temperature, however, is generally considered a less important driving factor in tropical forests compared to rainfall. The main effect of temperature on recovery is in higher elevation systems such as cloud forests, where the slow growth rate of trees can increase the time of recovery (Ewel 1980). Additionally, Zarin et al. (2001) found that growing season degrees (growing season length × growing season temperature) was a significant predictor of biomass across numerous old fields in the Brazilian Amazon, an area of little topographic variation, suggesting that temperatures may be an important predictor of recovery at larger scales.

A final important abiotic gradient is soil type. Much of the tropics are covered by oxisols and ultisols, which have low nutrient levels and high acidity. Some areas have more fertile, volcanic soils, such as andisols and inceptisols, although these soils commonly have low available phosphorus. Some authors (Moran et al. 2000; Zarin et al. 2001) have noted that over large spatial scales, differences in soil texture and fertility more strongly affect the recovery of biomass than previous land use (discussed later). For example, Moran et al. (2000) studied forest recovery across a range of land uses (pasture, swidden, mechanized agriculture) at five locations in Colombia and Brazil. He found that interregional variations in biomass and stand height were best explained by soil fertility, whereas within single locations previous land use explained most of the variation. Soil patterning at smaller scales can also affect species distributions. For example, Herrera and Finegan (1997) found that the different abundances of two common tree species, *Vochysia ferruginea* and *Cordia alliodora*, reflected differences in exchangeable acidity, slope, and magnesium.

Surrounding Landscape Mosaic

Gradients of climate and soil provide a template on which various anthropogenic factors act to influence neotropical forest recovery. A primary anthropogenic factor affecting succession in tropical old fields is the mosaic of surrounding land cover types, such as remnant forest, complex agroforests, shifting cultivation, pasture, or intensive agriculture (reviewed in Guariguata and Ostertag 2001; Holl 2002; Chazdon 2003, 2007). Although swidden agriculture (shifting cultivation) is important along some agricultural frontiers (Finegan and Nasi 2004), the intensity and scale of agriculture in the tropics is generally increasing. Therefore, the many regenerating sites embedded within agricultural landscapes are increasingly isolated from sources of seeds for recolonization.

Many studies in secondary growth habitats in the tropics demonstrate that seed rain and seedling establishment, particularly of large, animal-dispersed species, decline rapidly with increasing distance from the forest edge both in wet and dry forests (e.g., Aide and Cavelier 1994; Harvey 2000; Zimmerman et al. 2000; Mesquita et al. 2001). The scale over which this decline occurs ranges from within a few meters of the pasture edge up to 100 meters. Needless to say, many abandoned old fields are >100 meters from the forest edge, a distance at which there is generally minimal seed rain unless there is woody vegetation to attract seed dispersers. This isolation may be mediated by land use types, such as agroforestry, that facilitate movement of some seed dispersers within the agricultural matrix (Finegan and Nasi 2004; Harvey et al. 2004; Kupfer et al. 2004). Given that the area under secondary succession is increasing, successional forests may be the primary source of seeds of early-successional species in many recovering areas (Finegan and Nasi 2004). Nonetheless, there have been few studies comparing seed rain or vegetation recovery as a function of different surrounding land uses.

The lack of seed dispersal into abandoned old fields can affect both the rate of recovery as well as the successional trajectory, and a number of authors have recorded lower numbers of individuals and species establishing in old fields farther from forest (Mesquita et al. 2001; Chinea 2002; Ferguson et al. 2003). For example, Hooper et al. (2004) found that community composition varied substantially with distance from the forest edge in abandoned pastures in Panama. However, most of these past studies on the effect of distance to forest edge on vegetation community composition have been carried out over relatively short time periods (<5 years) and longer-term studies are critical to furthering our understanding of successional trajectories.

The surrounding landscape also has substantial effects on the faunal communities, which may affect not only seed dispersal, but other common tropical plant community mutualisms, such as pollination and herbivory. There is growing evidence that changes in mammal assemblages due to hunting, isolation, or fragmentation (Dirzo and Miranda 1990; Chapman and Chapman 1995) can be detrimental to the recruitment of many neotropical trees. Large-seeded species that are animal dispersed may be particularly at risk of disappearing from fragmented areas due to loss of dispersers (Cordeiro and Howe 2003). In addition, disruption of these animal communities can cause profound changes in seed fate (Dirzo and Miranda 1990), secondary seed dispersal (Forget 1993), and seedling recruitment (Benítez-Malvido 1998). Despite the likely impacts of the surrounding land use mosaic on seed predation and seedling herbivory, it has received much less study (Holl and Lulow 1997; Duncan and Duncan 2000; Jones et al. 2003) and is a ripe area for future research on vegetation dynamics in tropical old fields.

Type and Intensity of Previous Land Use

Old field succession is strongly influenced by the surrounding landscape as well as local land use history. Numerous studies show that the type of agriculture prior to abandonment and the disturbance intensity/frequency of that agriculture are primary factors affecting both the rate and trajectory of old field succession within a given soil type (reviewed in Guariguata and Ostertag 2001; Myster 2004; Chazdon 2007; chapter 4; table 6.1). Previous land use affects old field succession through a number of mechanisms, including differences in fire frequency, soil physical and chemical properties, and remnant vegetation (Aide et al. 1995; Myster 2004).

Numerous studies have compared the effects of different types of agriculture (e.g., swidden agriculture, coffee plantations, pasture), with the result that recovery of biomass, stem density, height, and species richness is generally fastest in swidden agriculture/shifting cultivation systems and lowest in abandoned pastures (table 6.1). Variable recovery rates were recorded for monoculture crops, such as coffee, and are dependent on management intensity (table 6.1). In some cases, differences in biomass across past land uses resulted from the amount of biomass remaining at the time of abandonment rather than the rate of recovery (Hughes et al. 2000). In contrast to most results, Zarin et al. (2001) found no differences in biomass related to past land uses (either pasture or crop cultivation) in a survey of eighty-five plots along a chronosequence across most of the Brazilian Amazon. This likely reflects the large scale of the study covering a broad climatic and soil gradient and high variability in intensity within each of the land use categories.

In most cases, the species composition recorded on land with different former uses is quite variable (e.g., Pascarella et al. 2000; Mesquita et al. 2001), suggesting different successional trajectories. For example, Aide et al. (1995) found that heavily used pastures are first colonized by a group of shrubs and treelets in the Rubiaceae, Melastomataceae, and Myrtaceae, rather than by pioneer species that typically colonize less intensively disturbed sites. Marcano-Vega et al. (2002) reported that species composition of former shade coffee plantations, sun coffee plantations, and pastures was starting to converge thirty to forty years after abandonment. More long-term studies are needed to determine whether initial differences in vegetation translate into different successional trajectories.

The only study comparing pasture recovery along a gradient of mechanization and intensity of previous use showed lower biomass and species richness in heavily used pastures, as well as different species composition in light, moderate, and heavily used pastures (Uhl et al. 1988). Repeated burning and grazing substantially reduces resprouting (Uhl 1987; Nepstad et al. 1996), which can affect the rate and direction of recovery. Hughes et al. (1999) reported an inverse correlation between the rate of accumulation of biomass following abandonment and the number of years the site had been used for agriculture (a mixture of pasture and cornfields).

Some studies have shown reduced biomass in sites with more cycles of clearing and cultivation in swidden agricultural systems (Hughes et al. 2000; Lawrence et al. 2004), whereas others report no effect on biomass (Steininger 2000; Gehring et al. 2005). Both Steininger and Gehring et al. note, however, that despite biomass similarities, more frequently cropped systems have a different species composition during the successional process than less intensively used sites.

FREQUENCY OF ANTHROPOGENIC FIRE

The relative frequency of fire in different land uses substantially affects the rate and direction of recovery. In wet tropical forest, fires do not occur naturally, and they are historically rare or nonexistent in most dry tropical forests (Janzen 2002; Vieira and Scariot 2006). Therefore, few tropical plants are adapted to fire, whereas most commonly planted pasture grasses resprout quickly after fire. Throughout the tropics anthropogenic fires are common, with burning following land clearing to remove brush and generate a short nutrient pulse. Moreover, fire frequency is increasing as rainfall decreases, due to regional deforestation, and biomass increases in early secondary growth areas (Cochrane et al. 1999; Nepstad et al. 1999).

Fire is a major obstacle to forest recovery as it kills most tropical forest seedlings, reduces resprouting, and strongly reduces the viability of soil seed banks (Uhl et al. 1988; Janzen 2002; Kennard 2002; Hooper et al. 2004). Fire also volatilizes a large amount of N (Davidson et al. 2004), which can reduce N availability. Zarin et al. (2005) compiled data on multiple sites in the Amazon basin to show that sites burned five or more times have 50% reduction in carbon accumulated in regenerating vegetation. Increasing fire frequency can substantially change the direction of the successional trajectory toward fire-adapted species (Hooper et al. 2004; Vieira and Scariot 2006). With increasing fire frequency, some tropical dry forest areas gradually shift toward savannah vegetation (Vieira and Scariot 2006). In dry forests in northwestern

TABLE 6.1

Effect of past land use on recovery of basal area/biomass, stem density, species richness, and vegetation community composition in neotropical old fields, ordered by region.

Authors	Comparison	Location	Basal area/biomass ↓ with ↑ intensity of land use	Stem density ↓ with ↑ intensity of land use	Species richness ↓ with ↑ intensity of land use	Different species composition	Comments
Marcano-Vega et al. 2002	Pasture vs. shade and sun coffee plantation	Puerto Rico				ns	Suggest that over time (30–40 years) the composition of secondary forest on different land use types is becoming more similar.
Pascarella et al. 2000	Pasture vs. coffee plantations	Puerto Rico	+		−	+	Higher biomass in coffee, but higher species richness in pasture.
Martínez-Ramos (unpub.)	Pastures vs. swidden	Mexico	+	+	−		Species richness higher in pasture than in swidden.
Méndez-Bahena 1999	Pasture vs. swidden	Chiapas, Mexico	+			+	
Ferguson et al. 2003	Pasture, swidden[a], agroforestry[b], monoculture	Guatemala	+	+	+		Swidden and agroforestry, higher biomass, stem density, and richness; monoculture and pasture, lower.
Griffith 2004	Pasture, swidden, agroforestry, monoculture	Nicaragua	+	+	+		Swidden, highest biomass, stem density, and richness; agroforestry (including cattle) and pasture, intermediate; monoculture, lowest

Study	Land use comparison	Location			Comments
Mesquita et al. 2001	Immediate regrowth[c] vs. pasture	Amazon, Brazil	+	+	
Moran et al. 2000	Swidden vs. mechanized agriculture and pasture	Amazon, Brazil, and Rio Negro Basin, Columbia	+	+	Canopy height taller in shifting cultivation than mechanized agriculture or pasture within a given soil type.
Steininger 2000	Cultivated for 1 year vs. >5 years of pasture usage	Amazon, Brazil	+		
Zarin et al. 2001	Pasture vs. crop cultivation	Amazon, Brazil	ns		Sites distributed over a larger scale than other studies, reflecting a wide range of pasture and crop use intensities. Soil texture and climate important predictors.

Note: + = effect of past land use in the direction stated; minus sign = effect of past land use the opposite direction than that stated; ns = no significant effect of past land use; blank = no data.

[a] Swidden agriculture refers to a shifting agricultural system where an area of land is cleared and then used for production of subsistence crops for one to three years.

[b] Agroforestry refers to a permanent agricultural system where crops are mixed with trees.

[c] Immediate regrowth refers to when a forest was cleared but never used for agriculture.

Costa Rica, fire control is the main "restoration" strategy, because in the absence of fire, forests recover quickly without intervention (Janzen 2002).

LAND USE EFFECTS ON SOIL PHYSICAL AND CHEMICAL CONDITIONS

As discussed previously, natural soil differences at multiple spatial scales affect the rate of biomass accumulation and height growth (e.g., Moran et al. 2000; Zarin et al. 2001; Thompson et al. 2002). Differences in compaction, texture, nutrient availability, and microbial composition certainly reflect natural underlying edaphic gradients but are also strongly influenced by past land use type and intensity.

Mechanized agriculture can heavily compact soils, which changes soil nutrient availability and impedes root growth, thereby slowing recovery (Buschbacher et al. 1988; Uhl et al. 1988). Likewise, intensively used soils may have altered microbial communities (Cleveland et al. 2003) and lack mycorrhizae (Janos 1996; Carpenter et al. 2001), which form associations with many tropical tree species. Previous land use can leave a legacy of differences in crop root exudates, which may affect soil microbes, decomposition, and seed germination (Myster 2004), but these effects have received little study. In the most highly disturbed soils, recovery can be extremely slow, and the system may not recover without intervention (Parrotta 1995; Carpenter et al. 2001; Montagnini 2001).

Other than the most severely degraded tropical old fields, however, the degree to which nutrient availability limits the rate and trajectory of succession is debated. Some support the position that soil nutrients have an important effect on succession (Chazdon 2003; Lawrence 2003; Ceccon et al. 2004), whereas others argue that nutrient limitation does not affect recovery (Harcombe 1980; Ganade and Brown 2002). Certainly, soils in neotropical old fields are generally low in nutrients, most commonly phosphorus, and they may have low pH and high levels of exchangeable aluminum, but so do most forest soils to which these species are adapted (Vitousek 1984). Several studies have shown that fertilization of neotropical old fields results in a short-term increase in the relative biomass of grasses and other ruderal, herbaceous species (Harcombe 1977; Uhl 1987; Gehring et al. 1999; Davidson et al. 2004). Some studies have demonstrated that fertilization with N and/or P may also enhance the biomass of tree species (Vieira et al. 1994; Gehring et al. 1999; Holl et al. 2000; Davidson et al. 2004), whereas other studies have reported no response of trees to fertilization (Uhl 1987; Nepstad et al. 1996; Holl 1999; Ganade and Brown 2002). Guariguata et al. (1997) attribute rapid recovery of forest in abandoned pasture in lowland rain forest in Costa Rica in part to fertile soils. These contradictory conclusions likely reflect both past land use differences as well as variation in soil types.

Vegetation Composition at Time of Abandonment

Previous land use strongly affects the vegetation composition of the old field at the time of abandonment. This vegetation may either inhibit, facilitate, or both inhibit and facilitate the establishment of later-successional species (sensu Connell and Slatyer 1977). African pasture grasses often form a dense monoculture up to 3 m tall in previously grazed areas (Holl 1999; Hooper et al. 2004; Jones et al. 2004), which can arrest succession, at least over the short term (e.g., ten to fifteen years) by forming a strong barrier to the establishment of woody seedlings (Aide et al. 1995; Holl et al. 2000). Likewise, dense ferns, vines, and some early-successional shrubs can inhibit recovery (e.g., Uhl et al. 1988; Zahawi and Augspurger 1999; Slocum et al. 2004). While in most cases the net effect of these ruderal species is to strongly inhibit establishment and growth of woody species, they may ameliorate stressful microclimate conditions in abandoned old fields, thereby facilitating seed germination, particularly in drier tropical forests (Holl 1999; Ganade and Brown 2002; Vieira and Scariot 2006; chapter 4). These results parallel research in the temperate zone showing that early-successional vegetation may simultaneously facilitate and inhibit later vegetation depending on the life history stage and environmental conditions of the site (e.g., Walker and Chapin 1986; Gill and Marks 1991).

In contrast, when past land use leaves a range of woody species, such as shade coffee, pastures with remnant trees, or living fences, these trees strongly facilitate recovery by increasing seed dispersal, ameliorating microclimatic conditions, and increasing soil nutrients (reviewed in Harvey et al. 2004). For example, Holl et al. (2000) found that the number of animal-dispersed seeds falling under remnant pasture trees was similar to the number in the forest and two orders of magnitude higher than in open pasture. These remnant trees may remain in the sites for a long time after abandonment, thereby affecting community composition (Marcano-Vega et al. 2002).

Conclusions and Implications for Restoration

Despite the fact that the study of tropical old field succession has received much less study than succession in the temperate zone, a number of common themes emerge from the growing number of studies of the recovery of

TABLE 6.2

Future directions in tropical old field succession and restoration

Priorities for future research

- Long-term data sets on tropical forest recovery in the same sites (rather than chronosequences), particularly evaluating the effect of different restoration treatments on successional trajectories.
- Studies of tropical forest succession and restoration replicated across large spatial scales to separate site and landscape factors affecting recovery and to allow for general recommendations for restoration.
- Investigation of the effect of the surrounding landscape mosaics on forest regeneration, in particular, plant–animal interactions (e.g., seed dispersal, seed predation, seedling herbivory).
- Experiments testing when in restoration to introduce later successional species that would naturally colonize over extremely slow time scales. Requires an understanding of the physiological tolerances of these species.

Recommendations for restoration

- Tailor restoration strategies to the degree of degradation from past land use and the resilience of the system. Requires some baseline data on factors limiting natural recovery.
- Focus on introducing later-successional, dispersal-limited species in sites where pioneer species establish quickly.
- Prioritize restoration sites with diverse sources of plant propagules nearby.
- Consider long-term effects on succession of species introduced as part of restoration efforts.

neotropical forest in old fields (table 6.2). It is clear that the rate of recovery is quite fast in many neotropical old fields, with a number of studies reporting 75%–100% of forest biomass regenerating within twenty to thirty years after abandonment (Guariguata et al. 1997; Marcano-Vega et al. 2002; Chazdon 2003; Ruiz et al. 2005). In other sites, particularly with aggressive ruderal vegetation, recovery may be arrested for at least a number of years (Aide et al. 1995; Zahawi and Augspurger 1999; Holl et al. 2000).

As I argue earlier, much of the variance in the rate and trajectory of recovery can be explained by three broad categories of variables (figure 6.2): abiotic conditions, surrounding landscape, and previous land use type and intensity. Unfortunately, few past studies of regeneration and restoration have compared the relative importance of factors at different scales on forest regeneration, as most have been conducted on small plots at single sites (reviewed in Holl 2002; Meli 2003). So, it is often impossible to discern which of these parameters are most important over a large scale, although the rate of recovery is almost certainly due to a combination of factors at a range of scales.

Restorationists have the greatest possibility to influence the second and third set of variables. Restoration projects are often opportunistic, but given the large areas of degraded agricultural lands in Central and South America and the limited resources available, acquisition of former agricultural lands for conservation should be prioritized in areas where there are sources of propagules from remnant forests, buffer strips, or living fences in the landscape (Harvey et al. 2004). Such areas would require less effort to reintroduce the diverse set of tropical forest species. In areas lacking nearby sources of propagules, more resources should be invested in providing structure, such as patches of trees (Zahawi and Augspurger 2006) to attract seed dispersers.

Likewise, it is important to tailor restoration strategies to the degree of degradation from past land use and the resilience of the system. In some resilient systems, the only intervention necessary to jump-start recovery is to remove ongoing stresses such as fire and grazing (Janzen 2002). In fact, in some cases, more aggressive restoration efforts may be detrimental to recovery. For example, Sampaio et al. (forthcoming) have shown that planting seedlings in pastures in Brazilian dry forests where extensive resprouting occurs decreases stem density because planting seedlings kills existing roots. In cases where recovery is slow, the most broadly effective strategy is to plant a suite of fast-growing tree species, which attract seed dispersers, improve soil structure and soil nutrient availability, shade out pasture grasses, and provide intermediate light conditions (Montagnini et al. 1995; Loik and Holl 1999; Jones et al. 2004; Lamb et al. 2005). It is important to consider, however, the effect of these species on the successional trajectory, as the composition of overstory species that are planted can affect the density and richness of seedlings that establish. These species also may inhibit growth of naturally establishing seedlings, which will affect long-term forest structure (Parrotta 1995; Cusack and Montagnini 2004; Silver et al. 2004).

The rate of recovery depends not only on the surrounding landscape and past disturbance regime but also on the parameter measured. In nearly every study of neotropical old field succession, biomass and species richness have recovered much faster than species composition (Finegan 1996). Specifically, many studies of secondary tropical forests ranging in age from thirty to sixty years show that these forests lack some species, in particular, large-seeded, mature-forest species (Aide et al. 2000; Chinea 2002; Martínez-Garza and Howe 2003; chapter 4). This is likely due to dispersal limitation, and these species may colonize with more time. In addition, many tropical trees regenerate under the existing forest canopy, and this advance regeneration is largely lacking in old fields. It may be that if these species are not present at the beginning of succession they may not establish until the forest matures to the point of demonstrating gap dynamics (Finegan 1996; Chazdon 2007).

Numerous authors have suggested that the most effective restoration strategy for tropical old fields that are rapidly colonized by pioneer species is to plant selected, mature-forest species that do not readily recolonize (Martínez-Garza and Howe 2003; Lamb et al. 2005; chapter 4), in order to more quickly restore the full complement of species. Some of these species have shown high survival in recently abandoned pasture (Loik and Holl 1999; Hooper et al. 2002), but more research is needed on when in the successional process these species should be introduced to maximize survival and growth.

Deterministic Versus Stochastic Succession

As discussed earlier, ecologists have long debated whether ecosystem succession is deterministic given the environmental conditions, or whether it is stochastic and may result in different endpoints. There is no question that highly diverse tropical systems succeed to a more diverse set of endpoints than is commonly reported in studies of succession in temperate old fields (Chazdon 2007). There is often a predictable set of pioneer species that colonize early in succession, but longer-term old field studies (Finegan and Delgado 2000; Capers et al. 2005), as well as studies of recovery from hurricanes (Vandermeer et al. 2004), commonly show strong divergence in species composition on apparently similar sites.

There are several possible explanations for these alternative successional trajectories. First, there may be subtle environmental gradients that explain differences in species composition, so in fact succession is deterministic dependent upon environmental conditions. Second, research in less disturbed tropical forests suggests that seedling establishment is strongly dispersal limited, meaning that species composition is highly dependent on what propagules arrive at the site (Webb 1972; Dalling et al. 1998; Hubbell et al. 1999). High variation in the composition of seeds arriving at a given site is expected given the large number of potential species, common episodic fruiting in the tropics, and the lack of advance regeneration. In sites with a high degree of resprouting, succession may be more deterministic given that many species are likely to regenerate in the same location.

The fact that there may be multiple alternative endpoints for tropical old field succession means that it will be more difficult for restorationists to achieve a specific desired state, at least with respect to species composition. If a restoration project is aimed at providing certain services, such as erosion control or carbon sequestration, this may not matter. But, if conservation of certain species is desired, then it may be necessary to introduce later-successional species, as discussed earlier. As Chazdon (2007) highlights, many reference forest remnants have been subjected to human intervention over time, so it is difficult to define what is a realistic endpoint for tropical forest restoration.

Initial Floristics Versus Relay Floristics

A final theoretical question I posed at the outset was whether tropical forest succession more closely follows an initial floristics or relay floristics model. In forests where a large percentage of the species resprout, many of the species that were there prior to disturbance will be present early in succession and will likely remain until later in succession. Although there are no detailed studies of the long-term dynamics of resprouts, resprout seedlings have higher survival than those established from seed (Ceccon et al. 2004). Drier tropical forests, where drought stress limits seedling survival (Vieira and Scariot 2006), seem to have few successional stages (Ewel 1977; Kennard 2002). Many wet forest studies, in which most of the dominant species are animal dispersed, show two to four stages of dominant species during succession (Finegan 1996; Tucker et al. 1998; Guariguata and Ostertag 2001; Peña-Claros 2003). The question that remains is whether those species all establish in sites within the first few years and simply mature at different rates or whether they colonize at different times (Finegan 1996; Peña-Claros 2003). Because there have been no studies of tropical old fields that have begun monitoring at the time of abandonment and continued for multiple decades, it is a difficult question to answer. Chazdon (2007) suggests that quite a number of species establish early on in succession, but there are rarer species that do not establish until later. From a restoration perspective, a better understanding of when different species establish during succession and what the physiological tolerances of species are to different abiotic conditions is critical to determine when in the succession process target species should be introduced.

Acknowledgments

Discussions with my graduate students and participants in the National Center for Ecological Analysis and Synthesis working group on "Biodiversity and conservation value of agricultural landscapes of Mesoamerica" were influential in developing this chapter. I appreciate helpful comments on the manuscript from R. Cole, V. Cramer, D. Vieira, and two anonymous reviewers.

References

Aide, T. M., and J. Cavelier. 1994. Barriers to lowland tropical forest restoration in the Sierra Nevada de Santa Marta, Colombia. *Restoration Ecology* 2:219–29.

Aide, T. M., J. K. Zimmerman, L. Herrera, M. Rosario, and M. Serrano. 1995. Forest recovery in abandoned tropical pasture in Puerto Rico. *Forest Ecology and Management* 77:77–86.

Aide, T. M., J. K. Zimmerman, J. B. Pascarella, L. Rivera, and H. Marcano-Vega. 2000. Forest regeneration in a chronosequence of tropical abandoned pastures: Implications for restoration ecology. *Restoration Ecology* 8:328–38.

Arroyo-Mora, J. P., G. A. Sanchez-Azofeifa, B. Rivard, J. C. Calvo, and D. H. Janzen. 2005. Dynamics in landscape structure and composition for the Chorotega region, Costa Rica, from 1960 to 2000. *Agriculture Ecosystems and Environment* 106:27–39.

Benítez-Malvido, J. 1998. Impact of forest fragmentation on seedling abundance in a tropical rain forest. *Conservation Biology* 12:380–89.

Benítez-Malvido, J., M. Martínez-Ramos, and E. Ceccon. 2001. Seed rain vs. seed bank, and the effect of vegetation cover on the recruitment of tree seedlings in tropical successional vegetation. In *Life forms and dynamics in tropical forests*, ed. G. Gottsberger and Sigrid Liede, 185–203. J. Cramer, Berlin.

Buschbacher, R., C. Uhl, and E. A. S. Serrao. 1988. Abandoned pastures in eastern Amazonia. 2. Nutrient stocks in the soil and vegetation. *Journal of Ecology* 76:682–99.

Campos, J. 2001. Management of goods and services from neotropical forest biodiversity: Diversified forest management in Mesoamerica. In *Assessment, conservation and sustainable use of forest biodiversity*, 5–16. CBD Technical Series No. 3, Montreal.

Capers, R. S., R. L. Chazdon, A. R. Brenes, and B. V. Alvarado. 2005. Successional dynamics of woody seedling communities in wet tropical secondary forests. *Journal of Ecology* 93:1071–84.

Carpenter, F. L., S. P. Mayorga, E. G. Quintero, and M. Schroeder. 2001. Land-use and erosion of a Costa Rican ultisol affect soil chemistry, mycorrhizal fungi and early regeneration. *Forest Ecology and Management* 144:1–17.

Ceccon, E., S. Sanchez, and J. Campo. 2004. Tree seedling dynamics in two abandoned tropical dry forests of differing successional status in Yucatan, Mexico: A field experiment with N and P fertilization. *Plant Ecology* 170:277–85.

Chapman, C. A., and L. J. Chapman. 1995. Survival without dispersers: Seedling recruitment under parents. *Conservation Biology* 9:675–78.

Chazdon, R. L. 2003. Tropical forest recovery: Legacies of human impact and natural disturbances. *Perspectives in Plant Ecology Evolution and Systematics* 6:51–71.

———. 2007. Dynamics of tropical secondary forests. In *Chance and determinism in tropical forest succession*, ed. S. A. Schnitzer and W. P. Carson. Blackwell Publishing, Oxford, UK.

Chinea, J. D. 2002. Tropical forest succession on abandoned farms in the Humacao Municipality of eastern Puerto Rico. *Forest Ecology and Management* 167:195–207.

Clement, R. M., and S. P. Horn. 2001. Precolumbia land-use history in Costa Rica: A 3000-year record of forest clearance, agriculture and fires from Laguna Zoncho. *Holocene* 11:419–26.

Clements, F. E. 1916. *Plant succession and indicators*. Henry Wilson, New York.

Cleveland, C. C., A. R. Townsend, S. K. Schmidt, and B. C. Constance. 2003. Soil microbial dynamics and biogeochemistry in tropical forests and pastures, southwestern Costa Rica. *Ecological Applications* 13:314–26.

Cochrane, M. A., A. Alencar, M. D. Schulze, C. M. Souza, D. C. Nepstad, P. Lefebvre, and E. A. Davidson. 1999. Positive feedbacks in the fire dynamic of closed canopy tropical forests. *Science* 284:1832–35.

Connell, J. H., and R. O. Slatyer. 1977. Mechanisms of succession in natural communities and their role in community stability and organization. *American Naturalist* 111:1119–41.

Cordeiro, N. J., and H. F. Howe. 2003. Forest fragmentation severs mutualism between seed dispersers and an endemic African tree. *Proceedings of the National Academy of Sciences* 100:14052–56.

Cusack, D., and F. Montagnini. 2004. The role of native species plantations in recovery of understory woody diversity in degraded pasturelands of Costa Rica. *Forest Ecology and Management* 188:1–15.

Dalling, J. W., S. P. Hubbell, and K. Silvera. 1998. Seed dispersal, seedling establishment and gap partitioning among tropical pioneer trees. *Journal of Ecology* 86:674–89.

Davidson, E. A., C. J. R. De Carvalho, I. C. G. Vieira, R. D. Figueiredo, P. Moutinho, F. Y. Ishida, M. T. P. Dos Santos, J. B. Guerrero, K. Kalif, and R. T. Saba. 2004. Nitrogen and phosphorus limitation of biomass growth in a tropical secondary forest. *Ecological Applications* 14:S150–S163.

Dirzo, R., and A. Miranda. 1990. Contemporary neotropical defaunation and forest structure, function, and diversity: A sequel. *Conservation Biology* 4:444–47.

Duncan, R., and V. Duncan. 2000. Forest succession and distance from forest edge in an afro-tropical grassland. *Biotropica* 31:33–41.

Egler, F. E. 1954. 1. Initial floristic composition: A factor in old field vegetation development. *Vegetatio* 4:412–17.

Ewel, J. 1980. Tropical succession: Manifold routes to maturity. *Biotropica* 12:2–7.

Ewel, J. J. 1977. Differences between wet and dry successional tropical ecosystems. *Geo-Eco-Trop* 1:103–17.

Ferguson, B. G., J. Vandermeer, H. Morales, and D. M. Griffith. 2003. Post-agricultural succession in El Peten, Guatemala. *Conservation Biology* 17:818–28.

Finegan, B. 1996. Pattern and process in neotropical secondary rain forests: The first 100 years of succession. *Trends in Ecology and Evolution* 11:119–24.

Finegan, B., and D. Delgado. 2000. Structural and floristic heterogeneity in a 30-year-old Costa Rican rain forest restored on pasture through natural secondary succession. *Restoration Ecology* 8:380–93.

Finegan, B., and R. Nasi. 2004. The biodiversity and conservation potential of shifting cultivation landscapes. In *Agroforestry and biodiversity conservation in tropical landscapes*, ed. G. Schroth, G. A. B. da Fonseca, C. A. Harvey, C. Gascon, H. L. Vasconcelos, and A. N. Izac, 153–97. Island Press, Washington, DC.

Forget, P. M. 1993. Postdispersal predation and scatterhoarding of *Dipteryx panamensis* (Papilionaceae) seeds by rodents in Panama. *Oecologia* 94:255–61.

Ganade, G., and V. K. Brown. 2002. Succession in old pastures of Central Amazonia: Role of soil fertility and plant litter. *Ecology* 83:743–54.

Gehring, C., M. Denich, M. Kanashiro, and P. L. G. Vlek. 1999. Response of secondary vegetation in eastern Amazonia to relaxed nutrient availability constraints. *Biogeochemistry* 45:223–41.

Gehring, C., M. Denich, and P. L. G. Vlek. 2005. Resilience of secondary forest regrowth after slash-and-burn agriculture in Central Amazonia. *Journal of Tropical Ecology* 21:519–27.

Geist, H. J., and E. F. Lambin. 2002. Proximate causes and underlying driving forces of tropical deforestation. *Bioscience* 52:143–50.

Gill, D. S., and P. L. Marks. 1991. Tree and shrub seedling colonization of old fields in central New York. *Ecological Monographs* 61:183–205.

Gleason, H. A. 1917. The structure and development of the plant association. *Bulletin of the Torrey Botanical Club* 44:463–81.

Grau, H. R., T. M. Aide, J. K. Zimmerman, J. R. Thomlinson, E. Helmer, and X. M. Zou. 2003. The ecological consequences of socioeconomic and land-use changes in post-agriculture Puerto Rico. *Bioscience* 53:1159–68.

Griffith, D. M. 2004. *Succession of tropical rain forest along a gradient of agricultural intensification: Patterns, mechanisms and implications for conservation.* PhD dissertation, University of Michigan, Ann Arbor.

Guariguata, M. R., R. L. Chazdon, J. S. Denslow, J. M. Dupuy, and L. Anderson. 1997. Structure and floristics of secondary and old-growth forest stands in lowland Costa Rica. *Plant Ecology* 132:107–20.

Guariguata, M. R., and R. Ostertag. 2001. Neotropical secondary forest succession: Changes in structural and functional characteristics. *Forest Ecology and Management* 148:185–206.

Harcombe, P. A. 1977. The influence of fertilization on some aspects of succession in a humid tropical forest. *Ecology* 58:1375–83.

———. 1980. Soil nutrient loss as a factor in early tropical secondary succession. *Biotropica* 12 (supp.):8–15.

Harvey, C. A. 2000. Windbreaks enhance seed dispersal into agricultural landscapes in Monteverde, Costa Rica. *Ecological Applications* 10:155–173.

Harvey, C. A., N. I. Tucker, and A. Estrada. 2004. Live fences, isolated trees, and windbreaks: Tools for conserving biodiversity in fragmented tropical landscapes. In *Agroforestry and biodiversity conservation in tropical landscapes*, ed. G. Schroth, G. A. B. da Fonseca, C. A. Harvey, C. Gascon, H. L. Vasconcelos, and A. N. Izac, 261–89. Island Press, Washington, DC.

Hecht, S. B., S. Kandel, I. Gomes, N. Cuellar, and H. Rosa. 2006. Globalization, forest resurgence, and environmental politics in El Salvador. *World Development* 34:308–23.

Herrera, B., and B. Finegan. 1997. Substrate conditions, foliar nutrients and the distributions of two canopy tree species in a Costa Rican secondary rain forest. *Plant and Soil* 191:259–67.

Holl, K. D. 1999. Factors limiting tropical rain forest regeneration in abandoned pasture: Seed rain, seed germination, microclimate, and soil. *Biotropica* 31:229–41.

———. 2002. Tropical moist forest restoration. In *Handbook of ecological restoration*, Vol. 2., ed. A. J. Davy and M. Perrow, 539–58. Cambridge University Press, Cambridge, UK.

Holl, K. D., M. E. Loik, E. H. V. Lin, and I. A. Samuels. 2000. Tropical montane forest restoration in Costa Rica: Overcoming barriers to dispersal and establishment. *Restoration Ecology* 8:339–49.

Holl, K. D., and M. E. Lulow. 1997. Effects of species, habitat, and distance from edge on post-dispersal seed predation in a tropical rainforest. *Biotropica* 29:459–68.

Hooper, E., R. Condit, and P. Legendre. 2002. Responses of 20 native tree species to reforestation strategies for abandoned farmland in Panama. *Ecological Applications* 12:1626–41.

Hooper, E. R., P. Legendre, and R. Condit. 2004. Factors affecting community composition of forest regeneration in deforested, abandoned land in Panama. *Ecology* 85:3313–26.

Howe, H. F., and J. Smallwood. 1982. Ecology of seed dispersal. *Annual Review of Ecology and Systematics* 82:201–28.

Hubbell, S. P., R. B. Foster, S. T. O'Brien, K. E. Harms, R. Condit, B. Wechsler, S. J. Wright, and S. L. De Lao. 1999. Light-gap disturbances, recruitment limitation, and tree diversity in a neotropical forest. *Science* 283:554–57.

Hughes, R. F., J. B. Kauffman, and D. L. Cummings. 2000. Fire in the Brazilian Amazon. 3. Dyamics of biomass, C, and nutrient pools in regenerating forests. *Oecologia* 124:574–88.

Hughes, R. F., J. B. Kauffman, and V. J. Jaramillo. 1999. Biomass, carbon, and nutrient dynamics of secondary forests in a humid tropical region of Mexico. *Ecology* 80:1892–1907.

Janos, D. P. 1996. Mycorrhizas, successsion, and the rehabilitation of deforested lands in the humid tropics. In *Fungi and environmental change*, ed. J. C. Frankland, N. Magan, and G. M. Gadd, 129–62. Cambridge University Press, Cambridge, UK.

Janzen, D. H. 2002. Tropical dry forest: Area de Conservacion Guancaste, northwestern Costa Rica. In *Handbook of ecological restoration*, Vol. 2., ed. by A. J. Davy and M. Perrow, 559–83. Cambridge University Press, Cambridge, UK.

Jones, E. R., M. H. Wishnie, J. Deago, A. Sautu, and A. Cerezo. 2004. Facilitating natural regeneration in *Saccharum spontaneum* (L.) grasslands within the Panama Canal watershed: Effects of tree species and tree structure on vegetation recruitment patterns. *Forest Ecology and Management* 191:171–83.

Jones, F. A., C. J. Peterson, and B. L. Haines. 2003. Seed predation in neotropical premontane pastures: Site, distance, and species effects. *Biotropica* 35:219–25.

Kennard, D. K. 2002. Secondary forest succession in a tropical dry forest: Patterns of development across a 50-year chronosequence in lowland Bolivia. *Journal of Tropical Ecology* 18:53–66.

Kupfer, J. A., A. L. Webbeking, and S. B. Franklin. 2004. Forest fragmentation affects early successional patterns on shifting cultivation fields near Indian Church, Belize. *Agriculture Ecosystems and Environment* 103:509–18.

Lamb, D., P. D. Erskine, and J. D. Parrotta. 2005. Restoration of degraded tropical forest landscapes. *Science* 310:1628–32.

Lawrence, D. 2003. The response of tropical tree seedlings to nutrient supply: Meta-analysis for understanding a changing tropical landscape. *Journal of Tropical Ecology* 19:239–50.

Lawrence, D., H. F. M. Vester, D. Perez-Salicrup, J. R. Eastman, B. L. Turner II, and J. Geoghegan. 2004. Integrated analysis of ecosystem interactions with land-use change: The southern Yucatan peninsular region. In *Ecosystems and land-use change*, ed. R. S. DeFries, G. P. Asner, and R. A. Houghton, 277–92. American Geophysical Union, Washington, DC.

Lockwood, J. L. 1997. An alternative to succession. *Restoration and Management Notes* 15:45–50.

Loik, M. E., and K. D. Holl. 1999. Photosynthetic responses to light for rainforest seedlings planted in abandoned pasture, Costa Rica. *Restoration Ecology* 7:382–91.

Marcano-Vega, H., T. M. Aide, and D. Baez. 2002. Forest regeneration in abandoned coffee plantations and pastures in the Cordillera Central of Puerto Rico. *Plant Ecology* 161:75–87.

Martínez-Garza, C., and H. F. Howe. 2003. Restoring tropical diversity: Beating the time tax on species loss. *Journal of Applied Ecology* 40:423–29.

Mayaux, P., P. Holmgren, F. Achard, H. Eva, H. Stibig, and A. Branthomme. 2005.

Tropical forest cover change in the 1990s and options for future monitoring. *Philosophical Transactions of the Royal Society B* 360:373–84.

Meli, P. 2003. Tropical forest restoration: Twenty years of academic research. *Interciencia* 28:581–89.

Méndez-Bahena, A. 1999. Sucesión secundaria de la selva húmeda y conservación de recursos naturales en Máquez de Comillas, Chiapas. MS thesis, Universidad de Michoacana de San Nicolás de Hidalgo, Michoacan, Mexico.

Mesquita, R. C., G. K. Ickes, G. Ganade, and G. B. Williamson. 2001. Alternative successional pathways in the Amazon Basin. *Journal of Ecology* 89:528–37.

Montagnini, F. 2001. Strategies for the recovery of degraded ecosystems: Experiences from Latin America. *Interciencia* 26:498–506.

Montagnini, F., A. Fanzeres, and S. Guimaraes da Vinha. 1995. The potentials of 20 indigenous tree species for soil rehabilitation in the Atlantic forest region of Bahia, Brazil. *Journal of Applied Ecology* 32:841–56.

Moran, E. F., E. S. Brondizio, J. M. Tucker, M. C. Da Silva-Forsberg, S. McCracken, and I. Falesi. 2000. Effects of soil fertility and land-use on forest succession in Amazonia. *Forest Ecology and Management* 139:93–108.

Myster, R. W. 2004. Post-agricultural invasion, establishment, and growth of neotropical trees. *Botanical Review* 70:381–402.

Nepstad, D. C., C. Uhl, C. A. Pereira, and J. M. Cardoso da Silva. 1996. A comparative study of tree establishment in abandoned pasture and mature forest of eastern Amazonia. *Oikos* 76:25–39.

Nepstad, D. C., A. Verissimo, A. Alencar, C. Nobre, E. Lima, P. Lefebvre, P. Schlesinger, C. Potter, P. Moutinho, E. Mendoza et al. 1999. Large-scale impoverishment of Amazonian forests by logging and fire. *Nature* 398:505–8.

Parrotta, J. A. 1995. Influence of overstory composition on understory colonization by native species in plantations on a degraded tropical site. *Journal of Vegetation Science* 6:627–36.

Pascarella, J. B., T. M. Aide, M. I. Serrano, and J. K. Zimmerman. 2000. Land-use history and forest regeneration in the Cayey Mountains, Puerto Rico. *Ecosystems* 3:217–28.

Peña-Claros, M. 2003. Changes in forest structure and species composition during secondary forest succession in the Bolivian Amazon. *Biotropica* 35:450–61.

Ruiz, J., M. C. Fandino, and R. L. Chazdon. 2005. Vegetation structure, composition, and species richness across a 56-year chronosequence of dry tropical forest on Providencia Island, Colombia. *Biotropica* 37:520–30.

Sampaio, A. B., K. D. Holl, and A. Scariot. Forthcoming. Regeneration of seasonal deciduous forest tree species in long-used pastures in Central Brazil. *Biotropica*.

Sarmiento, F. O. 1997. Landscape regeneration by seeds and successional pathways to restore fragile tropandean slopelands. *Mountain Research and Development* 17:239–52.

Silver, W. L., L. M. Kueppers, A. E. Lugo, R. Ostertag, and V. Matzek. 2004. Carbon sequestration and plant community dynamics following reforestation of tropical pasture. *Ecological Applications* 14:1115–27.

Silver, W. L., R. Ostertag, and A. E. Lugo. 2000. The potential for carbon sequestration through reforestation of abandoned tropical agricultural and pasture lands. *Restoration Ecology* 8:394–407.

Slocum, M. G., T. M. Aide, J. K. Zimmerman, and L. Navarro. 2004. Natural regeneration of subtropical montane forest after clearing fern thickets in the Dominican Republic. *Journal of Tropical Ecology* 20:483–86.

Steininger, M. K. 2000. Secondary forest structure and biomass following short and extended land-use in central and southern Amazonia. *Journal of Tropical Ecology* 16:689–708.

Tansley, A. G. 1935. The use and abuse of vegetational concepts and terms. *Ecology* 16:284–307.

Temperton, V. M., and R. J. Hobbs. 2004. The search for ecological assembly rules and its relevance to restoration ecology. In *Assembly rules and restoration ecology*, ed. V. M. Temperton, R. J. Hobbs, T. Nuttle, and S. Halle, 34–54. Island Press, Washington, DC.

Thompson, J., N. Brokaw, J. K. Zimmerman, R. B. Waide, E. M. Everham III, D. J. Lodge, C. M. Taylor, D. Garcia-Montiel, and M. Fluet. 2002. Land use history, environment, and tree composition in a tropical forest. *Ecological Applications* 12:1344–63.

Tucker, J. M., E. S. Brondizio, and E. F. Moran. 1998. Rates of forest regrowth in eastern Amazonia: A comparison of Altamira and Bragantina regions, Para State, Brazil. *Interciencia* 23:64–73.

Uhl, C. 1987. Factors controlling succession following slash-and-burn agriculture. *Journal of Ecology* 75:377–407.

Uhl, C., R. Buschbacher, and E. A. S. Serrao. 1988. Abandoned pastures in eastern Amazonia. 1. Patterns of plant succession. *Journal of Ecology* 76:663–81.

Uhl, C., D. Nepstad, R. Buschbacher, K. Clark, B. Kauffman, and S. Subler. 1989. Disturbance and regeneration in Amazonia: Lessons for sustainable land-use. *Ecologist* 19:235–40.

Vandermeer, J., I. G. De La Cerda, I. Perfecto, D. Boucher, J. Ruiz, and A. Kaufmann. 2004. Multiple basins of attraction in a tropical forest: Evidence for nonequilibrium community structure. *Ecology* 85:575–79.

Vesk, P. A., and M. Westoby. 2004. Sprouting ability across diverse disturbances and vegetation types worldwide. *Journal of Ecology* 92:310–20.

Vieira, D. L. M., and A. Scariot. 2006. Principles of natural regeneration of tropical dry forests for restoration. *Restoration Ecology* 14:11–20.

Vieira, D. L. M., A. Scariot, A. B. Sampaio, and K. D. Holl. 2006. Tropical dry-forest regeneration from root suckers in Central Brazil. *Journal of Tropical Ecology* 22: 353–57.

Vieira, I. C. G., C. Uhl, and D. Nepstad. 1994. The role of the shrub *Cordia multispicata* Cham. as a "succession facilitator" in an abandoned pasture, Paragominas, Amazonia. *Vegetatio* 115:91–99.

Vitousek, P. M. 1984. Litterfall, nutrient cycling, and nutrient limitation in tropical forests. *Ecology* 65:285–98.

Walker, L. R., and F. S. I. Chapin. 1986. Physiological controls over seedling growth in primary succession on an Alaskan floodplain. *Ecology* 67:1508–23.

Webb, L. J., J. G. Tracey, and W. T. Williams. 1972. Regeneration and pattern in the subtropical rain forest. *Journal of Ecology* 60:675–96.

Young, T. P., J. M. Chase, and R. T. Huddleston. 2001. Community succession and assembly: Comparing, contrasting and combining paradigms in the context of ecological restoration. *Ecological Restoration* 19:5–18.

Zahawi, R. A., and C. K. Augspurger. 1999. Early plant succession in abandoned pastures in Ecuador. *Biotropica* 31:540–52.

———. 2006. Tropical forest restoration: Tree islands as recruitment foci in degraded lands of Honduras. *Ecological Applications* 16:464–78.

Zarin, D. J., E. A. Davidson, E. Brondizio, I. C. G. Vieira, T. Sa, T. Feldpausch, E. A. Schuur, R. Mesquita, E. Moran, P. Delamonica, M. J. Ducey, G. C. Hurtt, C. Salimon, and M. Denich. 2005. Legacy of fire slows carbon accumulation in Amazonian forest regrowth. *Frontiers in Ecology and the Environment* 3:365–69.

Zarin, D. J., M. J. Ducey, J. M. Tucker, and W. A. Salas. 2001. Potential biomass accumulation in Amazonian regrowth forests. *Ecosystems* 4:658–68.

Zimmerman, J. K., J. B. Pascarella, and T. M. Aide. 2000. Barriers to forest regeneration in an abandoned pasture in Puerto Rico. *Restoration Ecology* 8:350–60.

Chapter 7

Patterns and Processes of Old Field Reforestation in Australian Rain Forest Landscapes

PETER D. ERSKINE, CARLA P. CATTERALL,
DAVID LAMB, AND JOHN KANOWSKI

Rain forests were widely cleared in Australia during the nineteenth and twentieth centuries to establish agriculture in areas thought to have high soil fertility and to be relatively drought free. Some of these areas were abandoned soon after clearing because of inherent low soil productivity or inappropriate farming methods. More recent changes to agricultural trade policies have lead to the breakup of some farming lands and a decline in several of the key industries in these rain forest landscapes. Over time it is likely that more of these farming areas will change land uses or be abandoned.

In the last few decades there has also been an increased public awareness that rain forests have been overcleared. Several organizations and government agencies have begun to reforest these landscapes using a variety of methods and have had varying levels of success. Understanding what controls flora and fauna colonization in the old fields of Australian rain forest landscapes is essential for guiding effective and efficient restoration programs. However, research conducted into these processes and patterns is patchy at best, and there has been little synthesis. The focus of this chapter is thus to summarize vegetation development on abandoned agricultural lands in rain forest regions, by examining both autogenic (self-directed) successional processes and those that are actively managed (such as tree planting or weed removal). We concentrate on Australia's humid tropics (wet tropics) and subtropics, because this is where most of the historic rain forest still exists today; the rain forests in the two regions show many floristic and faunal and some climatic and edaphic similarities (e.g., existence on basalt plateaus); the land use history in the two regions has been similar; and the current issues in both regions are similar. Furthermore, most relevant research into rain forest development on old fields has been conducted in these regions.

119

Characteristics and History of Australian Rain Forests

Rain forests possess structural, floristic, and physical characteristics that distinguish them from other Australian vegetation types. They are generally characterized by a closed canopy of broad-leaved tree species in combination with vines, lianes, epiphytes, palms, strangler figs, and trees with buttress roots; they also have a humid understory microclimate (Webb et al. 1976). However, there is continuous variation within and between rain forests across Australia, and this has resulted in sophisticated classification schemes based on structural/physiognomic characteristics (Webb et al. 1976; Webb 1978; Lynch and Neldner 2000). For simplicity, in this chapter we refer only to the broad categories of tropical and subtropical rain forests, so that overall trends are apparent, rather then focusing on particular differences between rain forest types.

Australian rain forests differ from rain forest in many other parts of the world because of their unique flora and fauna and their broad latitudinal distribution, and because they are spatially patchy and often surrounded by fire-prone sclerophyll vegetation (Webb and Tracey 1981; Barlow and Hyland 1988). Rain forest in the wet tropics are naturally less fragmented than those in other areas, but many areas of contiguous rain forests naturally occurred in the subtropics; for example, those associated with the volcanic-derived soils around Mt. Warning. The basalts underlying much of the Atherton Tablelands in the wet tropics support extensive stands of rain forest.

Human Settlement and Land Use

Australian Aborigines have occupied rain forest areas for at least 40,000 years, but there is little evidence that they cleared forest or practiced any sort of cultivation (Horsfall and Hall 1990). They did obtain a variety of products from the rain forest and used fire as part of their land management practices, but their impact on these forests is complex, not well understood, and controversial (Bowman 2000; Hill et al. 2000).

Immediately prior to European settlement, rain forest vegetation covered around 4% of the continent (Webb and Tracey 1981). One of the first resources that early settlers exploited from subtropical and tropical rain forests was red cedar, *Toona ciliata*, a species so highly prized that it became known as "red gold" (Valder 1987). The luxuriant growth of rain forests suggested to the early timber cutters and surveyors that the soils would support a range of agricultural activities. In many areas, the rain forest did occur on rich soils, and a number of government schemes encouraged people to clear forests and follow agrarian pursuits. Lowland tropical and subtropical rain forests were widely cleared, and sugarcane became the dominant agricultural activity in these areas. Large areas of upland rain forests were also cleared, and dairying was the dominant industry established in these landscapes. Estimates suggest that over this period of settlement, nearly six million hectares of rain forest were removed to access these soil resources (Frawley 1991). The results of rain forest clearing fell far short of early expectations as many areas were never productive because of the steep terrain, inherent poor soil fertility, or nutrient leaching by high rainfall (Frawley 1991). Subsequently, many cleared rain forest soils were abandoned after agricultural pursuits failed. Nevertheless there was government focus on agricultural expansion and intensification up until the 1960s aided by new technologies, such as phosphorous fertilizers, refrigeration, transport networks, and bulk processing (Webb 1966).

In comparison to many other parts of the world, deforestation and agricultural development began only recently (mostly 80–150 years ago), and because of the restricted areas of rain forest land in Australia, only relatively small areas have been abandoned for more than one hundred years, with this occurring less and later in the wet tropics.

Recent Socioeconomic and Land Use Changes

During the last few decades of the twentieth century, rain forest conservation emerged as a significant political issue, with debate over whether state governments should continue harvesting timber from rain forests, or whether they should be protected from consumptive uses. By 2000 this debate had been essentially settled and most large tracts of government-owned rain forest were in some form of conservation reserve (Catterall et al. 2004). As a consequence of government deregulation and removal of tariff protection from world commodity prices, there has also been a widespread decline in dairying and sugarcane production in both the tropical and subtropical rain forest regions. Government assistance packages were provided to help farmers convert to other industries or to leave their land. This in turn has resulted in more land being taken up by nonagricultural settlers interested in conservation and restoration, or by people generally uninterested in farming. Several government programs aimed at reforesting these lands have come and gone since 1990, but a new window of opportunity currently exists for reforestation though managing natural regrowth in these changing landscapes. Understanding how to use this opportunity relies upon a number of factors, among which knowledge of rain forest successional processes on old fields and appropriate timing and placement of management interventions are crucial.

Autogenic Plant Succession

There are three main types of large-scale successions that occur in Australian rain forest landscapes. The first occurs when fields that have been cleared and farmed for a significant period of time are abandoned. The second type is that which occurs when large areas have been disturbed in some way but are then able to regenerate more or less immediately. A third type of rain forest succession, although not within an old field situation, occurs when changes in regional fire regimes, such as decreased fire frequency since European settlement, allow rain forests to expand across ecotones with eucalypt forests (Russell-Smith et al. 2004).

Successional patterns are different in each of these three situations. One of the reasons for this is in the nature of the starting point. Old field sites tend to have very few native rain forest species present, and all seed must be dispersed into the site (Hopkins 1990). In this situation the more common pioneers tend to have a dominant role in the early-successional stages. Many old fields, but especially those in the tropical lowlands, also have a dense cover of introduced grasses and herbs, which makes seedling establishment difficult. By contrast, successions commencing after a recent disturbance usually have many of the original species still on the site as either seedlings (from the original subcanopy seedling pool), seed (in soil seed banks), or old stumps or roots. Many pioneer species may also be present but usually play a less-dominant role because of the density of the existing seedling pool. Invasive exotic plants are rarely dominant. In these cases the successional process is shortened, and species commonly found in later-successional stages may be present relatively early in the recovery process.

Role of Pioneers and Other Early Colonists

A relatively small group of species acts as pioneers. These include tree species from genera such as *Solanum*, *Omalanthus*, *Duboisia*, *Mallotus*, and *Trema*. Other early colonists include longer-lived species typical of early secondary succession, from genera such as *Acacia* and *Alphitonia*. Species of these two groups are found in both the tropics and subtropics. All are fast growing and shade intolerant. All have seed that is widely dispersed by birds or other animals, and all are commonly represented in soil seed banks. Although they are common, they can differ in their relative abundance. For example, *Acacia* is often the most dominant species in old field successions, perhaps because of its persistent seeds, taller stature, and greater longevity (up to eighty years) (Hopkins and Graham 1984). It is especially dominant in the early stages of rain forest successions in old fields in drier areas (e.g., <1,000 mm annual rainfall) or on degraded soils (Maggs and Hewett 1993; Hopkins et al. 1996). It is much less common in successions occurring after disturbances, such as storms, or in the successional development into eucalypt forest.

It is not clear whether any of these species are facilitators in the sense implied by Connell and Slatyer (1977). *Acacia* and *Trema* are potential nitrogen fixers, but few studies have found evidence of significant rates of fixation in many rain forest successions. However such fixation may be more significant in degraded old field sites than in other rain forest sites that are already relatively fertile. Fleshy-fruited pioneer species may attract birds that carry the seed of later-successional species, although the importance of fleshy fruited-ness has been questioned by some authors (Willson and Crome 1989; Toh et al. 1999). That is, a nonpioneer species could perhaps fulfill the same function. What these pioneers can do, however, is grow quickly and shade out grasses and other weeds, thereby improving the seed bed for other rain forest plants. Most have relatively short life spans (<20 years), so they eventually create a patchwork of small canopy gaps that assist later colonists to join the succession. However, there are some situations where dense, even-aged populations of *Acacia* appear to temporarily inhibit colonization by later arrivals, particularly at sites with low soil fertility (Hopkins et al. 1996). This stage may last for forty to eighty years in the tropics and subtropics.

Successional Pathways

After the initial colonization stage, a variety of species may enter the succession. An indication of the range is shown in table 7.1, which shows the numbers of species and seeds in the seed rain in a twenty-year-old successional site dominated by *Acacia* compared with that in undisturbed rain forest. The seed rain at the young successional site has representatives from a variety of successional stages and life-forms, including a large number of vines. It also has a significant number of species representative of a more mature successional phase (10% of total species caught in seed traps), even though the average seed rain density of these is low. Large year-to-year variations can occur in seed production (Connell and Green 2000), and the identities of these species and their relative abundances in the seed rain necessarily depend on the identities of species flowering and seeding in a particular year, the availability of seed dispersers able to move across the landscape, and the distance of the site from intact forest.

Large numbers of seeds of light-demanding, early-successional species that arrive in seed rain are unable to germinate and develop beneath a closed canopy. These species enter the soil seed bank and may remain dormant for

TABLE 7.1

Numbers of species and seed in seed rain over a twelve-month period in subtropical rainforest in southern Queensland

Life-form	20-year-old forest			Undisturbed forest		
	Total species (%)		Seeds/m^2	Total species (%)		Seeds/m^2
Pioneer	12	(16)	692[a]	4	(6)	10
Longer-lived secondary	5	(7)	2	4	(6)	30
Primary	8	(10)	16	18	(28)	2,033[b]
Shrubs	7	(9)	105	6	(9)	2
Vines	20	(26)	121	17	(26)	62
Herbs	15	(20)	32	10	(15)	9
Unknown	9	(12)	2	6	(9)	1
Total	76	(100)	970	65	(100)	2,148

Source: Data from Abdulhadi 1989.
Note: Based on traps each having a 0.4 m^2 area.
[a] *Acacia* = 90% of total.
[b] *Pseudoweinmannia* = 98% of total.

TABLE 7.2

Seed in soil seed banks in subtropical rain forest sites. All sites were within 5 km of each other. Undisturbed site 1 was adjacent to the two-year site, and undisturbed site 2 was adjacent to the 20, 50 and 60 sites.

Age (years) since disturbance	Number of seeds/m^2	Number of species
Undisturbed[1]	564	45
2	348	34
Undisturbed[2]	603	61
20	2,082	73
50	1,694	85
60	947	74

Source: Data from Abdulhadi 1989.

many years until dormancy is broken and germination is triggered by some form of disturbance (Hopkins and Graham 1984). Table 7.2 shows how this seed bank is initially depleted, when seed germinate following a disturbance, but then increases over time in both size and diversity. Some seed appears able to remain dormant in this seed bank for more than fifty years, although the population appears to decline after this time (Abdulhadi 1989). However, in the absence of further large-scale disturbances, the size of the pool, though not the diversity of species it contains, declines after this time.

Processes Inhibiting or Diverting Successions

Successions can be either inhibited at an early stage or diverted at a later stage once reforestation has occurred. One species that appears able to inhibit successions in some situations is the scrambling shrub *Lantana camara* (Webb et al. 1972; Fensham 1996). Once established on old fields, it is able to prevent many new colonists from establishing, and there is evidence in the subtropics of successions being blocked in this way for several decades (Webb et al. 1972). In the subtropics, there has been some concern that *Cinnamomum camphora*, which can form dense groves, may likewise exclude other species from colonizing old field sites (Natural Resources and Mines 2005), although this has been questioned by some researchers (Neilan et al. 2006).

Established successions can also be diverted by new disturbances. One common means by which this happens is when a forest is affected by storms. In this case competitive relationships in the succession between vines and trees are altered to favor vines. Hegarty (1988) found vines could be divided into those commonly found in early-successional stages and those more commonly found in later stages. The former have tendrils, while those from later-successional stages tend to have a higher proportion of twiners. Storms create conditions favoring species with tendrils that tend to form monospecific canopies over the top of the tree canopies. Recurrent storms along the coastal ranges in north Queensland have disrupted forest canopies and led to vine tangles enveloping remnant trees, forming "climber towers" (Webb 1958). In an analogous fashion, both native and exotic vines frequently isolate trees and small remnants within cleared agricultural land in both the tropics and subtropics, and this process may also deflect the progress of old field regrowth (Dunphy 1991; Big Scrub Rainforest Landcare Group 2005).

Fires can also divert successions. This is most common in situations where rain forest is advancing into open eucalypt forests or abandoned pastures dominated by fire-prone grasses such as blady grass, *Imperata cylindrica*. These sites are susceptible to fires until such time as a developing layer of woody rain forest species is able to form a closed canopy and shade out grasses. This new canopy will also create a new microclimate that hastens fuel decomposition and increases humidity, both of which decrease the fire hazard. However, the fire frequency of a site is dependent on a number of factors, including the composition of the plant species recruited; accumulated fuel loads; the length of the dry season or drought periods; and the activity of arsonists (Ash 1988). Moreover, fuel loads are changing with the introduction and spread of species such as molasses grass (*Melinus minutiflora*) and Guinea grass (*Urochloa maxima*), which are both highly flammable and form dense mats. These species grow rapidly and have the ability to prevent seedling regeneration and rain forest successional processes.

Rain Forest Restoration Frameworks in Former Rain Forest Landscapes

There are many pathways by which forest cover has been actively restored to former rain forest landscapes in Australia's humid tropics and subtropics, the number and variety increasing greatly since the 1980s. Most projects can be grouped into three broad categories: ecological restoration planting, timber plantations, and regrowth management (table 7.3). These pathways vary widely in terms of the proponents' main goals, the method of reforestation, subsequent management requirements, and the financial costs involved, as well as landscape context. At least in the short term, they also vary in terms of outcomes for biodiversity, including the development of forest structure, floristic composition, and use by fauna (tables 7.3 and 7.4). Active reforestation has often occurred on land converted from rain forest to pasture, bananas, and sugarcane, typically for periods of several decades to more than a century.

Ecological Restoration Planting

Most ecological restoration planting projects are less than two decades old. Their proliferation since the 1990s was mainly a result of government-sponsored schemes to encourage revegetation works by private landholders, community groups, and government agencies (Lamb et al. 1997; Catterall et al. 2004; Tucker et al. 2004). Ecological restoration planting projects have also been aided by improvements in knowledge and techniques, with reports and guidebooks published in both the subtropics (Phillips 1991; Kooyman 1996; Horton 1999; Big Scrub Rainforest Landcare Group 2005) and tropics (Goosem and Tucker 1995). A number of replantings have also been funded by private landholders, especially in the subtropics, and others as part of environmental mitigation within development projects. Together with the development of mixed-species farm forestry, these initiatives have stimulated the emergence of a fledgling industry of revegetation contractors, in both tropics and subtropics.

The specific methods used to establish ecological restoration plantings vary considerably, some using a lower-diversity mix of early-successional species (e.g., *Omalanthus*, *Alphitonia* spp.), and others, a high number of species, including later-successional plants (e.g., see table 7.4). However, all methods share the goal of rapidly overcoming two factors considered to limit rain forest succession within cleared land: a depleted soil seed bank and competition from vigorously growing, introduced pasture grasses and legumes (and associated fire risks). Typically, the pasture is sprayed with herbicide and

TABLE 7.3

Main modes of active forest restoration practiced on cleared land in former rain forest landscapes of Australia's humid tropics and subtropics

Typical attributes	Ecological restoration plantings (2, 9, 17, 22)	Timber plantations (2, 8, 17)	Regrowth management (1, 2)
Definition	Areas in which seedlings or seeds are planted with the goal of rapidly creating native rain-forestlike vegetation.	Areas in which seedlings of timber trees are planted, with a primary goal of timber production.	Areas in which autogenic succession occurs following land abandonment, which are managed to produce rain-forestlike vegetation.
Land-use and landscape context (2, 23, 32)	Public or private land, often low agricultural potential, e.g., on steep banks of waterways, along drainage lines.	Shift over time from public land cleared for plantations toward private land, formerly used for agriculture. Land often of marginal agricultural value (moderately steep slopes, often poorer soils).	Mainly private land (including ex-pastures, bananas, cane plantations). Usually on land of very low agricultural value; often on banks of waterways, steep rocky slopes, poorer soils.
Main goal (2, 6, 32)	Ecosystem reinstatement, linking or enlarging remnant forest patches to aid in biodiversity and wildlife conservation, streambank stabilization, water quality improvement.	Timber production, windbreaks, scenic amenity, carbon sequestration, provision of wildlife habitat, streambank stabilization, water quality improvement.	Most regrowth has been initiated when agricultural production has been reduced or abandoned; management goals (where they exist) resemble those for ecological restoration.
Vegetation initially established (2, 3, 17, 23, 24)	Dense plantings of seedlings (mostly around 3,000/ha), of a high diversity of plants (20–50 species). Mainly rain forest plants native to the region, including many that are fleshy fruited.	More widely spaced plantings of seedlings (mostly between 400 and 1,000/ha), of one to several tree species whose timber has commercial value. Often a large proportion of exotic and wind-dispersed species.	Varies with context. Initial regrowth on former agricultural land is often dominated by invasive exotic plants, many of which are fleshy fruited.
Species selection and techniques	The initial species and functional mixtures, and establishment techniques, vary considerably. Mostly a mix of pioneer and later successional species, established by planting seedlings; and a few direct seeding trials. (5, 9, 22)	A wide range of approaches. Includes large-scale industrial monospecies plantations, and small-scale monospecies or mixed-species, cabinet timber plantations on private land, often as part of agroforestry. (8, 24)	A few publications have discussed possible approaches, most of which are focused on the use or management of invasive exotic plants. (1, 12, 36)

TABLE 7.3
(Continued)

Typical attributes	Ecological restoration plantings (2, 9, 17, 22)	Timber plantations (2, 8, 17)	Regrowth management (1, 2)
Historical development (2, 3, 16, 23, 24)	Most efforts began in the 1980s, following increased public awareness of environmental and conservation issues.	Most monospecific plantations of native rain forest trees were trialed and established during the 1920s–60s. Promotion and development of small-scale, mixed-species plantations has occurred since the 1980s.	Small land areas were abandoned soon after first clearing (1860s–1950s), increasing after declines in the dairy industry from the 1970s. Management of regrowth for restoration has mostly been considered since the 1990s.
Spatial configuration and extent	Patches, mainly small (<5 ha); often linear strips, <50 m wide, along streams or adjacent to remnants. Total area small (e.g., ca. 10 km² in wet tropics (7); probably no more than twice that in total.	Industrial plantations cover tens to hundreds of ha. Small-scale plantations are often <5 ha, sometimes linear strips. Total area much larger than for ecological replanting (>500 km² in total). (19, 23)	Regrowth varies widely in size and configuration; aggregate area unknown, but exceeds 200 km² in at least one landscape (28, 29). Most interventions have involved small-scale management of exotic plants; total area very small (<1 km²).
Structural development	A high canopy foliage cover (>70%) typically achieved within 3–5 years; by around 10 years many structural features are broadly similar to rain forest (litter depth, fine woody debris, gross stem densities), although large debris, large stems, and special rain forest life-forms are scarce. (10, 15, 27)	Canopy closure occurs more slowly, especially in more widely spaced plantations, and if eucalypts are common; structural attributes of rain forest develop more slowly than in ecological restoration plantings and may be limited by management (15).	Developmental rates inadequately documented; regrowth ca. 20–40 years dominated by exotic trees has high canopy foliage cover, height, stem densities and litter layer, but lacks large debris, large stems, and special rain forest life-forms. (28)

Ongoing management	Most management restricted to first 3–5 years (initial herbicide, heavy mulching, selective weed control; closed canopy suppresses grass); exclusion of grazing livestock.	Intense weed control until canopy closure, including herbicide and/or slashing. Trees subsequently pruned and thinned. Cattle grazing may be used to reduce grass and understory cover.	Initial trials and suggestions for areas in which early regrowth is dominated by invasive exotic plants include selective thinning; overstory removal; enrichment planting. (29, 36)
Financial costs ($AUD, ca. 2000)	Establishment $20,000–$25,000/ha Maintenance often low after ca. 3 years.	Establishment $4,000–$8,000/ha Maintenance moderate. Financial returns from thinning during the rotation and at final harvest. (20)	Establishment negligible Maintenance varies with site conditions, $3,000–$4,000/ha + (R. Woodford pers. comm.).
Floristic recruitment	Seeds and seedlings of a wide range of native rain forest plants may recruit, often assisted by frugivores and influenced by seed type and landscape context. (31, 34)	Seedlings of a wide range of native rain forest plants may occur, often assisted by frugivores, and influenced by seed type, site management, and landscape context. (16, 19)	Seedlings of a range of native rain forest plants may recruit, often assisted by frugivores, and possibly influenced by seed type and landscape context. (28, 30, 35)
Outcomes for fauna	A wide range of rain forest animals colonize within 5 years. By 10 years there can often be moderate similarity to rain forest in species composition and high similarity in some functional measures. However, rain forest specialists are rare. (10, 13, 14, 18, 27)	A limited range of rain forest animals colonize by 10 years; greater colonization by generalists or open-habitat species. In old (40–70 years), lightly managed plantations adjacent to rain forest, fauna species are very similar to rain forest, although some specialists are rare.	Largely unknown, but well-developed regrowth (including weedy regrowth) is used by a wide range of rain forest birds. Even well-developed, native, autogenic rain forest regrowth often lacks rain forest specialists. Fauna other than birds little studied. (4, 11, 21, 25, 28)

Sources: 1. Big Scrub Rainforest Landcare Group 2005; 2. Catterall et al. 2004; 3. Catterall et al. 2005; 4. Crome et al. 1994; 5. Doust 2004; 6. Emtage et al. 2001; 7. Erskine 2002; 8. Erskine et al. 2005; 9. Goosem and Tucker 1995; 10. Grimbacher and Catterall 2007; 11. Hausmann 2004; 12. Harden et al. 2004; 13. Jansen 1997; 14. Jansen 2005; 15. Kanowski et al. 2003; 16. Kanowski, Catterall, and Wardell-Johnson 2005; 17. Kanowski, Catterall et al. 2005; 18. Kanowski et al., 2006; 19. Keenan et al. 1997; 20. Keenan et al. 2005; 21. Kitching et al. 2000; 22. Kooyman 1996; 23. Lamb et al. 1997; 24. Lamb et al. 2001; 25. Moran et al. 2004a; 26. Moran et al. 2004b; 27. Nakamura et al. 2003; 28. Neilan et al., 2006; 29. Scanlon 2000; 30. Toh et al. 1999; 31. Tucker and Murphy 1997; 32. Tucker et al. 2004; 33. Wardell-Johnson et al. 2005; 34. White et al. 2004; 35. Willson and Crome 1989; 36. Woodford 2000.
Note: Statements are overviews drawn from references cited here and elsewhere in this chapter and supplemented by the authors' experience and unpublished theses. Note that there is large variation within each restoration mode, and the goals and methods of different modes may intergrade in practice.

TABLE 7.4

Top five tree or shrub genera in rank abundance, in the canopy of rain forest (F), and of three modes of reforestation

Genus	Family	Status[a]	Dispersal[b]	F-t	F-s	RG-t	RG-s	CT-t	CT-s	ER-t	ER-s
Acacia	Mimosaceae	S	B/O				4		3	3	
Acronychia	Rutaceae		B			2					
Alphitonia	Rhamnaceae	S	B				5	4		4	4
Araucaria	Araucariaceae		W/O						3		
Argyrodendron	Sterculiaceae		W	1	1						
Castanospermum	Fabaceae		O		2						
Chukrasia	Meliaceae	X	W					4			
Cinnamomum	Lauraceae	X	B				1				
Commersonia	Sterculiaceae	S	O								1
Darlingia	Proteaceae		W							5	
Elaeocarpus	Elaeocarpaceae		B					3	3	1	
Endiandra	Lauraceae		B	1	3						
Eucalyptus	Myrtaceae	N	W					2			2
Ficus	Moraceae		B	5	5						4
Flindersia	Rutaceae		W	4				1	1	2	3
Glochidion	Euphorbiaceae	S	B			1					

Genus	Family	Status[a]	Dispersal vector[b]				
Grevillea	Proteaceae		W			2	
Guioa	Sapindaceae	S	B	5		3	
Ligustrum	Oleaceae	X	B			2	
Macaranga	Euphorbiaceae	S	B	5			
Omalanthus	Euphorbiaceae	S	B				5
Pittosporum	Pittosporaceae	S	B				4
Polyscias	Araliaceae	S	B	5			4
Psidium	Myrtaceae	X	B	2			
Rhodamnia	Myrtaceae	S	B	2			
Rhodosphaera	Anacardiaceae		B			3	
Syzygium	Myrtaceae	S	B	3	3		

Notes: RG = regrowth, CT = cabinet timber plantations, ER = ecological restoration planting, in humid basaltic landscapes in tropical (t) and subtropical (s) Australia. Five to ten replicate sites for each site type/region. Rank abundance calculated from the total frequency of occurrence separately for each site type and region, from data collected in five, 5 m radius plots per site (e.g., *Acronychia* = second most abundant genus within tropical RG sites, but not one of the five most abundant genera in other site types). Canopy is defined as the upper 30% of vegetation, >2 m above the ground. For further details see Catterall et al. (2004), Wardell-Johnson et al. (2005).
[a] Status: X = exotic; N, not typical of rain forest; S, usually considered to be typically pioneers or early successional (generalized to genus level based on species in the regions).
[b] Dispersal vector = predominant mode of seed dispersal (B = birds or bats; W = wind; O = other).

mulched, and seedlings planted at a high density (1–2 m apart), so that the young trees soon form a closed canopy and shade out grasses and weeds. The planted trees also serve as attractants or perch sites for frugivorous fauna, which disperse the seeds of rain forest plants. Due to the recent establishment of most ecological restoration plantings, there have been few comparative studies of how their vegetation development varies with planting technique (but see Kooyman 1996; Florentine and Westbrooke 2004). However, the high-density planting technique is generally very effective in both suppressing grass growth and enabling the recruitment of native rain forest plants (Kanowski et al. 2003; Florentine and Westbrooke 2004; White et al. 2004; Kanowski, Catterall, and Wardell-Johnson 2005).

A major constraint on the use of the ecological restoration plantings to initiate rain forest restoration in degraded landscapes is their high cost (table 7.3). As a result, the aggregate area of achieved restoration is small. For example, federal government funding of around $4.2 million for ecological restoration planting in the wet tropics bioregion during 1997–99, with matching input from the community, allowed projects to be established over 4.4 km^2 (Catterall et al. 2004). This is less than 0.2% of the area of cleared rain forest land in the region (>2,000 km^2) (Kanowski et al. 2003).

Timber Plantations

Timber plantations may provide an alternative means of catalyzing the regeneration of rain forest over larger areas of former pasture or agricultural land (Lamb 1998), although the ecological outcomes will be contingent on plantation designs, management, and harvesting practices (Catterall et al. 2005; Kanowski, Catterall, and Wardell-Johnson 2005). The variety of designs used for timber plantation projects in the study region have varied even more widely than has been the case for ecological restoration plantings. In the 1920s–80s, monospecific timber plantations (mainly using the native rain forest tree *Araucaria cunninghamii*) were established by government agencies, through conversion of publicly owned rain forest, without an intervening pasture phase and adjacent to areas of remnant forest. These plantations constitute the oldest form of deliberate reforestation in the study landscapes and provide some information on the potential for plantations to catalyze regeneration of rain forest plants over the long term (Keenan et al. 1997). However, extrapolating from these plantations to the situation where plantations are established on land used for pasture for many decades and/or on sites located far from existing forest is problematical (Kanowski et al. 2003; Kanowski, Catterall, and Wardell-Johnson 2005).

Since the 1980s, there has been an increasing promotion of mixed-species, multipurpose plantations on private land, for example, to provide both biodiversity and production (timber, cattle grazing) outcomes. These mixed-species cabinet timber plantings have often comprised a variable mixture of rain forest, eucalypt, and exotic tree species (e.g., see table 7.4), planted in relatively small plots. Their establishment has been encouraged by government subsidy, notably the Community Rainforest Reforestation Program (CRRP) during the 1990s.

Timber plantations are typically established at a lower stocking rate than ecological restoration plantings (table 7.3), which means canopy closure may be delayed for some years, limiting the ability of younger plantations to provide sufficient shade to suppress pasture grasses. In some agroforestry plantations, cattle graze these grasses until the canopy closes. Furthermore, management of plantations for rapid timber growth requires pruning, thinning, and understory suppression, which may limit the recruitment of rain forest plants. On the other hand, given suitable management and proximity to seed sources, timber plantations can facilitate the recruitment of rain forest plants to cleared land over the longer term (Parrotta et al. 1997; Lamb 1998; Kanowski et al. 2003). The potential tradeoffs between production and the development of biodiversity in timber plantations depend both on local factors, such as establishment methods, initial composition, and management, and on context factors, such as patch size, proximity to native forest, and the nature of the surrounding landscape (Catterall et al. 2005; Wardell-Johnson et al. 2005; Kanowski, Catterall, and Wardell-Johnson 2005).

Managed Regrowth

Management intervention to alter the trajectory of natural regrowth offers a third potential restoration pathway, although exploration of options and mechanisms for this is in its infancy. In the study regions, autogenic second-growth rain forest has until recently been restricted to relatively small areas, initially because of a high intensity of land use, and even after abandonment, because of the resistance of long-established pasture to colonization by rain forest plants (Toh et al. 1999; Florentine and Westbrooke 2004). However, a number of invasive exotic tree and shrub species are able to establish in pasture (e.g., species of *Solanum, Lantana, Cinnamomum, Ligustrum*), and are often dominant in weedy regrowth that began in areas abandoned since the mid-twentieth century (table 7.4; also see earlier section, "Processes Inhibiting or Diverting Succession"). These plants bear fleshy fruits and are dispersed by birds and bats, promoting their rapid spread (Gosper et al. 2005).

Once established at moderate density, they provide improved conditions for rain forest plant recruitment: shade that suppresses the pasture grasses, and habitat for frugivorous fauna that disperse the seeds of rain forest plants (Willson and Crome 1989; Neilan et al. 2006).

When economic conditions cause the decline of a dominant rural industry, regrowth dominated by invasive exotic plants has the potential to develop over large areas. For example, the decline of the dairy and banana industries in the subtropics in the mid- to late twentieth century was followed by the rapid spread of the camphor laurel (*Cinnamomum camphora*). By the year 2000, regrowth dominated by camphor laurel occupied around 25% of the former Big Scrub region (750 km^2), a landscape in which more than 99% of the original rain forest had been cleared for pasture (Neilan et al. 2006). This regrowth both provides fauna habitat and allows the regeneration of rain forest seedlings, which appear poorly adapted to recruiting in pastures. In this situation, tolerance of the initial weedy regrowth, followed by management targeted at reducing any suppression of the growing rain forest seedlings by the *C. camphora* overstory, may prove a cost-effective restoration technique over large areas (Scanlon 2000; Neilan et al. 2006).

Such suggestions are controversial, as exotic species are poorly regarded by many conservationists, often for good reasons. For example, the control of exotic plants has been an important part of rain forest rehabilitation and ecological restoration plantings. Exotic vines in particular are often the target of rain forest restoration activities, because they can smother small rain forest patches, especially where the canopy is broken (Harden et al. 2004). It is possible that, in the longer term, the many small patches and narrow strips of ecological plantings and lightly managed timber plantations may acquire an unwanted bird-dispersed flora of aggressive vines and other exotic plants, which in the absence of maintenance may overgrow them (Catterall et al. 2004; Wardell-Johnson et al. 2005). In extensively cleared and long-modified landscapes, it may be necessary to accept that exotic plants and animals may play both undesirable and useful roles in successional processes, both of which can be managed through informed intervention.

Our ability to make decisions about desirable actions for all forms of forest restoration is limited by the lack of knowledge about long-term trajectories under different starting conditions, management regimes, and landscape contexts. In the study regions, it has been difficult to explore this space-for-time substitution, because most situations of forest redevelopment on old fields have been under way for less than a few decades. Government subsidy for reforestation activities has not been matched with investment in research and monitoring aimed at extracting lessons for future endeavors, and there are no ongoing, long-term, scientific monitoring studies.

Has Knowledge About Old Field Succession Influenced Reforestation Methodologies?

The considerable amount of research into successional development has made only modest contributions to the way in which various forms of reforestation have been carried out. While increased knowledge of the autecology of individual species in old field successions has clearly been beneficial, the lack of knowledge about assembly rules, functional interactions between species during successions, or successional trajectories has meant most restoration practices have developed as a consequence of separate and more targeted trial and error by practitioners, coupled with limited empirical studies about restoration. Studies of old field succession have not revealed the consequences of using different species numbers, combinations, patterns of relative abundance, or total densities when initiating an ecological restoration planting.

Both the diversity of plantation establishment techniques, which have been used in recent years, and the empirical studies of restoration have shown that it is not necessary to replicate the successional sequence normally seen in old field successions of the past. In fact, this may not be desirable if the goal is to accelerate the rate of development of rain forest and, furthermore, may not be feasible in the face of increasing numbers of invasive exotic plants. It seems unlikely that further studies of old field successions will reveal how to accelerate successional development on badly degraded sites. In hindsight, perhaps this type of knowledge was never going to be gained from passive observational studies. Perhaps the real opportunities lie in more manipulative experiments in old fields, such as excluding certain species or creating canopy gaps of various sizes and adding new species to the developing communities. Enrichment trials of this type have been carried out after logging, but not in many old field successions in Australia.

Animal–Plant Interactions and Their Roles in Reforestation

Many types of rain forest fauna have greatly declined because of clearing for agriculture, and reforestation has a potential to reverse these trends. Fauna are also important to the developmental trajectory of the vegetation, because the reproductive success, recruitment, survival, and growth of rain forest plants are strongly influenced by the roles that animals play in pollination, seed dispersal, and the predation of seeds and seedlings (Kanowski et al. 2004).

It might be expected that variation in fauna during reforestation would correspond with variation in floristic diversity and composition, and this should be the case for some herbivores and frugivores (Kitching 2000; Moran et al. 2004a). However, the physical structure of the vegetation appears to be the main determinant of habitat suitability for many rain forest–dependent animals (Grimbacher and Catterall 2007; Catterall et al. 2007). Canopy cover, litter depth, and specific attributes, such as dead timber, trunk crevices, and areas of vine tangle or high stem density, are structural components whose development is not strongly tied to either the diversity or the identity of plant species. However, the effects of plant diversity, floristic composition, and physical structure are difficult to separate in practice, because they usually develop together during vegetation succession (Kanowski, Catterall et al. 2005).

Ecological restoration planting sites acquire rain forest fauna at a faster rate than do timber plantations, at least in part because they more rapidly form a closed canopy (table 7.3). Regrowth dominated by exotic plants is also capable of providing many of the structural features of rain forest (Kanowski et al. 2003; Neilan et al. 2006). However, the resources and specific structures associated with older-growth forest develop slowly, and both closed-canopy plantations and advanced regrowth at several decades of age still lack the full complement of rain forest–dependent fauna (Crome 1994; Kitching 2000; Moran et al. 2004a, 2004b; Neilan et al. 2006). Strategic management interventions may accelerate faunal development. For example, selective thinning could foster spatial heterogeneity, promote the growth of large trees, and add large woody debris. Alternatively, fungal inoculation might encourage tree-hollow development.

Spatial patterns of vegetation at the landscape scale will also affect the ability of fauna to colonize and use reforested areas (Catterall et al. 2004). Spatial attributes such as isolation from remnant rain forest have been explored in a limited number of studies (Tucker and Murphy 1997; Kanowski, Catterall et al. 2005; Neilan et al. 2006; Grimbacher and Catterall 2007), and so far suggest that the effects of distance are most apparent within the first few hundred meters from a rain forest boundary.

Seed dispersal by vertebrate frugivores is a key process in the development of rain forest on old fields, because most Australian rain forest plants bear fleshy fruits that are dispersed by birds and bats (Crome 1990). There are many (twenty to thirty) species of birds and fruit bats in the study regions that regularly eat fruit (table 7.5). Many methods for active ecological restoration of vegetation are designed to make explicit use of this vertebrate-assisted seed dispersal, in part by planting fleshy-fruited species that are attractive to a wide range of frugivores (Goosem and Tucker 1995; Tucker and Murphy 1997; Toh et al. 1999; Florentine and Westbrooke 2004). Frugivorous birds show large interspecific differences in the ecological traits that determine their likely seed dispersal roles, and these are correlated with their dependence on unfragmented rain forest (Moran et al. 2004a, 2004b). As a result, relatively smaller-sized fruits or seeds are likely to be more frequently dispersed than larger seeds

TABLE 7.5

Birds and bats regularly dispersing fruit in rain forest landscapes of tropical and subtropical Australia

Taxon	Genera	Number of species	
		Tropics	Subtropics
Birds			
Casuariidae (cassowaries)	*Casuarius*	1	
Columbidae (pigeons, doves)	*Ducula*	1	
	Lopholaimus	1	1
	Ptilinopus	3	3
Cuculidae (cuckoos)	*Eudynamys*	1	1
	Scythrops	1	1
Meliphagidae (honeyeaters)	*Meliphaga*	3	1
	Lichenostomus	2	1
	Xanthotis	1	
Campephagidae (cuckoo-shrikes, trillers)	*Coracina*	2	1
	Lalage	1	1
Oriolidae (orioles, figbirds)	*Oriolus*	1	1
	Sphecotheres	1	1
Artamidae (currawongs)	*Strepera*	1	1
Paradisaeidae (birds of paradise)	*Ptiloris*	1	1
Ptilonorhynchidae (bowerbirds)	*Ailuroedus*	1	1
	Prionodura	1	
	Ptilonorhynchus	1	1
	Scenopoeetes	1	
	Sericulus		1
Dicaeidae (flowerpeckers)	*Dicaeum*	1	1
Zosteropidae (white-eyes)	*Zosterops*	1	1
Sturnidae (starlings)	*Aplonis*	1	
Mammals			
Pteropodidae (fruit bats, flying foxes)	*Nyctimene*	1	1
	Pteropus	2	2

Source: Data from Catterall et al. 2004; Moran et al. 2004b.
Note: Includes major and mixed frugivores as per Moran et al. 2004b, but not seed-grinders or minor frugivores.

(and over longer distances) in fragmented landscapes and regrowth areas (Moran et al. 2004a, 2004b; Gosper et al. 2005). Consequently, small-seeded plants may tend to dominate recruits to regrowth patches (Neilan et al. 2006).

Landscape-scale food resources, although poorly understood, are likely to be important in supporting frugivore populations, and hence in maximizing seed recruitment to reforested areas (Date et al. 1991; Neilan et al. 2006). The mode of active forest restoration also plays a role: timber plantations are often dominated by wind-dispersed (dry-fruited) plants (table 7.4) and, at least while young, tend to be visited less frequently by rain forest frugivores than are ecological restoration plantings (Kanowski, Catterall, and Wardell-Johnson 2005). Conversely, wind-dispersed rain forest plants may be less likely to colonize reforested sites farther from rain forest (Tucker and Murphy 1997; White et al. 2004). These considerations have led to suggestions that the focus of enrichment plantings during postestablishment and regrowth management should be biased toward large-seeded and wind-dispersed plants (Tucker and Murphy 1997; Tucker et al. 2004). Large-seeded plants are also likely to survive relatively well when introduced to sites through direct seeding (Doust 2004).

Old Field Development in Rain Forest Landscapes of Australia's Humid Tropics and Subtropics

A variety of successional processes have occurred in the old fields of Australian rain forest landscapes. How these sites have developed has been dependent upon a number of factors, including the type and duration of the agriculture phase; the underlying soil type; the condition of the soil at abandonment; the presence of seed banks; the location of the landscape in the old field; and the size and composition of remnant rain forest patches in the landscape. Although there has been little management of rain forest regrowth in Australian old fields, a variety of active restoration pathways (e.g., ecological restoration planting, timber plantations, direct seeding) have been developed by different individuals and organizations. How these restored forests develop depends on the techniques used, the starting conditions, and the landscape context. Intensive replanting has been successful in recreating forestlike ecosystems, which contain rain forest fauna, although these restored sites may be of less immediate use to specialized rain forest species, especially endemics, and may not be useful for some biota because of the small patch size and/or isolation. Invasive plants are of increasing importance, although they may either inhibit or assist forest development, and they create dilemmas about when to remove them and when to tolerate their presence.

Opportunities for Rain Forest Restoration

With the relatively short history of land abandonment in Australian rain forest landscapes, we cannot use space-for-time methods to infer longer-term trajectories of rain forest successions. Large-scale restoration is desirable in heavily cleared areas (e.g., Big Scrub, northern New South Wales) or the existing remnants will degrade further due to the time-lagged effects of past clearing. However, current restoration practices will be able to cover only very small areas due to prohibitively high costs of labor and materials. Nevertheless, with the wealth of the Australian population and the motivation of committed organizations and individuals, restoration efforts will continue. Recent socioeconomic changes (e.g., those within the dairy and sugar industries) have also helped create an opportunity for rain forest restoration by potentially freeing up land, but the longer-term outcomes of these demographic and industry changes will depend on what other possible land uses may be adopted in the future.

Creative manipulation of autogenic weedy regrowth appears to be a good opportunity to achieve restoration outcomes. The future successional trajectories of these areas are very uncertain, especially in the face of climate change and in the context of large-scale land use change, and with increases in plant (and probably fauna) invasions. These factors also make it very unlikely that either restoration or managed autogenic successions will "return" the forests to a pre-European species composition even though they could possibly resemble them in terms of a set of functional criteria, such as physical structure, species richness, and variety of life-forms. Timber plantations with the native hoop pine (*Araucaria cunninghamii*) and exotics (e.g., *Pinus*) may also jump-start forest development, and this capacity may be greatly enhanced by the use of some fleshy-fruited species in plantations (Tucker et al. 2004). Indeed, biodiversity concerns have been integrated into an investigation of hoop pine plantation expansion on the Atherton Tablelands and will include monitoring of wildlife usage and water quality as part of the project. However, plantations with some exotic species need to be established carefully, as they do not always create suitable habitat for rain forest species, and the plantation trees themselves may become environmental weeds (Kanowski, Catterall, and Wardell-Johnson 2005).

Conclusion

Because of the relative recency of both land clearing and revegetation in Australia, coupled with a changing environmental context that includes species invasions and global climate change, it is not possible to reliably predict either the future trajectories of old field regrowth or the results of management efforts. In these circumstances, large-scale experimental management, coupled with monitoring, are needed to provide a better understanding of patterns and processes involved in rain forest development on old fields (Catterall et al. 2005; Neilan et al. 2006).

Better inventories of old field regrowth (both autogenic and actively restored) throughout these regions are also needed to provide a baseline for effective treatment and manipulation of these areas. Standard vegetation mapping techniques used by land management agencies have not been designed to distinguish different types of old field regrowth. Therefore, new mapping criteria need to be developed, especially if exotics species are present.

These same information gaps also apply more generally to other vegetation types in tropical and subtropical Australia (e.g., dry rain forests, brigalow, sclerophyll forests), within which extensive land clearing has taken place since the 1960s. In subcoastal and inland regions, much of this recently cleared land has had only a short (or no) intervening agriculture phase prior to the commencement of regrowth. Inland regions also contain vast areas of land that have never been cleared or cultivated, but whose vegetation has been severely impacted by more than a century of heavy livestock grazing. In recent years there have been increasing initiatives to remove or reduce livestock densities. Understanding the trajectories of vegetation development and options for ecological restoration in these situations presents many parallels to the questions involved in old field regrowth.

Acknowledgments

The authors thank Grant Wardell-Johnson, Stephen McKenna, and Nigel Tucker (plant data); Cath Moran and Terry Reis (frugivore data); and Susan Doust, Don Butler, and Jessie Wells for discussions about rain forest old fields.

REFERENCES

Abdulhadi, R. 1989. Sub-tropical rain forest seed dynamics. PhD thesis, University of Queensland, Brisbane, Australia.

Ash, J. 1988. The location and stability of rainforest boundaries in north-eastern Queensland, Australia. *Journal of Biogeography* 15:619–30.

Barlow, B. A., and B. P. M. Hyland. 1988. The origins of the flora of Australia's wet tropics. *Proceedings of the Ecological Society of Australia* 15:1–17.

Big Scrub Rainforest Landcare Group. 2005. *Subtropical rainforest restoration: A practical manual and data source for landcare groups, land managers and rain forest regenerators.* Bangalow, New South Wales, Australia.

Bowman, D. M. J. S. 2000. *Australian rain forests. Islands of green in a land of fire.* Cambridge University Press, Cambridge, UK.

Catterall, C. P., J. Kanowski, and G. Wardell-Johnson. 2007. Biodiversity and new forests: Interacting processes, prospects and pitfalls of rainforest restoration. In *Living in a dynamic tropical forest landscape*, ed. N. Stork and S. Turton. Blackwell Publishing, Oxford.

Catterall, C. P., J. Kanowski, D. Lamb, D. Killin, P. D. Erskine, and G. Wardell-Johnson. 2005. Trade-offs between timber production and biodiversity in rainforest tree plantations: Emerging issues from an ecological perspective. In *Reforestation in the tropics and subtropics of Australia using rainforest tree species*, ed. P. Erskine, D. Lamb, and M. Bristow, 206–21. Rainforest CRC and RIRDC, Canberra, Australia.

Catterall, C. P., J. Kanowski, G. Wardell-Johnson, H. C. Proctor, T. Reis, D. A. Harrison, and N. I. J. Tucker. 2004. Quantifying the biodiversity values of reforestation: Perspectives, design issues and outcomes in Australian rainforest landscapes. In *Conservation of Australia's forest fauna*, ed. D. Lunney, 359–93. Royal Zoological Society of New South Wales, Sydney, Australia.

Connell, J. H., and P. T. Green. 2000. Seedling dynamics over thirty-two years in a tropical rain forest tree. *Ecology* 81:568–84.

Connell, J. H., and R. O. Slatyer. 1977. Mechanisms of succession in natural communities and their role in community stability and organization. *American Naturalist* 111:1119–44.

Crome, F., J. Isaacs, and L. Moore. 1994. The utility to birds and mammals of remnant riparian vegetation and associated windbreaks in the tropical Queensland uplands. *Pacific Conservation Biology* 1:328–43.

Crome, F. H. J. 1990. Rainforest successions and vertebrates. In *Australian tropical rainforest: Science, values, meaning*, ed. L. J. Webb and J. Kikkawa, 53–64. CSIRO, Melbourne, Australia.

Date, E. M., H. A. Ford, and H. E. Recher. 1991. Frugivorous pigeons, stepping stones and weeds in northern New South Wales. In *Nature Conservation 2: The role of corridors*, ed. D. A. Saunders and R. J. Hobbs, 241–44. Surrey Beatty and Sons, Sydney, Australia.

Doust, S. 2004. Seed and seedling ecology in the early stages of rain forest restoration. PhD thesis, University of Queensland, Brisbane, Australia.

Dunphy, M. 1991. Rainforest weeds of the Big Scrub. In *Rainforest remnants*, ed. S. Phillips, 85–93. New South Wales National Parks and Wildlife Service, Hurstville, New South Wales.

Emtage, N. F., S. R Harrison, and J. L. Herbohn. 2001. Landholder attitudes to and participation in farm forestry activities in sub-tropical and tropical eastern Australia. In *Sustainable farm forestry in the tropics*, ed. S. R. Harrison and J. L. Herbohn, 195–210. Edward Elgar, Cheltenham, UK.

Erskine, P. 2002. Land clearing and forest rehabilitation in the wet tropics of north Queensland, Australia. *Ecological Management and Restoration* 3:135–37.

Erskine, P. D., D. Lamb, and M. Bristow. 2005. Reforestation with rainforest trees: Challenges ahead. In *Reforestation in the tropics and subtropics of Australia using rainforest tree species*, ed. P. D. Erskine, D. Lamb, and M. Bristow, 264–72. RIRDC, Canberra, and Rainforest CRC, Cairns, Australia.

Fensham, R. J. 1996. Land clearance and conservation of inland dry rainforest in north Queensland, Australia. *Biological Conservation* 3:289–98.

Florentine, S. K., and M. E. Westbrooke. 2004. Evaluation of alternative approaches to rainforest restoration on abandoned pasturelands in tropical north Queensland, Australia. *Land Degradation and Development* 15:1–13.

Frawley, K. 1991. Past rainforest management in Queensland. In *The rainforest legacy*, ed. G. Werren and P. Kershaw, 85–105. Australian Government Publishing Service, Canberra, Australia.

Goosem, S., and N. Tucker. 1995. Repairing the rainforest : Theory and practice of rainforest re-establishment in North Queensland's wet tropics. Wet Tropics Management Authority, Cairns, Australia.

Gosper, C. R., C. D. Stansbury, and G. Vivian-Smith. 2005. Seed dispersal of fleshy-fruited invasive plants by birds: Contributing factors and management options. *Diversity and Distributions* 11:549–58.

Grimbacher, P. S., and C. P. Catterall. 2007. How much do site age, habitat structure and spatial isolation influence the restoration of rainforest beetle diversity? *Biological Conservation* 135:107–18.

Harden, G. J., M. D. Fox, and B. J. Fox. 2004. Monitoring and assessment of restoration of a rainforest remnant at Wingham Brush, NSW. *Austral Ecology* 29:489–507.

Hausman, F. 2004. The utility of linear riparian rainforest for vertebrates on the Atherton and Evelyn Tablelands, North Queensland. MPhil thesis, Griffith University, Brisbane, Australia.

Hegarty, E. E. 1988. Canopy dynamics of lianes and trees in subtropical rain forest. PhD thesis, University of Queensland, Brisbane, Australia.

Hill, R. S., P. Griggs, and B. B. N., Inc. 2000. Rainforests, agriculture and aboriginal fire-regimes in wet tropical Queensland, Australia. *Australian Geographical Studies* 38:138–57.

Hopkins, M. S. 1990. Disturbance: The forest transformer. In *Australian tropical rain forests: Science, values, meaning*, ed. L. J. Webb and J. Kikkawa, 40–52. CSIRO, Melbourne, Australia.

Hopkins, M. S., and A. W. Graham. 1984. Viable soil seed banks in disturbed lowland tropical rainforest sites in North Queensland. *Australian Journal of Ecology* 9:71–79.

Hopkins, M. S., P. Reddell, R. K. Hewett, and A. W. Graham. 1996. Comparison of root and mycorrhizal characteristics in primary and secondary rainforest on a metamorphic soil in north Queensland, Australia. *Journal of Tropical Ecology* 12:871–85.

Horsfall, N., and J. Hall. 1990. People and the rainforest: An archaeological perspective. In *Australian tropical rainforest: Science, values, meaning*, ed. L. J. Webb and J. Kikkawa, 33–39. CSIRO, Melbourne, Australia.

Horton, S., ed. 1999. *Rainforest remnants: A decade of growth*. NSW NPWS, Sydney, Australia.

Jansen, A. 1997. Terrestrial invertebrate community structure as an indicator of the success of a tropical rainforest restoration project. *Restoration Ecology* 5:115–24.

Jansen, A. 2005. Avian use of restoration plantings along a creek linking rainforest patches on the Atherton Tablelands, north Queensland. *Restoration Ecology* 13:275–83.

Kanowski, J., C. P. Catterall, A. J. Dennis, and D. A. Westcott, eds. 2004. *Animal-plant interactions in rainforest conservation and restoration*. Cooperative Research Centre for Tropical Rainforest Ecology and Management, Cairns, Australia.

Kanowski, J., C. P. Catterall, H. Proctor, T. Reis, N. I. J. Tucker, and G. W. Wardell-Johnson. 2005. Rainforest timber plantations and animal biodiversity in tropical and subtropical Australia. In *Reforestation in the tropics and subtropics of Australia using rain forest tree species*, ed. P. Erskine, D. Lamb, and M. Bristow, 183–205. Rainforest CRC and RIRDC, Canberra, Australia.

Kanowski, J., C. P. Catterall, and G. W. Wardell-Johnson. 2005. Consequences of broad-scale timber plantations for biodiversity in cleared rainforest landscapes of tropical and subtropical Australia. *Forest Ecology and Management* 208:359–72.

Kanowski, J., C. P. Catterall, G. W. Wardell-Johnson, H. Proctor, and T. Reis. 2003. Development of forest structure on cleared rainforest land in eastern Australia under different styles of reforestation. *Forest Ecology and Management* 183:265–80.

Kanowski, J., T. Reis, C. P. Catterall, and S. Piper. 2006. Factors affecting the use of reforested sites by reptiles in cleared rainforest landscapes in tropical and subtropical Australia. *Restoration Ecology* 14:67–76.

Keenan, R., D. Doley, and D. Lamb. 2005. Stand density management in rainforest plantations. In *Reforestation in the tropics and subtropics of Australia using rainforest tree species*, ed. P. D. Erskine, D. Lamb and M. Bristow, 141–60. RIRDC, Canberra, and Rainforest CRC, Cairns, Australia.

Keenan, R., D. Lamb, O. Woldring, T. Irvine, and R. Jensen. 1997. Restoration of plant biodiversity beneath tropical tree plantations in northern Australia. *Forest Ecology and Management* 99:117–31.

Kitching, R. L., A. G. Orr, L. Thalib, H. Mitchell, M. S. Hopkins, and A. W. Graham. 2000. Moth assemblages as indicators of environmental quality in remnants of upland Australian rain forest. *Journal of Applied Ecology* 37:284–97.

Kooyman, R. M. 1996. *Growing rainforest: Rainforest restoration and regeneration. Recommendations for the humid sub-tropical region of northern New South Wales and south east Queensland.* Greening Australia, Queensland, Brisbane, Australia.

Lamb, D. 1998. Large-scale ecological restoration of degraded tropical forest lands: The potential role of timber plantations. *Restoration Ecology* 6:271–79.

Lamb, D., R. Keenan, and K. Gould. 2001. Historical background to plantation development in the tropics: a north Queensland case study. In *Sustainable farm forestry in the tropics*, ed. S. R. Harrison and J. L. Herbohn, 9–20. Edward Elgar, Cheltenham, UK.

Lamb, D., J. A. Parrotta, R. Keenan, and N. Tucker. 1997. Rejoining habitat fragments: Restoring degraded rainforest lands. In *Tropical forest remnants: Ecology, management, and conservation of fragmented communities*, ed. W. F. Laurance and R. O. Bierregaard Jr., 366–85. University of Chicago Press, Chicago.

Lynch, A. J. J., and V. J. Neldner. 2000. Problems of placing boundaries on ecological continua: Options for a workable national rainforest definition in Australia. *Australian Journal of Botany* 48:511–30.

Maggs, J., and B. Hewett. 1993. Organic C and nutrients in surface soils from some primary rainforests, derived grasslands and secondary rainforests on the Atherton Tableland in north east Queensland. *Australian Journal of Soil Research* 31:343–50.

Moran C., C. P. Catterall, R. J. Green, and M. F. Olsen. 2004a. Functional variation among frugivorous birds: Implications for rainforest seed dispersal in a fragmented subtropical landscape. *Oecologia* 141:584–95.

———. 2004b. Fates of feathered fruit eaters in fragmented forests. In *Conservation of Australia's forest fauna*, ed. D. Lunney, 699–712. Royal Zoological Society of New South Wales, Sydney, Australia.

Nakamura, A., H. Proctor, and C. Catterall. 2003. Using soil and litter arthropods to assess the state of rainforest restoration. *Ecological Management and Restoration* 4:S20–S26.

Neilan W., C. P. Catterall, J. Kanowski, and S. McKenna. 2006. Do frugivorous birds assist rainforest succession in weed dominated old field regrowth of subtropical Australia? *Biological Conservation* 129:393–407.

Natural Resources and Mines. 2005. Camphor laurel: Pest fact sheet. Department of Natural Resources and Mines, Queensland Government, Brisbane, Australia.

Parrotta, J. A., J. W. Turnbull, and N. Jones. 1997. Introduction. Catalyzing native forest regeneration on degraded tropical lands. *Forest Ecology and Management* 99:1–7.

Phillips S., ed. 1991. *Rainforest remnants.* NSW NPWS, Sydney, Australia.

Russell-Smith, J., P. J. Stanton, P. J. Whitehead, and A. Edwards. 2004. Rain forest invasion of eucalypt-dominated woodland savanna, Iron Range, north-eastern Australia. 1. Successional processes. *Journal of Biogeography* 31:1293–1303.

Scanlon, T. 2000. Camphor laurel kit: Everything you need to know about camphor and its control. North Coast Weeds Advisory Committee, Bellingen, New South Wales, Australia.

Toh, I., M. Gillespie, and D. Lamb. 1999. The role of isolated trees in facilitating tree seedling recruitment at a degraded sub-tropical rainforest site. *Restoration Ecology* 7:288–97.

Tucker, N. I. J., and T. M. Murphy. 1997. The effects of ecological rehabilitation on vegetation recruitment: Some observations from the wet tropics of north Queensland. *Forest Ecology and Management* 99:133–52.

Tucker, N. I. J., G. W. Wardell-Johnson, C. P. Catterall, and J. Kanowski. 2004. Agroforestry and biodiversity: Improving the outcomes for conservation in tropical north-eastern Australia. In *Agroforestry and biodiversity conservation in tropical landscapes,* ed. G. Schroth, G. Fonseca, C. A. Harvey, C. Gascon, H. Vasconcelos, and A. M. N. Izac, 431–52. Island Press, Washington, DC.

Valder, J. 1987. *Red cedar. The tree of Australia's history.* Reed Books, Sydney, Australia.

Wardell-Johnson, G. W., J. Kanowski, C. P. Catterall, S. McKenna, S. Piper, and D. Lamb. 2005. Rainforest timber plantations and the restoration of plant biodiversity in tropical and subtropical Australia. In *Reforestation in the tropics and subtropics of Australia using rainforest tree species,* ed. P. Erskine, D. Lamb, and M. Bristow, 162–78. Rainforest CRC and RIRDC, Canberra, Australia.

Webb, L. J. 1958. Cyclones as an ecological factor in tropical lowland rainforest in north Queensland. *Australian Journal of Botany* 6:220–28.

———. 1966. The rape of the forests. In *The great extermination. A guide to Anglo-Australian cupidity, wickedness and waste,* ed. A. J. Marshall, 156–205. Heinemann, London.

———. 1978. A general classification of Australian rainforests. *Australian Plants* 9:349–63.

Webb, L. J., and J. G. Tracey. 1981. Australian rainforest: Patterns and change. In *Ecological biogeography of Australia,* ed. A. Keast, 605–94. W. Junk, The Hague, The Netherlands.

Webb, L. J., J. G. Tracey, and W. T. Williams. 1972. Regeneration and pattern in the subtropical rainforest. *Journal of Ecology* 60:675–95.

———. 1976. The value of structural features in tropical forest typology. *Australian Journal of Ecology* 1:3–28.

White, E., N. Tucker, N. Meyers, and J. Wilson. 2004. Seed dispersal to revegetated isolated rainforest patches in north Queensland. *Forest Ecology and Management* 192:409–26.

Willson, M. F., and F. H. J. Crome. 1989. Patterns of seed rain at the edge of tropical Queensland rain forest. *Journal of Tropical Ecology* 5:301–8.

Woodford, R. W. 2000. Converting a dairy farm back to a rainforest water catchment. *Ecological Management and Restoration* 1:83–92.

Chapter 8

Succession on the Piedmont of New Jersey and Its Implications for Ecological Restoration

SCOTT J. MEINERS, MARY L. CADENASSO,
AND STEWARD T. A. PICKETT

The revegetation of disturbed ground through succession is one of the most fundamental processes studied by plant ecologists. In fact, the early history of plant ecology was dominated by attempts to understand the dynamic nature of vegetation though studying succession (e.g., the work of Cowles, Clements, Gleason). In a modern context, the manipulation of succession represents a dynamic approach to the creation and management of restored systems (Luken 1990). Vegetation development during restoration is analogous to successional dynamics and can benefit directly from the integration of successional theory into restoration ecology. Frederick Clements was an early champion of utilizing natural processes in restoration. His perspective, from "Ecology in the Public Service," is summarized in this statement:

> From the very nature of the climax and succession, development is immediately resumed when the disturbing cause ceases, and in this fact lies the basic principle of all restoration or rehabilitation. (1949a)

Our perspective on succession, and on vegetation dynamics in general, is a hierarchical one (Pickett et al. 1987; Pickett and Cadenasso 2005). This perspective recognizes three separate coarse-scale drivers of ecological change that revolve around variation in (1) site availability, (2) species availability, and (3) species performance. Within each of these major categories are nested specific determinants of community structure and composition. For example, landscape connectivity may alter species availability (Haddad et al. 2003), and herbivory may alter relative species performance. Successional dynamics are driven by changes in one or more of these variants. Similarly, ecological restoration depends on the manipulation of these independent causes of change. We define succession as the change in species composition or community architecture of a site over time (Pickett and Cadenasso 2005). This broad definition avoids some of the debate associated with narrower definitions and can encompass the variety of outcomes and dynamics suggested by the hierarchical perspective. Here we present data from a long-term experimental study of succession, specifically focusing on drivers of successional change and their application to restoration.

History of the Buell–Small Succession Study

In 1958, Murray Buell, Helen Buell, and John Small initiated an experimental study of succession at the Hutcheson Memorial Forest Center (HMFC), New Jersey, USA. This site (40°30′ N, 74°34′ W) is located within the piedmont of the Mid-Atlantic region and is operated by Rutgers University. The study had two main objectives (Pickett et al. 2001): (1) determine whether space-for-time substitution gives a reliable account of the successional dynamics of the area, and (2) determine whether Egler's (1954) initial floristic composition hypothesis of succession merited support. Space-for-time substitution studies were commonly used to infer long-term successional dynamics within a short study period (Pickett 1989) but had not been specifically tested for accuracy of prediction.

The ten fields of the Buell–Small Succession Study (BSS) range in size from 0.4 to 1.0 ha. Eight of these fields are along the edge of an old-growth forest, while the remaining two are separated from the old growth by the other fields (figure 8.1). Beginning in 1958, fields were abandoned as pairs in alternate years until 1966. The experimental structure of the study generated fields that differed in season of abandonment (fall or spring), final crop (hay

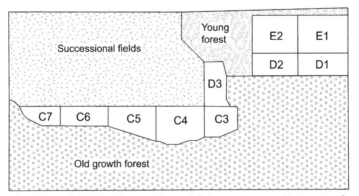

Figure 8.1. Map of the Hutcheson Memorial Forest, Rutgers University, showing locations of the Buell–Small Succession Study (BSS) fields.

field or row crops—sorghum and soybeans), and soil treatment at abandonment (plowed or vegetation left intact). Within each of ten fields, forty-eight plots were arranged in a regular pattern that varied with the shape of the field. At each sampling, the percentage cover of all species present was visually estimated in each permanently marked 0.5 × 2.0 m plot. Data collection occurred every year, after the release from agriculture, until 1979, when sampling switched to alternate years, though 2005.

Site History and Conditions

The area surrounding the HMFC was farmed continuously since 1701, when Dutch farmers settled the region. The Piedmont region of central New Jersey is characterized by soils derived from Triassic red shales that tend to dry out in late summer and are not very fertile (Ugolini 1964). However, the Mettler's Farm, which became the HMFC, was reputed to be well managed and relatively fertile (Pickett et al. 2001). Records indicate that cow manure was applied to the fields regularly, but that in 1963 amendments shifted to chemical fertilizers.

The climate of the area is characterized as mesic continental. The mean annual temperature is 11.5 °C, with average daily high temperatures of −1.2 °C in January and 23.7 °C in July. Annual rainfall averages 1,180 mm and is evenly distributed throughout the year. Monthly precipitation ranges from an average of 76 mm in February to 123 mm in July. While average site conditions are mesic, fluctuations in rainfall combined with the tendency of the soil to dry out have resulted in periods of drought that have influenced plant growth and vegetation dynamics (Buell et al. 1961; Small 1961; Bartha et al. 2003).

Target Community Composition

The old-growth forest of HMFC is a mixed-oak deciduous forest. When originally described, upland areas were dominated by three oak species, *Quercus alba*, *Q. velutina*, and *Q. rubra*, which accounted for nearly half of the stems and more than half of the basal area (Monk 1961b). Other species present included, in order of abundance, *Carya ovalis*, *Fraxinus americana*, and *Acer rubrum*. The understory of the upland was described as having a nearly continuous subcanopy of *Cornus florida* and a shrub layer dominated by *Viburnum acerifolium* (Monk 1957). Lowland portions of the site were dominated by a canopy of *Acer rubrum*, *Fraxinus americana*, and *Nyssa sylvatica*, with a shrub layer composed largely of *Lindera benzoin* and

Viburnum dentatum. The understory also supported a diversity of herbaceous species (Monk 1957; Monk 1961b; Frei and Fairbrothers 1963; Davison and Forman 1982).

Though the forest is referred to as old growth, it is dynamic and has experienced variation over time. Before European settlement, frequent surface fires occurred at roughly ten-year intervals (Buell et al. 1954). Fires were set by the indigenous people of the area until shortly after European colonization, with the last fire occurring in 1711 (Buell et al. 1954). The disturbance regime of the forest also includes periodical damage by severe winds that accompany hurricanes. The most recent of these was in 1950 and resulted in 299 trees being blown down (Frei and Fairbrothers 1963). Contemporary challenges to the forest include high deer (*Odocoileus virginiana*) densities and the persistent invasion of several exotic plants.

Tree regeneration is remarkably low at the site, particularly for oaks and hickories (*Carya* spp.). However, *Acer saccharum* and *Fagus grandifolia*, both characteristic of mesic forests, and the exotic *Acer platanoides* still appear to regenerate. This suggests that the forest may be transitioning from an oak-dominated forest to a more mesic forest type (Monk 1961a). The once dominant *Viburnum acerifolium* understory has also declined steadily over time (Davison and Forman 1982). These changes in forest composition may be related to increased deer density or the loss of fire from the system. In 1973 an outbreak of gypsy moths (*Lymantria dispar*) led to widespread oak defoliation and a resulting increase in light-demanding species in the subcanopy and understory (Moulding 1977; Davison and Forman 1982). The herbaceous understory of the forest has also been severely altered, first by grazing pressure from deer, then from apparent displacement by the exotic grass *Microstegium vimineum*, which has formed a nearly continuous carpet throughout much of the old-growth forest.

Successional Trends Over Time

The successional patterns of the Buell–Small Succession Study (BSS) represent a straightforward progression of life-forms from short-lived herbaceous species to longer-lived woody species (figure 8.2). Annuals, such as *Ambrosia artemisiifolia*, *Bromus* spp., and *Erigeron annuus*, are abundant only the first one or two years after abandonment. Biennials and short-lived perennials, such as *Aster pilosus*, *Daucus carota*, and *Verbascum blattaria*, quickly replace these annuals but are in turn displaced by longer-lived perennials, such as *Calystegia sepium*, *Fragaria virginiana*, and *Solidago* spp. Shrubs and lianas dominate the system approximately fifteen to thirty years postabandon-

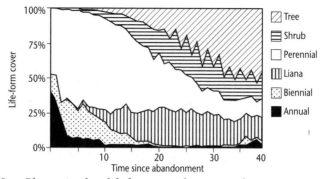

FIGURE 8.2. Change in plant life-forms over forty years of succession expressed as a percentage of total plant cover. Data are summed across all ten BSS fields.

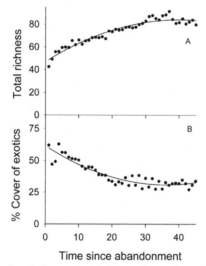

FIGURE 8.3. Successional change in (a) the total number of species, and (b) the percentage cover of exotic species, averaged across the ten BSS fields.

ment. The vast majority of these two life-forms is made up of two exotics, *Lonicera japonica* (liana) and *Rosa multiflora* (shrub). Native lianas common in the site include *Parthenocissus quinquefolia*, *Toxicodendron radicans*, and *Vitis* spp.; native shrubs include several species of *Cornus*, *Rhus*, and *Rubus*. Tree cover begins to dominate the site thirty years postabandonment and consists of native taxa such as *Acer* spp., *Juniperus virginiana*, *Quercus* spp., and *Cornus florida*.

While the number of species occurring in each 1 m² permanent sampling plot has not changed over time (Meiners et al. 2002), the total number of species in each field has continued to increase over time (figure 8.3a). This suggests that the number of species that can coexist at small spatial scales is limited. Plot richness averaged twelve to thirteen species during the first thirty-five years of succession (Meiners et al. 2002), which is well within the range of species richness documented for a variety of herbaceous communities at the 1 m^2 scale (Gross et al. 2000). Compositional variation among plots leads to a much larger species pool across each field, and this pool becomes larger as succession proceeds. Clearly, successional communities do not become closed to new species as envisioned by Egler (1954). However, most common species become established relatively early in succession, as predicted.

After forty years of succession, it appears that the total richness of the BSS fields may be beginning to decline slightly. If this trend continues, the BSS may provide support for the intermediate disturbance hypothesis, which predicts maximum diversity at intermediate stages of succession, when the community contains a mix of early- and late-successional species (Loucks 1970; Connell 1978). As succession proceeds past this maximum, early-successional species are selectively lost from the community because they depend on disturbance or high resource availability for persistence (Bazzaz 1996). Though greatly reduced in abundance, most early-successional species have persisted as occasional occupants of small openings in the canopy or have appeared as a pulse following droughts. It is likely that at least some of the species that have disappeared from one or more of the BSS fields persist in the plots as dormant seeds and may reappear if the proper conditions arise. The occasional reappearance of *Ambrosia artemisiifolia* is an example of this kind of dynamic.

Exotic species are important components of successional systems worldwide (Inouye et al. 1987; Rejmánek 1989; Bastl et al. 1997). This is largely due to the dependence of many exotic species on disturbance for establishment. As many of the exotic species that occur early in succession are agricultural weeds, it is not surprising that the cover of exotic species starts high but becomes proportionately lower as succession proceeds (figure 8.3b). Exotic taxa are also disproportionately represented in early-successional lifeforms. Exotics make up 53% of both annual and biennial species, 29% of perennials, 28% of shrubs, 21% of lianas, and 19% of trees. Because trees and other woody taxa dominate the later stages of succession, the decline in exotic cover over time may be partially driven by a lack of suitable species within the local pool rather than an inherent advantage to being native. Furthermore, most of the exotic tree species in the BSS are shade intolerant and limited to open areas.

Repeatability of Successional Sequences

Central to the utility of successional processes in restoration is whether succession is predictable in space and time. If succession leads to variable endpoints, then the outcome of a restoration may also be variable. We can directly address the predictability of succession in time by comparing the BSS data to a space-for-time substitution study conducted within the same area (Bard 1952). We can also examine the consistency of successional dynamics across the ten BSS fields to address the predictability of succession in space.

Gily Bard's thesis work (directed by Murray Buell) described successional patterns from the Piedmont of New Jersey by sampling twenty-six sites of various ages and inferring successional transitions from one stage to the next (a chronosequence approach). Because all study areas occurred within 8 km of the old-growth forest of the HMFC and used similar sampling methodology (1 m^2 vegetation plots), Bard's study provides an excellent comparison for the experimental BSS data. The general transition of life-forms that Bard described is repeated in the BSS data (figure 8.2). However, there are some significant departures that illustrate the inherent variability of successional systems.

Bard's (1952) chronosequence study describes the perennial C4 grass *Schizachyrium scoparium* as one of the major stages of succession, dominating vegetation for nearly thirty years. However, this species is relatively unimportant within the BSS fields. At its peak, it never occurred in more than 20% of the plots and rarely made up more than 50% of the cover of any individual plot. The relative unimportance of this species in succession at HMFC may be caused by two factors. First, seeds of *S. scoparium* may not have arrived in sufficient numbers because of the landscape position of the fields (Pickett 1983). Bordered on one side by old-growth forests and the other by active and recently abandoned farmland, few reproductive individuals of this species were likely within dispersal range. Second, the lack of *S. scoparium* in the BSS fields may reflect local site conditions. In other successional systems, *S. scoparium* is often associated with infertile sites (Billings 1938). Though the soils of the area are often nutrient poor (Ugolini 1964), the farm that became the BSS was considered relatively fertile. Preexisting site conditions may have placed *S. scoparium* at a competitive disadvantage to species adapted to exploit abundant soil resources.

The second significant deviation from Bard's successional projection is the abundance of shrubs and lianas. Both of these life-forms were found in low abundance in Bard's study (Bard 1952). Within the BSS, shrubs and lianas dominated the middle stages of succession (figure 8.2), though the majority of this cover consisted of two exotic invaders, *Lonicera japonica* and *Rosa multiflora*. Both of these species were present in Bard's chronosequence, but only in minimal amounts. The initiation of the BSS study coincided with the expansion of these species in the landscape and reflects a change in regional species pools. Together, these two relative newcomers to the local flora have dramatically altered succession by changing the structure of midsuccessional communities. Both of these species lead to decreases in local species richness through the inhibition of plant establishment (Meiners et al. 2001; Yurkonis and Meiners 2004). These recent invasions highlight the dynamic nature of local and regional species pools and the potential for rapid changes in successional trajectories in response to invasion.

The composition and dynamics of fields within the BSS also show dramatic spatial variability, though this depends on the scale of observation. The abundance of life-forms is much more predictable across the site than the abundance of any individual species, though there is considerable variation among life-forms. Annuals and biennials tend to vary dramatically among fields, probably reflecting the variation in abandonment conditions. In contrast, the cover of perennials, lianas, shrubs, and trees is quite repeatable among fields over the first forty years of succession. The lack of predictability for individual species probably reflects dispersal limitation that generates spatial variation for even the most abundant species. When a species first appears in the successional community, it is very patchy, leading to great variation among fields. As species spread and increase in cover, this among-field variation decreases. Interestingly, when a species begins to decline, it also does so in a patchy manner, generating higher among-field variability again.

Fields also differ in the timing and rate of tree colonization, though all have eventually become tree dominated. Fields in the lowland portions of the site developed into forests dominated by *Acer rubrum*, mirroring the composition of the lowland old-growth forest. Fields located within the upland have developed a more diverse canopy, including *Quercus* spp., *Carya* spp., *Cornus florida*, and *Fraxinus americana*. Compositionally, tree species vary dramatically among fields with little convergence over time. For example, the largest-seeded species at the site, *Juglans nigra*, is abundant in only one of the ten fields. This probably reflects dispersal limitation and suggests that a mature individual may have been near that field at the time of abandonment.

The difficulties in predicting successional outcomes and trajectories described here have several implications for restoration:

- The past may not be the key to the future. Looking to the surrounding landscape for indications of what may be expected in a successional system is a critical first step, but may not reflect what will occur within a restoration. Variation in site conditions may lead to different transitional community types and potentially different successional trajectories. Furthermore, target communities may themselves be dynamic, responding to contemporary biotic and abiotic influences.

- Identify local species pools. These pools will determine the species that are available within the regenerating community, wanted and unwanted alike. Desired species that are not within the surrounding landscape may need to be added to the species pool. Likewise, heavily invaded landscapes are likely to generate heavily invaded successional systems. While the landscape surrounding a restoration often cannot be altered, management of invasive species can be included in restoration.

- Variation is natural. While spatial heterogeneity is often great within successional systems, compositional heterogeneity becomes an issue only when it involves missing life-forms or species characteristic of late-successional stages. This condition may represent problems within the site that are likely to persist without active remediation. Early-successional species are typically transient and so will need to be managed only if they are of conservation value.

Factors Affecting Tree Regeneration

Because succession in the BSS is leading to a closed-canopy forest, this study has relevance to the restoration of forested systems. Factors that regulate the establishment and growth of trees may have the most direct impact on the rate of succession and potentially the success of a restoration. Tree regeneration at HMFC has been reviewed elsewhere (Myster 1993), but it is worth revisiting specifically from a restoration perspective. This key transition evokes several of the important drivers of succession (Pickett et al. 1987), which are evident in the BSS.

The experimental variation in agricultural history led initially to differences in the vegetation composition and rate of succession among fields (Myster and Pickett 1990). Fields that differed in agricultural treatments— row crops versus hay fields—exhibit the greatest variation in community structure and dynamics. Hay fields, planted in *Dactylis glomerata*, had distinctly different vegetation dynamics for the first eight years after abandonment (Myster and Pickett 1990). Because competition for water between tree seedlings and grasses has been proposed as a major limitation to tree establishment in herbaceous communities (Davis et al. 1998), agricultural history may be expected to influence tree regeneration. However, a comparison of the rate of increase in tree cover between row crops and hay fields does not suggest that grasses inhibit trees within the BSS (figure 8.4). Hay fields actually appear to be colonized more quickly, though this may be related to the small size of the hay fields relative to the row crop fields. However, some portions of the row crop fields have remained open for more than forty years, suggesting dispersal is not the sole limiting factor. In general, grasses are relatively unimportant in this system and do not persist. While herbaceous community composition and dynamics were strongly affected by agricultural history, tree regeneration appears to have been unaffected.

A clear driver of the rate and composition of tree regeneration is the proximity of fields to seed sources. Trees colonized fields adjacent to the old-growth forest much more quickly than those separated from the forest by another field (figure 8.4). There are also compositional differences in these fields. Fields adjacent to the old-growth forest supported a more diverse forest canopy that contains wind-, bird-, and mammal-dispersed species. The isolated fields became dominated by *Juniperus virginiana*, a bird-dispersed conifer, and have remained so for more than forty years. Fields adjacent to the forest were colonized early by *J. virginiana*, but these have largely been displaced by more shade-tolerant tree species (Quinn and Meiners 2004). Succession to deciduous forest has been slowed in the isolated fields by the

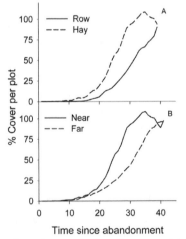

FIGURE 8.4. Influence of previous agricultural treatment (a) and proximity to seed source (b) on the rate of tree colonization of BSS fields. See figure 8.1; data plotted in panel (a) are from hay (C6 and C7) and row crop (C4 and C5) fields adjacent to the old-growth forest. Data plotted in panel (b) are from hay fields adjacent to (C6 and C7), and isolated from (E1 and E2), the old-growth forest.

relative inability of later-successional tree species to colonize the fields, and they have remained dominated by *J. virginiana*. Dispersal limitation can also be seen within fields as tree seedling densities initially declined with distance from the forest edge (Myster and Pickett 1992). This spatial pattern disappeared as the fields become structurally more complex (McDonnell and Stiles 1983; McDonnell 1986), bringing more vertebrate-dispersed species into the fields (Myster and Pickett 1992).

Though not directly assessed in the BSS, herbivory on tree seedlings has been extensively studied at HMFC. Seedlings of palatable trees tend to be clustered closer to the forest edge than nonpalatable species, suggesting that herbivores limit the establishment of these species within fields (Myster and Pickett 1992). Overall, mammalian herbivores severely limit the survival and growth of trees (Myster and McCarthy 1989; McCarthy 1994; Meiners et al. 2000; Cadenasso et al. 2002). Invertebrates have also been shown to regulate tree establishment (Facelli 1994; Meiners et al. 2000), but these tend to be more important in the regeneration of small-seeded trees (Myster and McCarthy 1989). In addition, seed predators consume large portions of the seeds dispersed into successional areas of HMFC (McCarthy 1994; Manson and Stiles 1998; Meiners and LoGiudice 2003), potentially limiting recruitment (DeSteven 1991; Gill and Marks 1991).

Despite the overall impact of herbivory on tree regeneration, there is considerable spatial variability in the intensity of herbivore activity. This spatial variation may be driven by large, landscape-level features or local heterogeneity in the environment. For example, leaf litter is heterogeneously deposited within successional areas (Facelli and Carson 1991). Insect herbivores concentrate their activity in areas containing leaf litter (Facelli 1994), compounding the direct impacts of litter on seedling emergence (Myster 1994). At larger spatial scales, seed predators respond to the presence of forest edges, generating heterogeneity in predation pressure (Myster and McCarthy 1989; Manson and Stiles 1998; Meiners and LoGiudice 2003). In addition to spatial variation, there may be considerable temporal variation in herbivore activity (Manson and Stiles 1998; McCormick and Meiners 2000; Meiners et al. 2000). Taken together, this variation may generate temporal and spatial refugia that allow tree regeneration to occur periodically. For example, seed predator activity varies dramatically among years and is concentrated close to forest edges. Seeds dispersed away from the edge in years with low predation risk would have a much greater chance of establishment than seeds produced in other years or dispersed to other locations.

The observations of patterns and the experimental assessment of processes in the transition between herb and shrub dominance to the establishment and dominance by trees in the BSS yield insights for restoration. Management directives for the restoration of forested systems include the following:

- Tree regeneration appears independent of site history. Within the BSS, the influences of past agricultural treatments had little influence on tree regeneration. This appears contradictory to many studies that find residual effects of past land use. In systems where agricultural history generates persistent variation in vegetation or where grasses dominate, competition with herbaceous vegetation may become limiting to tree regeneration.
- Dispersal limits establishment. Trees are overall the poorest dispersers in a sere and may require the most direct intervention to ensure adequate regeneration. This is particularly true for species with poor dispersal ability in fragmented systems or for large sites. Lack of propagules from late-successional tree species may arrest succession at an earlier stage.
- Herbivory is a rate-limiting process. The density of herbivores at HMFC clearly limits tree regeneration. Sites with high densities of seed predators or herbivores may require protection of seeds and seedlings or the management of herbivores to ensure adequate and timely regeneration. Temporal and spatial heterogeneity in herbivore activity may provide opportunities for establishment in an otherwise hostile environment.

Exotic Species Invasions

Exotic species are common in both successional and restored communities and are often problematic (Walker and Smith 1997; Parker et al. 1999; Byers et al. 2002). From a successional perspective, species invasions become a problem only if they inhibit or alter successional trajectories or if they alter the composition of the target community. In most successional systems, and most plant communities in general, there are relatively few exotic species that warrant concern and potentially require management. Within the BSS, 117 exotic species have been recorded, compared to 216 native species. Of these invaders, only 12 became abundant enough to examine for impacts on the richness of local (1 m^2) communities. Of those, the invasions of only 4— *Elytrigia repens, Trifolium pratense, Lonicera japonica*, and *Rosa multiflora*— depressed richness (Meiners et al. 2001; Yurkonis and Meiners 2004). However, abundance of an invader does not necessarily lead to impacts on community structure. *Hieracium caespitosum*, a perennial invader that peaked ten years after abandonment, had no impacts on richness despite covering an average 30% of each plot at its peak.

Most of the exotic species recorded within the BSS are short-lived species associated with agriculture and do not represent serious management concerns. As successional communities develop, these exotic species naturally leave the system. *Elytrigia repens* and *Trifolium pratense* become successional dominants for very brief periods. The two most abundant exotics at the site, *Lonicera japonica* and *Rosa multiflora*, also decline with tree canopy closure. These two species are considered to be regionally problematic management concerns. However, in this site they did not prevent the trees that are slowly shading them out from establishing. Fields abandoned when these invaders had become more prevalent in the landscape may have become invaded sufficiently to inhibit tree regeneration, requiring intervention.

Currently, two exotic species appear to be management concerns within the BSS: *Alliaria petiolata* and *Microstegium vimineum*. Both of these are invaders of forested systems (Cavers et al. 1979; Gibson et al. 2002). As such, they have the potential to alter the composition of the target community. Both of these species have spread throughout the forested portions of the BSS fields. *Alliaria petiolata* exudes chemicals with antifungal activity that reduces the ability of competing plants to acquire nutrients (Roberts and Anderson 2005). It appears to reduce the diversity of the herbaceous understory and can cover large areas quickly (McCarthy 1997). Likewise, *Microstegium vimineum* forms a thick continuous layer that reduces the diversity and cover of understory herbs (unpublished data).

Because these species are both (1) likely to persist and (2) have considerable impacts on the community, they represent serious management concerns. Any species that meets both of those criteria should be considered a management concern, while species meeting only one will have minimal or only transient effects. Based on information from other sites, it appears that several other species meet these criteria, though they have not yet become abundant in the BSS fields. These include *Acer platanoides*, *Berberis thunbergii*, *Lonicera maackii*, and *Rubus phoenocolasius*.

The directives from the BSS for the management of exotic species in restoration are clear:

- They're not all bad. The vast majority of exotic species, even those that become abundant, have no effects on the plant community and represent little to no threat to the successional development of the site.
- Be patient. Most invaders, because they are adapted to open and disturbed habitats, tend to disappear from systems as the successional

sorting of life-forms and species progresses. Management to remove early-successional invaders may hasten their decline, but should be considered a low priority. However, if management introduces new disturbed sites in an area, it may in fact prolong their occupancy.

- Focus management efforts wisely. Management of exotic species invasions should focus on species that inhibit key successional transitions (e.g., tree regeneration), or that can both persist and have an impact on the target community. Management of taxa that do not meet these criteria may provide aesthetic benefits but should not improve the ecological function or ultimate success of the restoration.

Conclusion

Successional dynamics are the result of a complex suite of processes that interact to generate spatial and temporal heterogeneity. Knowledge of the primary processes that influence succession can help land managers effectively manipulate systems toward restoration. The Buell–Small Succession Study data suggest several major implications for restoration: (1) Dispersal appears to be a primary driver of restoration success, as it has the potential to regulate the availability of both late-successional and exotic species. (2) Site history clearly controls early-successional community dynamics but may not have persistent impacts. (3) Exotic species of early-successional stages are relatively nonproblematic, while exotics that persist within the target community have the potential to dramatically impact the system. Finally, (4) herbivory, particularly on the seeds or seedlings of trees, may limit the rate of canopy closure and the success of forest restorations. The applicability of these restoration lessons for other sites must be assessed based on the similarity of target communities and successional drivers. By placing site-specific factors appropriately in the framework of site availability, species availability, and species performance, managers should be able to successfully manage and enhance successional communities. The necessary linkage between successional processes and remediation was affirmed by Clements:

> Nature's cooperation is essential to the success of the many present endeavors to undo man's destructiveness. Since this cannot be compelled, it must be won by understanding and insight. (1949b)

Acknowledgments

We are indebted to the project initiators and the people who, for decades, have devoted time to the collection of this data so that we might have a complete view of the successional dynamics of this system. The Buell–Small Succession Study has been funded by many sources throughout the years, most recently by a NSF-LTREB award (DEB-0424605), a USDA-NRI grant (99-35315-7695), and the Council on Faculty Research of Eastern Illinois University. We thank M. E. Horn, B. M. Nott, T. A. Rye, W. L. Stewart, and E. M. Tulloss for comments on the manuscript; and J. R. Klass and K. A. Yurkonis for assistance in figure preparation.

REFERENCES

Bard, G. E. 1952. Secondary succession on the piedmont of New Jersey. *Ecological Monographs* 22:195–215.

Bartha, S., S. J. Meiners, S. T. A. Pickett, and M. L. Cadenasso. 2003. Plant colonization windows in a mesic old field succession. *Applied Vegetation Science* 6:205–12.

Bastl, M., P. Kocr, K. Prach, and P. Pyšek. 1997. The effect of successional age and disturbance on the establishment of alien plants in man-made sites: An experimental approach. In *Plant invasions: Studies from North America and Europe*, ed. J. H. Brock, 191–201. Backhuys, Leiden, The Netherlands.

Bazzaz, F. A. 1996. *Plants in changing environments*. Cambridge University Press, Cambridge, UK.

Billings, W. 1938. The structure and development of old field shortleaf pine stands and certain associated physical properties of soil. *Ecological Monographs* 8:437–500.

Buell, M. F., H. F. Buell, and J. A. Small. 1954. Fire in the history of Mettler's woods. *Bulletin of the Torrey Botanical Club* 81:253–55.

Buell, M. F., H. F. Buell, J. A. Small, and C. D. Monk. 1961. Drought effect on radial growth of trees in the William L. Hutcheson Memorial Forest. *Bulletin of the Torrey Botanical Club* 88:176–80.

Byers, J. E., S. Reichard, J. M. Randall, I. M. Parker, C. S. Smith, W. M. Lonsdale, I. A. E. Atkinson, T. R. Seastedt, M. Williamson, E. Chornesky, and D. Hayes. 2002. Directing research to reduce the impacts of nonindigenous species. *Conservation Biology* 16:630–40.

Cadenasso, M. L., S. T. A. Pickett, and P. J. Morin. 2002. Experimental test of the role of mammalian herbivores on old field succession: Community structure and seedling survival. *Journal of the Torrey Botanical Society* 129:228–37.

Cavers, P. B., M. I. Heagy, and R. F. Kokron. 1979. The biology of Canadian weeds. 35. *Alliaria petiolata* (M. Bieb) Cavara and Grande. *Canadian Journal of Plant Science* 59:217–29.

Clements, F. E. 1949a. Ecology in the public service. In *Dynamics of vegetation*, ed. B. W. Allred and E. S. Clements, 246–78. H. W. Wilson, New York.

———. 1949b. Plant succession and human problems. In *Dynamics of vegetation*, ed. B. W. Allred and E. S. Clements, 1–21. H. W. Wilson, New York.

Connell, J. H. 1978. Diversity in tropical rain forests and coral reefs. *Science* 199:1302–10.

Davis, M. A., K. J. Wrage, and P. B. Reich. 1998. Competition between tree seedlings and herbaceous vegetation: Support for a theory of resource supply and demand. *Journal of Ecology* 86:652–61.

Davison, S. E., and R. Forman. 1982. Herb and shrub dynamics in a mature oak forest: A thirty-year study. *Bulletin of the Torrey Botanical Club* 109:64–73.

DeSteven, D. 1991. Experiments on mechanisms of tree establishment in old-field succession: Seedling emergence. *Ecology* 72:1066–75.

Egler, F. E. 1954. Vegetation science concepts. 1. Initial floristic composition, a factor in old-field vegetation development. *Vegetatio* 4:412–17.

Facelli, J. M. 1994. Multiple indirect effects of plant litter affect the establishment of woody seedlings in old fields. *Ecology* 75:1727–35.

Facelli, J. M., and W. P. Carson. 1991. Heterogeneity of plant litter accumulation in successional communities. *Bulletin of the Torrey Botanical Club* 118:62–66.

Frei, K. R., and D. E. Fairbrothers. 1963. Floristic study of the William L. Hutcheson Memorial Forest (New Jersey). *Bulletin of the Torrey Botanical Club* 90:338–55.

Gibson, D. J., G. Spyreas, and J. Benedict. 2002. Life history of *Microstegium vimineum* (Poaceae), an invasive grass in southern Illinois. *Journal of the Torrey Botanical Society* 129:207–19.

Gill, D. S., and P. L. Marks. 1991. Tree and shrub seedling colonization of old fields in central New York. *Ecological Monographs* 61:183–205.

Gross, K. L., M. R. Willig, L. Gough, R. Inouye, and S. B. Cox. 2000. Patterns of species density and productivity at different spatial scales in herbaceous plant communities. *Oikos* 89:417–27.

Haddad, N. M., D. R. Browne, A. Cunningham, B. J. Danielson, D. J. Levey, S. Sargent, and T. Spira. 2003. Corridor use by diverse taxa. *Ecology* 84:609–15.

Inouye, R. S., N. J. Huntly, D. Tilman, J. R. Tester, M. Stillwell, and K. C. Zinnel. 1987. Old-field succesion on a Minnesota sand plain. *Ecology* 68:12–26.

Loucks, O. 1970. Evolution of diversity, efficiency, and community stability. *American Zoologist* 10:17–25.

Luken, J. O. 1990. *Directing ecological succession*. Chapman and Hall, London.

Manson, R. H., and E. W. Stiles. 1998. Links between microhabitat preferences and seed predation by small mammals in old fields. *Oikos* 82:37–50.

McCarthy, B. C. 1994. Experimental studies of hickory recruitment in a wooded hedgerow and forest. *Bulletin of the Torrey Botanical Club* 121:240–50.

———. 1997. Response of a forest understory community to experimental removal of an invasive nonindigenous plant (*Alliaria petiolata*, Brassicaceae). In *Assessment and management of plant invasions*, ed. J. O. Luken, 117–30. Springer, New York.

McCormick, J. T., and S. J. Meiners. 2000. Season and distance from forest–old field edge affect seed predation by white-footed mice. *Northeastern Naturalist* 7:7–16.

McDonnell, M. J. 1986. Old field vegetation height and the dispersal pattern of bird-disseminated woody plants. *Bulletin of the Torrey Botanical Club* 113:6–11.

McDonnell, M. J., and E. W. Stiles. 1983. The structural complexity of old field vegetation and the recruitment of bird-dispersed plant species. *Oecologia* 56:109–16.

Meiners, S. J., S. N. Handel, and S. T. A. Pickett. 2000. Tree seedling establishment under insect herbivory: Edge effects and inter-annual variation. *Plant Ecology* 151:161–70.

Meiners, S. J., and K. LoGiudice. 2003. Temporal consistency in the spatial pattern of seed predation across a forest–old field edge. *Plant Ecology* 168:45–55.

Meiners, S. J., S. T. A. Pickett, and M. L. Cadenasso. 2001. Effects of plant invasions on the species richness of abandoned agricultural land. *Ecography* 24:633–44.

———. 2002. Exotic plant invasions over 40 years of old field succession: Community patterns and associations. *Ecography* 25:215–23.

Monk, C. D. 1957. Plant communities of the Hutcheson Memorial Forest based on shrub distribution. *Bulletin of the Torrey Botanical Club* 84:198–218.

———. 1961a. Past and present influences on reproduction in the William L. Hutcheson Memorial Forest, New Jersey. *Bulletin of the Torrey Botanical Club* 88:167–75.

———. 1961b. The vegetation of the William L. Hutcheson Memorial Forest, New Jersey. *Bulletin of the Torrey Botanical Club* 88:156–66.

Moulding, J. 1977. The progression of a gypsy moth invasion of a mature oak forest in New Jersey. *William L. Hutcheson Memorial Forest Bulletin* 4:5–11.

Myster, R. W. 1993. Tree invasion and establishment in old fields at Hutcheson Memorial Forest. *Botanical Review* 59:251–72.

———. 1994. Contrasting litter effects on old field tree germination and emergence. *Vegetatio* 114:169–74.

Myster, R. W., and B. C. McCarthy. 1989. Effects of herbivory and competition on survival of *Carya tomentosa* (Juglandaceae) seedlings. *Oikos* 56:145–48.

Myster, R. W., and S. T. A. Pickett. 1990. Initial conditions, history, and successional pathways in ten contrasting old fields. *American Midland Naturalist* 124:231–38.

———. 1992. Effect of palatability and dispersal mode on spatial patterns of trees in old fields. *Bulletin of the Torrey Botanical Club* 119:145–51.

Parker, I. M., D. Simberloff, W. M. Lonsdale, K. Goodell, M. Wonham, P. M. Kareiva, M. H. Williamson et al.1999. Impact: Toward a framework for understanding the ecological effects of invaders. *Biological Invasions* 1:3–19.

Pickett, S. T. A. 1983. The absence of an *Andropogon* stage in old-field succession at the Hutcheson Memorial Forest. *Bulletin of the Torrey Botanical Club* 110:533–35.

———. 1989. Space-for-time substitutions as an alternative to long-term studies. In *Long-term studies in ecology*, ed. G. E. Likens, 110–35. Springer-Verlag, New York.

Pickett, S. T. A., and M. L. Cadenasso. 2005. Vegetation dynamics. In *Vegetation ecology*, ed. E. van der Maarel, 172–98. Blackwell, Malden, MA.

Pickett, S. T. A., M. L. Cadenasso, and S. Bartha. 2001. Implications from the Buell-Small succession study for vegetation restoration. *Applied Vegetation Science* 4:41–52.

Pickett, S. T. A., S. L. Collins, and J. J. Armesto. 1987. Models, mechanisms, and pathways of succession. *Botanical Review* 53:335–71.

Quinn, J., and S. J. Meiners. 2004. Growth rates, survivorship, and sex ratios of *Juniperus virginiana* on the New Jersey piedmont from 1963 to 2000. *Journal of the Torrey Botanical Society* 131:187–94.

Rejmánek, M. 1989. Invasibility of plant communities. In *Biological invasions: A global perspective*, ed. J. A. Drake, 369–88. John Wiley and Sons, Chichester, UK.

Roberts, K. J., and R. C. Anderson. 2005. Effect of garlic mustard [*Alliaria petiolata* (Beib, Cavara and Grande)] extracts on plants and arbuscular mycorrhizal (AM) fungi. *American Midland Naturalist* 146:146–52.

Small, J. A. 1961. Drought response in William L. Hutcheson Memorial Forest, 1957. *Bulletin of the Torrey Botanical Club* 88:180–83.

Ugolini, F. 1964. Soil development on the red beds of New Jersey. *William L. Hutcheson Memorial Forest Bulletin* 2:1–34.

Walker, L. R., and S. D. Smith. 1997. Impacts of invasive plants on community and ecosystem properties. In *Assessment and management of plant invasions*, ed. J. O. Luken, 69–86. Springer, New York.

Yurkonis, K. A., and S. J. Meiners. 2004. Invasion impacts species turnover in a successional system. *Ecology Letters* 7:764–69.

Chapter 9

Succession and Restoration in Michigan Old Field Communities

KATHERINE L. GROSS AND SARAH M. EMERY

At the time of European settlement (1820s), the plant communities of southern Michigan, in the upper-Midwest of the United States, were a mixture of forest and grasslands, the distribution of which was determined by a combination of soil, topographic, and climatic drivers (Albert et al. 1988). A history of repeated glaciations and the moderating effects of being surrounded by three of the Great Lakes of North America resulted in a complex of community types that was more varied than in other states at similar latitude in North America. For example, Comer and others (1995) analyzed notes and maps made by the General Land Office Survey (GLOS) during the mid-1800s and determined that there were ten different natural communities in the area of the W. K. Kellogg Biological Station in southwest Michigan.

Grasslands and oak savannas were extensive in southern Michigan at the time of European settlement and covered approximately 194,000 hectares of the state (Chapman 1984). The prairie systems of southwest Michigan were part of the "prairie-peninsula" (Transeau 1935) that likely developed 4,000–8,000 years ago during an extended drier climatic period, which probably also promoted frequent wildfires (MDMVA 2001; Clark et al. 2002). Although records are scant, it is likely that even at the time of European settlement the extent of grassland communities in southern Michigan was influenced by native Americans (Mascouten and Potawatomi tribes), who commonly set fires to maintain open areas for hunting (Legge et al. 1995). Large-scale conversion of the vegetation came with the settlement of this area in the 1830s by European Americans, primarily from New England (Gray 1999).

European settlement resulted in the conversion of forest and grassland openings to small farms and villages, and farming (along with forestry) was the major economic activity of the area until the mid-1950s. Agricultural statistics for the area report that wheat, followed by corn and oats, were the most common row crops from 1850 to 1953. Since 1954, corn has become the dominant crop grown in this area, with soybeans the second most common crop (Tomecek and Robertson 1996). Few herbicides or pesticides were used in conventional agricultural practices in this area until after World War II. The increase in the availability and use of inorganic fertilizers and pesticides began in the 1950s, interestingly accompanying the transition to corn and soybeans as dominant row crops in the region. These human activities, together with fire suppression and the introduction of nonnative species (e.g., multiflora rose, *Rosa multiflora* Thunb. ex Murr., and garlic mustard, *Alliaria petiolata* [Bieb.] Cavara and Grande) resulted in increasing fragmentation and degradation of the remaining small areas of pristine habitat in the region. Rising land costs, changes in agricultural markets, and the loss of economic opportunities in rural Michigan over the past thirty years have caused a demographic shift from rural areas to midsized cities (Hathaway 1960; Frey 2005). Suburban sprawl has caused further loss of agricultural land in southern Michigan, creating a new set of challenges for efforts to restore and retain native communities in this area (Knight 1999). There is increasing concern about (and both legal and public support for) the restoration of natural areas in Michigan. However, there are only limited numbers of studies that can be used to guide restoration efforts in this region.

Successional Studies at the W. K. Kellogg Biological Station

The W. K. Kellogg Biological Station (KBS) has a rich history of research related to understanding the mechanisms of succession following the abandonment of agricultural lands. KBS is located in southwest Michigan (42°24′ N, 85°24′ W; elevation 288 m) and includes over 1,500 ha of land that reflect the matrix of crop fields, old fields, second-growth forests, wetlands, and small lake systems typical of this region (Burbank and Gross 1992; Tomecek and Robertson 1996). The climate is strongly influenced by the Great Lakes, with an average annual rainfall of 890 mm (about half falling as snow) and a mean annual temperature of 9.7 °C. The principal soils at KBS are Typic Hapludalfs (Austin 1979), and most of the region consists of fertile outwash plain deposits and sandy loam soils from the Wisconsin glaciation events 12,000 years ago (Foster 1992). The station has supported ecological research and teaching since the 1950s, with a growth in these activities in the mid-1960s corresponding with the hiring of a full-time resident director and research faculty.

An important but unfortunately little-known experiment that demonstrated the importance of land use history was initiated at KBS in the early 1960s (Davis and Cantlon 1969; Foster 1992). John Cantlon, then professor of botany at Michigan State University, initiated a replicated field experiment at KBS (figure 9.1, SF1) to examine the effect of past land use on successional trajectories. This study quantified the long-term effects of a previous experiment designed to evaluate alternative control methods (herbicide, fire) for *Agropyron repens* (*Elymus repens*, quack grass). There were also long-term effects of a walnut tree crop (where *Rhus typhina*, smooth sumac, were planted as nurse plants) on the successional pathways and community composition of this site (Werner 1972; Foster 1992). Analyses of the vegetation dynamics over the six years following abandonment (Holt 1972; Werner 1972), and twenty-five years after abandonment (Foster 1992), revealed the important role of past land use. In contrast to the standard view of successional pat-

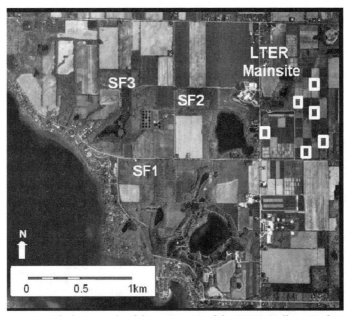

FIGURE 9.1. Aerial photograph of the main site of the W. K. Kellogg Biological Station (KBS) (source: Abrams Aerial Survey, 7 May 1993), with labels indicating locations of successional fields (SF1, SF2, SF3) used for observational studies of old field succession and the KBS Long Term Experimental Research project (LTER) experimental main site, where the replicate experimental successional fields are located (outlined in white). For further details see http://lter.kbs.msu.edu/experimental Design.html.

terns that predicts consistent transitions from old field, to shrubland, to climax deciduous forest (see Drury and Nesbit 1973 for a critical review of this paradigm for North America), Cantlon's work showed that the trajectory of succession was highly influenced by the land use prior to abandonment and initial patterns of colonization. Sites that had been planted to *Rhus typhina* had higher densities of tree seedlings than other areas (Werner and Harbeck 1982; Foster and Gross 1999). In contrast, areas that had been initially dominated by either *Agropyron repens* or *Poa* species (*P. compressa* or *P. pratensis*), most likely as a consequence of preabandonment herbicide trials done on the site, followed very different successional trajectories.

Subsequent short-term experimental studies in old fields in and around the KBS area have shown the important role that colonization dynamics and seed limitation, particularly traits related to seed dispersal, seed size, and seedling traits (Gross and Werner 1982; Emery 2005; Houseman and Gross 2006), competition from established vegetation (Foster and Gross 1997; Foster and Gross 1998; Foster 1999), and small-scale disturbance (Gross 1984; Goldberg and Gross 1988) can play in determining the arrival time and persistence of species in successional fields in this area. However, deriving generalizations about successional dynamics from these studies is difficult because they were all conducted at different times in old fields of various ages and land use histories. Several authors have commented on the limitation of chronosequences for interpreting successional dynamics because of the likely important (and typically unknown) effects that differences in land use history and spatial location can have on these processes (e.g., Pickett et al. 2001).

Unfortunately there are few studies in North America in which a replicated set of experimental plots has been established to follow long-term successional dynamics. The Buell–Small Succession Study established in 1958 by Murray Buell, Helen Buell, and John Small at the Hutcheson Memorial Forest associated with Rutgers University in New Jersey (USA) is a remarkable exception (see chapter 8).

The formation of the Long Term Ecological Research Project (LTER) at KBS let us address concerns about the general lack of replicated studies on old field succession by developing long-term monitoring of successional dynamics of old field communities. This has allowed us to test hypotheses regarding how ecological processes vary in old fields and to explore hypotheses about how plant communities develop in this region. In the following sections, we describe both experimental and observational studies from the KBS LTER and discuss how the results from each can contribute to a greater understanding of successional processes, which in turn can inform restoration of native grasslands and forests in this area.

Old Field Succession Studies at the KBS LTER Site

The KBS Long Term Ecological Research site (figure 9.1) was established in 1988 with a grant from the U.S. National Science Foundation. The project focuses on understanding ecological processes within row crop agriculture and comparing these agricultural systems to unmanaged old fields and forests (e.g., Robertson et al. 1997). The KBS LTER main site was established in a 42 ha field that had been cultivated in row crop agriculture for more than 100 years (mostly continuous corn) (Robertson et al. 1997). The entire site was planted to corn in 1987, followed by a winter rye (*Lolium*) cover crop, and then planted to soybeans in 1988. In spring 1989, treatment plots were established after plowing the entire site. Four treatments on the main site are different management systems for annual row crop agriculture (conventional to organic), and two are perennial crops (alfalfa and poplars) managed for biomass production. One treatment was established to examine early-successional, old field plant dynamics by abandoning plots after spring tillage in 1989 (six replicate, 1 ha plots; see Huberty et al. 1998 for details). To slow woody plant invasion and to keep the old field plots in a herb-dominated successional state, prescribed burns were initiated in 1997, and these plots are now burned in the spring two out of every three years.

Observational Studies of Successional Dynamics in Old Fields at KBS

In 1993, monitoring plots were established in three old fields at KBS that are representative of midsuccessional systems of the area to provide an additional reference base for comparative studies. These sites are sampled at a spatial and temporal scale identical to that of the experimental plots, facilitating comparisons between the younger, experimental plots and older communities. Both the observational and experimental old field systems of the KBS LTER reveal important insights about successional dynamics that can inform efforts to restore native grasslands in the region.

The three reference old fields of the KBS LTER site (figure 9.1) represent a range of prior land uses and times since abandonment, and are typical of the area, providing a general reference for long-term successional dynamics of old fields in the region (figure 9.2a). In each field, a 1 ha, permanent, sampling plot has been established. Plant community composition is determined annually from five 1 m^2 subplots within each large plot. Changes in plant species composition are based on summed species abundances across the five subplots, allowing for more accurate estimates of woody species abundances, as well as grass and forb species composition. A number of soil and ecosystem

(a)

(b)

FIGURE 9.2. (a)Typical Michigan old field plant community; (b) prescribed burn at the KBS LTER main site experimental successional fields, April 2004.

processes are monitored annually in these fields at the same spatial scale and temporal frequency as in the experimental site.

Successional Field 1 (figure 9.1, SF1) is located in Cantlon Field, a 3 ha field abandoned from agriculture in 1964. As described above, this field has a history of ecological studies dating from the 1960s (Davis and Cantlon 1969; Foster 1992). Cantlon Field was actively cultivated from the 1850s until 1964 in a wheat-alfalfa-corn rotation, except for a walnut tree crop and the herbicide application experiment described earlier. After abandonment, *Agropyron repens* and *Rhus typhina* quickly established as dominants, and by 1992 the field was dominated by these species and several exotic cool-season grasses, specifically *Bromus inermis* and *Phleum pratensis* (Burbank and Gross 1992; Foster 1992).

Successional Field 2 (SF2) is located in Louden Field, a 7 ha field abandoned from agriculture in 1948. There is little specific information on land use history of the field before 1911, though it was most likely in row crop cultivation. The field was fallow during 1911–15, and was a sheep pasture during 1915–42. A rotation of corn-oats-wheat was planted in the field during 1942–48, but no fertilizers or herbicides were applied during this time (Stergios 1970). In 1970, vegetation surveys showed that the field was dominated by a combination of *Poa compressa* and *Agropyron repens*, two exotic cool-season grasses (Stergios 1970). By the 1980s, the field was mostly composed of mixed grasses and *Solidago* spp. (Goldberg and Gross 1988). Woody species began to establish in the field in the 1970s (Werner and Harbeck 1982), and tree establishment and growth has continued to the point that the area is codominated by herbaceous plants, oaks (*Quercus* spp.), and black cherry trees (*Prunus serotina*) (Burbank and Gross 1992).

Successional Field 3 (SF3) is located in Turner Field, a 5 ha field last cultivated in a row crop rotation in 1963 (Gross and Werner 1982). Soils in the field are sandy and well drained. In 1992, when the LTER monitoring study was established, the field was dominated by several forb species, including *Hieracium* spp., *Aster pilosus, Solidago* spp., and *Centaurea maculosa*, and a mixture of cool-season grasses. *Prunus serotina*, a native weedy and woody colonizer, was also beginning to establish at this time (Burbank and Gross 1992).

Changes in the composition of the vegetation in each of these three fields have been regularly monitored annually since 1993 (except in 1998–99) as part of the KBS LTER sampling. The three fields differ markedly in aboveground production, species richness, composition, and the abundance of native and nonnative species (figure 9.3). SF3 has almost twice the nonnative species richness as the other two fields, while SF1 has very low native species richness compared with the other two sites. The ratio of native to nonnative species has consistently been higher in SF2 than the other two fields, potentially reflecting that it was abandoned from pasture rather than from row crops. Over a ten-year period of LTER monitoring, there were only small changes in species richness and composition in these fields; nonnative species richness declined slightly in SF3, while total species richness, especially of native species, increased in SF2 (figure 9.3).

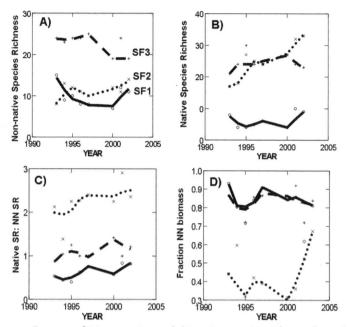

FIGURE 9.3. Patterns of (a) nonnative and (b) native species richness through time in successional old fields; (c) ratio of native to nonnative (NN) species in each field; (d) fraction of total aboveground biomass contributed by nonnative species. For all panels, SF1 is the solid line, SF2 is dotted, and SF3 is dashed. Fit lines are based on Locally Weighted Scatterplot Smoothing (LOWESS) to indicate general trends in the data.

This stability in composition of the plant community in SF1 is apparent in the comparison of functional composition over ten years, with most of the community consisting of perennial grasses and some perennial forbs (figure 9.4a), and a low turnover of species from year to year (figure 9.4b). SF2, and to a greater extent SF3, are much more dynamic, with large increases in woody biomass from 1993 to 2003 (figure 9.4a) and associated decreases in biomass of the herb community.

Though it is risky to generalize results from only three sites, differences in community vegetation patterns across fields can be attributed to past land use, successional age, soil fertility, and geographic location of these sites. For SF1, Foster (1992) found that weed control practices prior to abandonment (herbicides or fire) affected early-successional success of three dominant grasses (*Agropyron repens, Poa compressa, Poa pratensis*), which maintained dominance even through to 2003. In contrast, SF3 has seen large increases in woody species colonization, despite being abandoned at approximately the

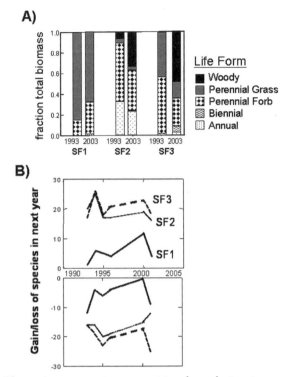

FIGURE 9.4. Change in community composition through time in successional old fields at the KBS. (a) Change in functional group biomass (average aboveground biomass for each field: SF1: 1993 = 458 g m^{-2}, 2003 = 310 g m^{-2}; SF2: 1993 = 175 g m^{-2}, 2003 = 125 g m^{-2}; SF3: 1993 = 253 g m^{-2}, 2003 = 41 g m^{-2}); (b) species turnover (absolute gain and loss of individual species from year to year) over ten years. SF1 = solid lines, SF2 = dotted lines, SF3 = dashed lines.

same time as SF1. Geographic location may partly explain why SF3 has higher woody biomass than SF1, as SF3 is surrounded by more wooded areas than SF1, which likely allows more opportunities for woody species to invade (Myster and Pickett 1993). SF2 is the oldest field of the three, and has the highest proportion of native species diversity and the lowest nonnative biomass. Interestingly, native diversity is still increasing in this field. Woody species biomass is also increasing, though is lower than SF3. Historically, this field had woodland on only one side (Stergios 1970), which may account for the lower occurrence of woody species in this field.

While differences in the successional trajectories of these three fields can be partially accounted for by prior land use, weed control, and surrounding community types, observational studies cannot identify mechanisms that drive differences in successional community composition and trajectory. The replicated, experimental, old field successional plots on the LTER main site help address this shortfall.

Successional Dynamics in Experimental Old Fields

Changes in plant community composition have been documented in the experimental old field plots of the KBS LTER main site (figure 9.1) since 1989. Native and nonnative species richness increased rapidly in the first five years after abandonment but stabilized over the next ten years (figure 9.5). The ratio of native to nonnative species has leveled out at around 1:1. Functional composition of the plant communities shows predictable changes from annual-dominated systems to perennial grass- and forb-dominated systems

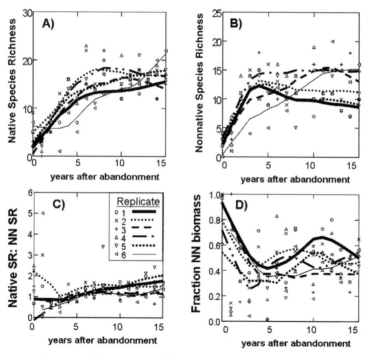

FIGURE 9.5. Patterns of (a) native and (b) nonnative species richness through time in the KBS LTER experimental successional plots; (c) ratio of native to nonnative (NN) species through time; (d) fraction of total biomass contributed by nonnative species. Total aboveground biomass varied between 300 and 800 g m^{-2} in the most recent five years of sampling. Fit lines are based on Locally Weighted Scatterplot Smoothing (LOWESS) to indicate general trends.

through time (see Huberty et al. 1998). We used Non-metric Multidimensional Scaling (NMS)(Legendre and Legendre 1998; McCune et al. 2002) to look at changes in species composition in these plots through time. NMS is a multivariate ordination technique that is similar to Principal Components Analysis but uses ranked distances between plots to estimate similarity to avoid assumptions about linearity or unimodality of the community data. Although the six replicate plots started with very different initial community composition, after five or six years they converged to very similar community assemblages (figure 9.6).

Other studies of community assembly and succession have demonstrated that initial community composition can have persistent effects on community diversity and functional structure (Drake 1990; Young et al. 2001). The

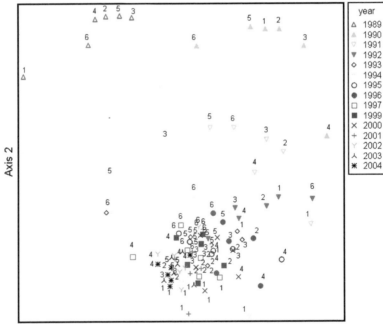

FIGURE 9.6. Nonmetric Multidimensional Scaling (NMS) ordination of plant community composition over sixteen years (1989–2004) in six replicate successional field experimental plots on the KBS LTER. Single species occurrences and species that had less than 1g biomass total over all years and plots were excluded from this analysis. Points are identified by replicate (plot) number (1–6); symbols indicate years. The closer two points are in the ordination, the more similar they are in community composition. The NMS ordination explained 73% of the variation in community composition among plots in two axes.

similarity in past land use and landscape distribution of these plots, and the use of prescribed burning to control woody plant invasions, may explain the convergence in community structure among replicates despite the initial differences in composition. Regular disturbances, such as fire, can be an important filter, overriding interactions among species (Hobbs and Huenneke 1992; Booth and Swanton 2002). Indeed, a comparison of community composition in 1996, before burning began, showed strong differences among replicates, especially in woody species invasions (Foster and Gross 1999), which have diminished since regular burning was initiated in 1997.

Old Field Succession and Restoration

While the old field successional studies at the KBS LTER site were not designed to address issues related to grassland restoration, these two approaches to studying successional trajectories offer insight into what management efforts could be applied to restore native grassland species and communities on abandoned agricultural land of the midwestern United States. These studies reinforce the importance of understanding the role of land use history, geographic location, disturbance regimes, and seed sources (native and nonnative) in structuring plant communities.

The marked differences in community composition among the three reference old fields suggest that variability in site location and history can lead to very different plant communities. This may be a consequence of differences in initial plant community composition due to cultivation practices or use of herbicides (Foster 1992), compounded by differences in seed sources in the surrounding landscape. Although change in species composition in later-successional old fields is typically slowed once perennials have established (Huston and Smith 1987), our data show that even forty to sixty years after abandonment, species composition in old field communities is very dynamic (figure 9.4). Invasion of woody species may account for this, as the establishment of a canopy by trees and shrubs can alter the abundance of understory herbaceous species (e.g., Werner and Harbeck 1982). Invasion rates of woody species can be influenced by landscape position and the structure of surrounding communities. The higher rate of tree invasion into SF3 likely is due to the proximity of wooded areas that surround this site; however, how this relates to the decline in nonnative (mostly herbaceous) species richness over the past ten years in this field in this site is not apparent (figure 9.3).

Despite large initial differences in plant species community composition, all six experimental old field succession plots converged on similar plant communities within ten years. This convergence may be due to (or facilitated by) the introduction of prescribed burning in the early 1990s, as these fields showed strong evidence of diverging prior to the establishment of the burning treatments that was related to differences in the establishment of woody species across replicates (Foster and Gross 1999). Both the season and frequency of burns on the LTER main site may be important explanatory variables for the convergence in community composition and relatively low species richness in the experimental plots. In the upper-midwestern United States, prescribed fires are often implemented in early spring, before the greening of vegetation (Howe 1995). These early-season burns often increase dominance of warm-season grasses, while suppressing nondominant forbs (Copeland et al. 2002). Frequent fires, such as the near-annual burning on the KBS LTER, may reinforce convergence of the plant communities. However, Collins (2000) showed that annual burning of tallgrass prairie systems in the United States caused strong directional change in plant communities compared with sites that were burned on four- or twenty-year cycles.

The almost-annual prescribed fires on the KBS LTER may be able to stabilize plant community composition by regulating nonnative species dynamics. Emery and Gross (2005) found that annual burning reduced the population growth rate of an invasive species, though alternate-year burning had no effect. Yet the reintroduction of fire into degraded grasslands or successional fields may not necessarily lead to the reestablishment of native grassland or savannah communities (Anderson et al. 2000; Suding et al. 2004; Suding and Gross 2006). Initial conditions and feedbacks created by the establishment of nonnative species or shifts in abundance of native species can create strong barriers to restoration efforts and send communities along alternate trajectories (Young et al. 2001; Suding et al. 2004).

It is interesting to note that nonnative species abundances in all replicates of the LTER experimental old fields appear to stabilize around 50% cover soon after field abandonment, suggesting that spread in dominance of nonnatives is not necessarily a consequence of initial invasion. One goal of restoration is to increase the native to nonnative species ratio in an area (Suding and Gross 2006), especially in sites where total elimination of nonnative species is impractical (D'Antonio and Meyerson 2002). It is encouraging that nonnative species cover has stabilized, though fire alone seems to be ineffective at further reducing cover of nonnatives in these plots. Some restoration studies have shown that variable fire regimes (e.g., alternating spring and summer burns) can increase native plant diversity (Copeland et al. 2002). Several other restoration studies have shown that fire needs to be coupled with native seed supplementation to effectively reduce abundance of nonnative species (e.g., Maret and Wilson 2005; Suding and Gross 2006).

Both observational and experimental studies of successional old fields in and around KBS provide evidence that supplementation of native seeds will be an important contributor to the establishment of native species and grassland restoration in this area (Houseman 2004; Suding and Gross 2006). Despite the reintroduction of frequent fire over the past decade, native species richness has stabilized at approximately fifteen to twenty species ha^{-1}. The reference old fields (SF1–3) in the area surrounding the KBS LTER rarely have more than thirty native species per 1 ha plot. Estimates of historic diversity of prairies in the Midwest region range close to 1,000 species (Ladd 1997). The decline and fragmentation of native grassland communities (now estimated at less than 1% of their original area) as a consequence of agriculture and other human activities may make it impossible for "natural processes" to promote the reestablishment of native species on abandoned fields. Competitive pressure from nonnative species that may respond positively to fire (see Suding and Gross 2006) may compound this problem unless native seed sources are supplemented as part of the management.

Conclusion

The patterns of succession observed over the last fifteen years in both the reference and experimental old field communities at the KBS LTER offer insight into how successional processes in the midwestern United States can guide grassland restoration efforts. Land use history, geographic location, disturbance regimes, fertility, and seed sources all play important roles in setting the trajectory of successional communities. If native savannah and grassland seed sources had not been lost from the regional species pool due to habitat loss and fragmentation, and if fire had been used as a management tool, the old field sites in the KBS area today might more closely resemble the native prairie–savannah vegetation that historically occurred in this area (Faber-Langendoen et al. 1995; Packard and Mutel 1997; Suding and Gross 2006). Currently, management appears to be effective at creating and maintaining stable, early-successional communities, despite differences in initial community composition. An interactive management approach, combining prescribed fire and native seed addition seems most promising for future restoration efforts.

Some ecologists argue that successful restorations will only be possible once we understand all components of a system (Bradshaw 1987). While our understanding of aboveground patterns and processes of succession is fairly strong, there is some indication that belowground processes may be critical missing links for successful restorations (Camill et al. 2004). Successional old fields can have reduced mycorrhizal diversity (Bever et al. 2003), reduced soil organic carbon and nitrogen (Kindscher and Tieszen 1998; Potter et al. 1999), and reduced soil invertebrate abundance (Hanel 2003) compared to intact prairie systems, which may explain the continued low species richness in the KBS successional systems. New and continuing research in these areas at the KBS LTER (e.g., Robertson et al. 2000; Buckley and Schmidt 2001; Grandy 2005) may offer additional insight into successful prairie restorations.

Acknowledgments

Research on the KBS LTER has been supported by grants from the Long Term Ecological Research program of the U.S. National Science Foundation (NSF). Other research on grassland restoration and diversity reported here were supported by grants from the Michigan Department of Military and Veterans Affairs, the NSF Ecology program and the A. W. Mellon Foundation. Numerous summer field and laboratory workers assisted in the collection of these data, but without the careful supervision and management by Carol Baker these long-term data could not be reliably analyzed and interpreted. This is KBS contribution number 1281.

REFERENCES

Albert, D. A., G. A. Reese, S. R. Crispin, M. R. Penskar, L. A. Wilsmann, and S. J. Ouwinga. 1988. *A Survey of Great Lakes Marshes in the Southern Half of Michigan's Lower Peninsula.* Report to the Michigan Department of Natural Resources, Land and Water Management Division. Michigan Natural Features Inventory, East Lansing, MI.

Anderson, R. C., J. E. Schwegman, and M. R. Anderson. 2000. Micro-scale restoration: A 25-year history of a southern Illinois barrens. *Restoration Ecology* 8:296–306.

Austin, F. R. 1979. *Soil survey of Kalamazoo County, Michigan.* USDA Soil Conservation Service and the Michigan Agricultural Experiment Station, East Lansing, MI.

Bever, J. D., P. A. Schultz, R. M. Miller, L. Gades, and J. D. Jastrow. 2003. Inoculation with prairie mycorrhizal fungi may improve restoration of native prairie plant diversity. *Ecological Restoration* 21:311–12.

Booth, B. D., and C. J. Swanton. 2002. Assembly theory applied to weed communities. *Weed Science* 50:2–13.

Bradshaw, A. D. 1987. Restoration: An acid test for ecology. In *Restoration ecology: A synthetic approach to ecological research*, ed. W. R. Jordan, M. E. Gilpin, and J. D. Aber, 23–29. Cambridge University Press, Cambridge, UK.

Buckley, D. H., and T. M. Schmidt. 2001. The structure of microbial communities in soil and the lasting impact of cultivation. *Microbial Ecology* 42:11–21.

Burbank, D. H., and K. L. Gross. 1992. *Vegetation of the W. K. Kellogg Biological Station.* Michigan State University, Agricultural Experiment Station, East Lansing, MI.

Camill, P., M. J. McKone, S. T. Sturges, W. J. Severud, E. Ellis, J. Limmer, C. B. Martin, R. T. Navratil, A. J. Purdie, B. S. Sandel, S. Talukder, and A. Trout. 2004. Community- and ecosystem-level changes in a species-rich tallgrass prairie restoration. *Ecological Applications* 14:1680–94.

Chapman, K. A. 1984. An ecological investigation of native grassland in southern lower Michigan. MA thesis. Western Michigan University, Kalamazoo, MI.

Clark, J. S., E. C. Grimm, J. J. Donovan, S. C. Fritz, D. R. Engstrom, and J. E. Almendinger. 2002. Drought cycles and landscape responses to past aridity on prairies of the northern Great Plains, USA. *Ecology* 83:595–601.

Collins, S. L. 2000. Disturbance frequency and community stability in native tallgrass prairie. *American Naturalist* 155:311–25.

Comer, P. J., D. A. Albert, H. A. Wells, B. L. Hart, J. B. Raab, D. L. Price, D. M. Kashian, R. A. Corner, and D. W. Schuen. 1995. *Michigan's native landscape, as interpreted from the General Land Office Surveys 1816–1856*. Michigan Natural Features Inventory, Lansing, MI.

Copeland, T. E., W. Sluis, and H. F. Howe. 2002. Fire season and dominance in an Illinois tallgrass prairie restoration. *Restoration Ecology* 10:315–23.

D'Antonio, C., and L. A. Meyerson. 2002. Exotic plant species as problems and solutions in ecological restoration: A synthesis. *Restoration Ecology* 10:703–13.

Davis, R. M., and J. E. Cantlon. 1969. Effect of size area open to colonization on species composition in early old-field succession. *Bulletin of the Torrey Botanical Club* 96:660–73.

Drake, J. A. 1990. The mechanics of community assembly and succession. *Journal of Theoretical Biology* 147:213–33.

Drury, W. H., and I. C. T. Nesbit. 1973. Succession. *Journal of the Arnold Arboretum, Harvard University* 54:331–68.

Emery, S. M. 2005. Population and community approaches to understanding invasion in grasslands. PhD dissertation. Michigan State University, East Lansing, MI.

Emery, S. M., and K. L. Gross. 2005. Effects of timing of prescribed fire on the demography of an invasive plant, spotted knapweed *Centaurea maculosa*. *Journal of Applied Ecology* 42:60–69.

Faber-Langendoen, D., R. Henderson, K. McCarty, S. Packard, and B. Pruka. 1995. *Midwest oak ecosystems recovery plan: A call to action*. Midwest Oak Savanna and Woodland Ecosystems Conference, Springfield, MO.

Foster, B. L. 1992. The role of land use history in structuring an old-field plant community. MS thesis. Michigan State University, East Lansing, MI.

———. 1999. Establishment, competition and the distribution of native grasses among Michigan old-fields. *Journal of Ecology* 87:476–89.

Foster, B. L., and K. L. Gross. 1997. Partitioning the effects of plant biomass and litter on *Andropogon gerardii* in old-field vegetation. *Ecology* 78:2091–2104.

———. 1998. Species richness in a successional grassland: Effects of nitrogen enrichment and plant litter. *Ecology* 79:2593–602.

———. 1999. Temporal and spatial patterns of woody plant establishment in Michigan old fields. *American Midland Naturalist* 142:229–43.

Frey, W. H. 2005. *Metro America in the new century: Metropolitan and central city demographic shifts since 2000*. Brookings Institution, Washington, DC.

Goldberg, D. E., and K. L. Gross. 1988. Disturbance regimes of midsuccessional old fields. *Ecology* 69:1677–88.

Grandy, A. S. 2005. Ecosystem consequences of soil aggregation following soil disturbance. PhD dissertation. Michigan State University, East Lansing, MI.

Gray, S. 1999. *History of land use in western Michigan.* 1999 KBS LTER All Investigators Meeting, Hickory Corners, MI.

Gross, K. L. 1984. Effects of seed size and growth form on seedling establishment of six monocarpic perennial plants. *Journal of Ecology* 72:369–87.

Gross, K. L., and P. A. Werner. 1982. Colonizing abilities of "biennial" plant species in relation to ground cover: Implications for their distributions in a successional sere. *Ecology* 63:921–31.

Hanel, L. 2003. Recovery of soil nematode populations from cropping stress by natural secondary succession to meadow land. *Applied Soil Ecology* 22:255–70.

Hathaway, D. E. 1960. Migration from agriculture: The historical record and its meaning. *American Economic Review* 50:379–91.

Hobbs, R. J., and L. F. Huenneke. 1992. Disturbance, diversity, and invasion: Implications for conservation. *Conservation Biology* 6:324–37.

Holt, B. R. 1972. Effect of arrival time on recruitment, mortality, and reproduction in successional plant populations. *Ecology* 53:668–73.

Houseman, G. R. 2004. Local and regional effects on plant diversity: The influence of species pools, colonizer traits and productivity. PhD dissertation. Michigan State University, East Lansing, MI.

Houseman, G. R., and K. L. Gross. 2006. Does ecological filtering across a productivity gradient explain variation in species pool-richness relationships? *Oikos* 115:148–54.

Howe, H. F. 1995. Succession and fire season in experimental prairie plantings. *Ecology* 76:1917–25.

Huberty, L. E., K. L. Gross, and C. J. Miller. 1998. Effects of nitrogen addition on successional dynamics and species diversity in Michigan old-fields. *Journal of Ecology* 86:794–803.

Huston, M., and T. Smith. 1987. Plant succession: Life history and competition. *American Naturalist* 130:168–98.

Kindscher, K., and L. L. Tieszen. 1998. Floristic and soil organic matter changes after five and thirty-five years of native tallgrass prairie restoration. *Restoration Ecology* 6:181–96.

Knight, R. L. 1999. Private lands: The neglected geography. *Conservation Biology* 13:223–24.

Ladd, D. 1997. Vascular plants of midwestern tallgrass prairies. In *The tallgrass restoration handbook*, ed. S. Packard and C. F. Mutel. Island Press, Washington, DC.

Legendre, P., and L. Legendre. 1998. *Numerical ecology.* Elsevier, Amsterdam.

Legge, J. T., P. J. Hickman, P. J. Comer, M. R. Penskar, and M. C. Rabe. 1995. *A floristic and natural features inventory of Fort Custer Training Center, Augusta, Michigan.* Michigan Natural Features Inventory, Lansing, MI.

Maret, M. P., and M. V. Wilson. 2005. Fire and litter effects on seedling establishment in western Oregon upland prairies. *Restoration Ecology* 13:562–68.

McCune, B., J. B. Grace, and D. L. Urban. 2002. *Analysis of Ecological Communities.* MjM Software Design, Gleneden Beach, OR.

MDMVA (Michigan Department of Military and Veteran Affairs). 2001. *Fort Custer Training Center integrated natural resources management plan and environmental assessment for fiscal years 2002–2006.* Michigan Department of Military and Veteran Affairs, Construction and Facilities Office, Environmental Section, Lansing, MI.

Myster, R. W., and S. T. A. Pickett. 1993. Effects of litter, distance, density and vegetation patch type on postdispersal tree seed predation in old fields. *Oikos* 66:381–88.

Packard, S., and C. F. Mutel. 1997. *The tallgrass restoration handbook*. Island Press, Washington, DC.

Pickett, S. T. A., M. L. Cadenasso, and S. Bartha. 2001. Implications from the Buell-Small Succession Study for vegetation restoration. *Applied Vegetation Science* 4:41–52.

Potter, K. N., H. A. Torbert, H. B. Johnson, and C. R. Tischler. 1999. Carbon storage after long-term grass establishment on degraded soils. *Soil Science* 164:718–25.

Robertson, G. P., K. M. Klingensmith, M. J. Klug, E. A. Paul, J. C. Crum, and B. G. Ellis. 1997. Soil resources, microbial activity, and primary production across an agricultural ecosystem. *Ecological Applications* 7:158–70.

Robertson, G. P., E. A. Paul, and R. R. Harwood. 2000. Greenhouse gases in intensive agriculture: Contributions of individual gases to the radiative forcing of the atmosphere. *Science* 289:1922–25.

Stergios, B. G. 1970. Seed dispersal, seed germination, and seedling establishment of *Hieracium aurantiacum* in an old field community. MS thesis. Michigan State University, East Lansing, MI.

Suding, K. N., and K. L. Gross. 2006. Modifying native and exotic species richness correlations: The influence of fire and seed addition. *Ecological Applications* 16:1319–26.

Suding, K. N., K. L. Gross, and G. R. Houseman. 2004. Alternative states and positive feedbacks in restoration ecology. *Trends in Ecology and Evolution* 19:46–53.

Tomecek, M. B., and G. P. Robertson. 1996. *Land use history for the Kellogg Biological Station and the surrounding area*. KBS LTER All Scientists Meeting, Hickory Corners, MI.

Transeau, E. N. 1935. The prairie peninsula. *Ecology* 16:423–37.

Werner, P. 1972. Effect of the invasion of *Dipsacus sylveestris* on plant communities in early old-field succession. PhD dissertation. Michigan State University, East Lansing, MI.

Werner, P. A., and A. L. Harbeck. 1982. The pattern of tree seedling establishment relative to staghorn sumac cover in Michigan old fields. *American Midland Naturalist* 108:124–32.

Young, T. P., J. M. Chase, and R. T. Huddleston. 2001. Community succession and assembly: Comparing, contrasting, and combining paradigms in the context of ecological restoration. *Ecological Restoration* 19:5–18.

Chapter 10

Old Field Succession in Central Europe: Local and Regional Patterns

KAREL PRACH, JAN LEPŠ, AND MARCEL REJMÁNEK

The geographical area considered in this chapter includes central European countries: Poland (P), the Czech Republic (CZ), Slovakia (SK), Hungary (H), and Romania (RO). Except the Czech Republic, which has been rather industrialized, these areas belong to traditionally agricultural countries. Nowadays, agriculture contributes to the gross national product by the following proportions (percentage of population employed in agriculture in parentheses): P, 2.9% (18.8%); CZ, 1.88% (4.0%); SK, 4.0% (4.7%); H, 4.0% (8.0%); RO, 11.4% (22.0%) (http://europa.eu.int/comm/agriculture).

From the late 1940s to the late 1980s, all of the countries belonged to the former Soviet bloc, and the so-called socialist economy had a substantial impact on agriculture, changing ownership, reshaping agricultural land, and using excessive amounts of fertilizers and pesticides, especially in CZ, SK, and H. In those years, the proportion of arable land increased, and many former grasslands (both hay meadows and pastures) were plowed up. Since the collapse of the communist totalitarian regimes in the late 1980s, many changes have occurred and are still in progress. These changes can be characterized by the decrease of agricultural productivity due to cheap imports, decrease in the use of fertilizers and pesticides, and decrease of arable land. For example, in the Czech Republic, under the communist rule, grasslands were reduced in area by 300,000 hectares through conversion to arable land, from an original extent of approximately one million hectares. Recently, about 230,000 hectares of grasslands have been reestablished. Unfortunately, the ecological value of the new grasslands is low due to the use of common, commercial seed mixtures. Spontaneous succession on the former arable land contributed and still contributes to the process (Prach et al. 2001) but much less than the intentional sowing of commercial mixtures. Other changes in agri-

180

cultural land use are expected in the countries accessed to the European Union in 2004 (Bogaerts et al. 2002).

The present occurrence of abandoned fields of different ages is determined by their history. Officially, nearly no arable land was abandoned between the late 1940s and late 1980s, but, in fact, various marginal parts of existing fields or small portions of arable land, which were difficult to access, were abandoned, especially in hilly regions. This was also the case in the area where an intensive study in former Czechoslovakia was conducted (Osbornová et al. 1990). Various abandoned fields between fifty and sixty years old can be found in former or still-existing military training areas, and these still await detailed investigation. Nowadays, there are many abandoned fields between one and fifteen years old, having been abandoned after the collapse of communism.

Quantitative data concerning spontaneous vegetation succession on abandoned fields are available from Poland (Falinski 1980; Symonides 1985, 1986), the Czech Republic (Prach 1981 and unpubl.; Lepš and Prach 1981; Lepš et al. 1982; Baumová 1985; Lepš 1987; Osbornová et al. 1990), Hungary (Csetcserits and Rédei 2001) and Romania (Ruprecht 2005, 2006). The aims of this chapter are to summarize the successional patterns over the geographical area considered and to provide a detailed view into mechanisms of succession in the intensively studied sites.

Variability of Vegetation Changes on a Landscape Scale

Three particular seres were distinguished in abandoned fields in the Bohemian Karst area near Prague, CR, as determined by site moisture: (a) dry, (b) mesic, and (c) wet (Osbornová et al. 1990). Besides vegetation dynamics and nutrient conditions, soil micromycetes, microarthropods, and small mammals were investigated (Osbornová et al. 1990). We reexamined the earlier data set, including more recent vegetation records made in the past twenty years in the intensively studied fields, if still available, and further justified the subdivision into the seres. In total, 108 vegetation records were processed by the indirect ordination method, DCA (Ter Braak and Šmilauer 2001 (figure 10.1a, 10.1b). The repeatedly recorded fields showed separate trajectories indicating successional development in the dry, mesic, and wet seres, respectively. Obviously, the soil moisture represents a continuum, and the classification into the seres is rather arbitrary. Nevertheless, it is useful for interpretation and generalization of the results, as well as for practical land management guidelines. The successional trajectories are clearly divergent from the initial stages, which are most similar among the seres, being formed by a common pool of weedy

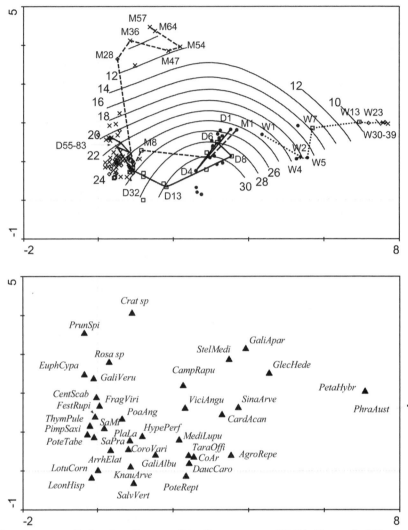

FIGURE. 10.1. Ordination diagram of the first two axes of DCA (Detrended Correspondence Analysis) analysis of all samples (a). DCA options used: log transformation of cover values, downweighting of rare species, detrending by segments. Individual old fields were categorized according to their age: one to five years, dots; six to fifteen years, squares; sixteen to thirty years, diamonds; over thirty years, crosses. Intensively studied plots are connected with lines and labeled D, M, and W for dry, mesic, and wet seres, respectively, and age. Repeated observations were done in fields D4 to D8, D55 to D83, M1 to M8, M28 to M64, W1 to W5, and W7 to W39. Isolines of number of species per plot were fitted using Generalized Additive Models. In the DCA analysis of species (b), only the most frequent species are displayed. The abbreviations are composed of four first letters of generic and species names, except SaMi = *Sanguisorba minor*, CoAr = *Convolvulus arvensis*, PlaLa = *Plantago lanceolata*, SaPra = *Salvia pratensis*.

species. The wet sere exhibits a trajectory that is quite different. The dry and mesic seres run more or less parallel up to approximately fifteen years, when shrubs start to expand in the mesic sere, which is accompanied by a retreat of all heliophilous species and later by the appearance of nitrophilous and other species typical of the woodland understory.

This pattern sharply contrasts with convergent succession observed among 130 abandoned fields of ages one to thirty-five years in Finland (Prach 1985). The difference may be explained by the environmental conditions and the species pool of late-successional species being more uniform in northern than in central Europe. Additionally, initial old field vegetation in Finland was largely determined by highly diverse agriculture practices. Convergence or divergence of successional seres is largely determined by the relative environmental variability among initial and late seral stages (Lepš and Rejmánek 1991). In general, the convergence in plant functional types seems to be more likely than the convergence in species composition (Fukami et al. 2005).

Trends in species diversity, expressed here by the number of species per plot, are as follows: (1) in the wet sere, there is a continuous decline in species richness, with the late seral stages represented by monodominant stands; (2) in the mesic sere, there is comparably high species diversity until the expansion of shrubs; while (3) in the dry sere, the high diversity remains until the oldest (probably stabilized) stages. This pattern corresponds with the expectation that under favorable site conditions, that is, mesic and wet in this case, a strong competitive dominant likely prevails. Diversity is highest under moderate stress that does not allow establishment of strong dominants (Mellinger and McNaughton 1975; Grime 2001). Under extreme site conditions, diversity is again low due to physiological constraints, but such sites do not usually belong to arable land.

Successional trends described here are based primarily on the quantitative data from intensively studied sites (Klaudisová 1974; Osbornová et al. 1990; Prach, unpubl.). Major patterns at the species level are illustrated by figure 10.1b. These results can be tentatively extrapolated to the territory of the Czech Republic and some neighboring areas, and are thus potentially useful in various restoration projects (Prach et al. 1999). Therefore, we provide additional remarks on similarities and differences in the three delineated seres.

Dry Sere

Variable and diverse annual weed flora, including rare and endangered species and those dependent largely on the previous management, prevail in the first years (see also Symonides 1986). Biennial *Carduus acanthoides* often dominates in the fourth year and is later replaced by perennial ruderal/weedy species, *Agropyron repens* and *Artemisia vulgaris*. After approximately twelve years, perennial grasses start to dominate, temporarily *Arrhenatherum elatius*, and later *Festuca rupicola* and *Poa angustifolia*. The latter two prevail after some twenty years and seem to maintain dominance with some fluctuations up to eighty-three years, that is, the age of the oldest documented abandoned fields. The replacement of the competitive grass *A. elatius* by the less competitive *F. rupicola* and *P. angustifolia* can be explained by the decreased concentration of available soil nutrients, namely phosphorus (Kovářová 1990; Rejmánek and Jeník 1978). Woody species (*Crataegus* sp. div., *Rosa* sp. div., *Acer campestre*) are present with a rather low cover up to ca. 10%. Species typical for (semi)natural, steppe-like grasslands gradually penetrate into the fields (figure 10.2). We assume that the shrubby grassland is probably a long-term stabilized stage. We have observed similar trends in drier habitats, especially on south-facing slopes in other drier and warmer parts of central Europe. Formation of seminatural wooded grasslands by spontaneous succession on dry sandy soils can be expected after approximately twenty years (Csetscerits and Rédei 2001).

Mesic Sere

The sere also starts with common annual weeds, with *Tripleurospermum maritimum*, *Stellaria media*, *Chenopodium album*, or *Echinochloa crus-galli* be-

FIGURE 10.2. Late successional "steppe-woodland" stage in an abandoned field of the dry sere. Bohemian Karst, Czech Republic, ca. 60 years after abandonment.

ing the most frequent. The next stage is usually dominated by *Agropyron repens*, which seems very typical for most of temperate Europe (Prach 1985). Its expansion depends largely on the previous field management. If the species was a common weed before the field abandonment, it already dominates by the second or third year. In other cases, it usually dominates between the fifth and tenth year. It may be accompanied by *Artemisia vulgaris* and/or *Cirsium arvense*, and together they often form a distinct ruderal stage (see also Schmiedeknecht 1995). Locally, the invasive alien *Solidago canadensis* can participate (Schmidt 1983, 1988; see also figure 16.3 in Rejmánek 1989), or the most expansive central European grass, *Calamagrostis epigeios*, can dominate (Rebele and Lehmann 2001). These vigorous clonal species, which form monotonous and long-persisting stands, are not desirable from the restoration point of view.

After approximately ten years, grasses typical for mesic site conditions, such as *Arrhenatherum elatius* (the most frequent dominant) and/or *Festuca pratensis*, *Dactylis glomerata* and *Poa pratensis* are usually present, unless *C. epigeios* or *S. canadensis* form a compact stand. *Holcus mollis* is typical for this stage of abandoned fields at higher altitudes. At the same time, an expansion of woody species usually starts. Thus, the grass-dominated communities are often only transitional. The most common woody species are *Crataegus* spp. (in lowlands and low uplands), *Prunus spinosa*, and *Rosa* spp. In the current central European landscapes the presence of the thorny shrubs seems to be favored by the high density of browsing animals such as red deer. *Crataegus* often forms dense, shrubby woodland, which may persist up to approximately 100 years of succession, depending upon its life span (Grime et al. 1988; authors' observations). The next development is difficult to predict because of the lack of old and representative stages. The oldest observed fields are now sixty-five-years old, and *Crataegus* still dominates, with *Fraxinus excelsior* starting in the understory (figure 10.3). Although the old, woody stages were studied in detail only in one area (Prach 1981; Osbornová et al. 1990), similar ones were observed in other nonmontane regions of the Czech Republic and Slovakia. On nutrient-rich mesic sites, a pioneer forest stage, dominated by *F. excelsior*, *Betula pendula*, and *Salix caprea*, was reported from Germany (Schmidt 1993, 1998).

Wet Sere

The initial plant community is usually formed by species typical for initial seral stages in successions at other wet sites, such as *Agrostis stolonifera*, *Juncus bufonius*, *Polygonum lapathifolium*, and *P. hydropiper*, mixed with

FIGURE 10.3. Shrubby successional stage of the mesic sere. *Crataegus* sp. div. dominate in the overstory, while *Fraxinus excelsior* starts in the understory. Bohemian Karst, Czech Republic, ca. 60 years after abandonment.

ordinary weeds of broad ecological amplitude also common to the other seres. The following development seems to be very variable site to site. In the study area, *Phragmites australis* dominated later in succession, while *Salix* spp. were observed elsewhere to establish quickly, sometimes even in the first years after abandonment. Documentation of this sere is rather fragmentary. The occurrence of wet fields in central Europe is exceptional. The succession usually occurs on wet margins of existing fields near streams, or if a drainage system has collapsed.

We believe that the trends described are typical for many areas of central Europe. Nevertheless, any extrapolations must be done with caution. Previous agricultural practice, different local propagule pools, grazing, and various random events can substantially modify successional trajectories (Olsson 1987; Blatt et al. 2005; Dovčiak et al. 2005, etc.). This is illustrated by the following example. Our data (Lepš and Matějka, unpubl. data) indicate that the establishment of willows (*Salix caprea* and *S. cinerea*) is restricted to a "successional window" (Johnstone 1986; Bartha et al. 2003) at a very early stage of succession (generally the first one or two years) when bare soil is present. This observation is based on the development of two adjacent fields in southern Bohemia. Under suitable conditions, the willows became established during the first year and after nearly twenty years dominated the whole plot. In the adjacent field, exhibiting the same environmental conditions, no willows became established, and the plot was dominated by competitive grasses and herbs, mostly *Calamagrostis epigeios*. Probably the most influential effect was the type of abandonment. Bare soil was available in a field plowed after the last harvest, whereas in the plot just harvested and abandoned, no bare soil was available and no willows became established (see also Schmidt 1998). There will also undoubtedly be differences in bare soil availability between different crops (see also Blatt et al. 2005). As willows have extremely short seed viability, and seedling survival is drastically reduced by drought stress (Karrenberg et al. 2002), their establishment in the first year is probably highly contingent, not only upon the state of the field at the time of abandonment, but also on the weather conditions, particularly during the time of seed release.

For the prediction of successional changes with the aim of site restoration it is often sufficient to use life-forms or ecological groups instead of species. The trends discerned from the intensively studied fields are presented in figure 10.4. Under relatively extreme site-moisture conditions, perennial graminoids prevail in the late-successional stages, while woody species prevail in the mesic sites. Biennials (monocarpic perennials) are more, but temporarily, frequent in both dry and wet than in mesic sites. These patterns seem to be common in most central European abandoned fields, and they fit the general expectations (Walker and del Moral 2003).

Sources of In-site Vegetation Variability in Space and Time

There are two main sources of within-site heterogeneity of vegetation: first, the variability in environmental conditions; and second, the variability in colonization potential. The study of the newly abandoned field of the dry sere demonstrated that the relative importance of those two factors changes quickly during the first four years (Lepš 1987). In the first year, the plant community was formed by plants germinating from the seed bank of the arable field. The germination was extremely sensitive to even a small gradient in soil conditions, and so was the composition of plant community. However, starting from the third year, the effect of the distance of propagule source started to become more and more important. The direction of the main gradient in species composition determined by indirect ordination corresponded in the first year of succession to the gradient in soil conditions. In the fourth year, it corresponded to the distance from the source of propagules in the neighboring permanent grassland (Lepš 1987). The relative importance of the two factors naturally depends on their range and steepness. As germination and seedling establishment are among the most sensitive processes in the life cycle,

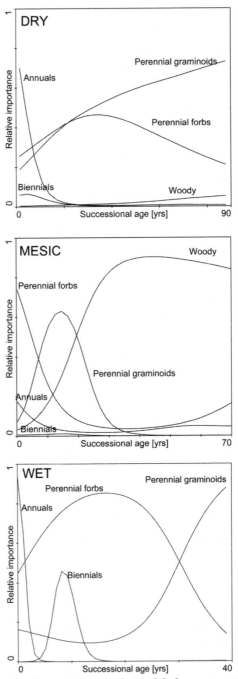

FIGURE 10.4. Changes in participation of main life-form categories during succession in dry, mesic, and wet seres, based on cover degrees. Note the different scales on the x-axis. Oldest stages were not available for the mesic and wet seres.

we can expect that the successional stages dependent on seed germination will be more sensitive to even very small environmental variation. The differences determined by the distance from propagule sources will be very important in the early stages of transient dynamics (Tilman 1988), that is, the stages constrained by dispersal limitation. It could be expected that in latter stabilized stages, the relative importance of environmental heterogeneity will increase again in situations where the heterogeneity will affect the competitive outcomes. Our recent study in an experimentally abandoned field (Jedlička and Lepš, unpubl. data) demonstrated another source of directional variability in the fields. In recent agriculture, the field margins are not as intensively treated as the rest of the field (e.g., the intensity of herbicide use is lower). It results in higher abundance of weeds in the margins of fields during the time of cultivation, and consequently higher concentration of propagules (seeds, but also rhizome fragments) at the field margins. This results in rather strong gradients in species composition perpendicular to field margins during the first two years after abandonment.

Mechanisms of Vegetation Change

In most cases, the seed bank of an arable field is abundant (Thompson et al. 1997). This seed bank depends on the type of management, the last crop, and so forth, and is sometimes accompanied by surviving plant parts enabling vegetative propagation (typically, *Agropyron repens* and *Cirsium arvense*). The decline of annual arable weeds is usually rapid, as they are not able to germinate within a dense cover of perennials. However, many of them survive in the seed bank long after being eliminated from the community. Wind-dispersed species are not very common in the first year, but some of them are able to colonize the community quickly. In particular, the wind-dispersed species from the Asteraceae family (both biennials and perennials) often attain dominance. We observed a rapid increase and then a sharp decline in the biennial *Carduus acanthoides*, whereas the decline in dominant perennials, typically *Artemisia vulgaris* in drier and *Taraxacum officinalis* in mesic conditions, was less pronounced. The decline in monocarpic plants was probably affected by allelopathy. Allelopathic effects in our system were demonstrated under laboratory conditions only; however, the effects on germination were rather strong, including the auto-inhibition of *Artemisia* (Soukupová 1990). The colonization by shrubs differs considerably, depending on the type of shrub dispersal. Whereas the establishment of wind-dispersed species (*Salix* spp.) seems to be limited to the successional "window" at very early stages with bare ground present, animal-dispersed species usually enter later when food, shelter, and/or perching opportunities are available.

Changes in Species Richness at a Fine Scale and Their Mechanisms

Changes in species richness are determined by two counteracting processes: the arrival of new colonizers and competitive exclusion. The trend in species richness differs according to the scale at which species richness is recorded (figure 10.5). Consequently, we characterized each studied stage of the dry sere by its species area curve (estimated usually on the basis of three replicated nested quadrats of increasing size, from 0.01 m² to 64 m²). The curve was characterized by $S = cA^z$, where S is number of species, A is area (in m²), and c and z are parameters estimated by regression analysis after double log transformation. Whereas c corresponds to estimated number of species per unit area (i.e., per 1 m²), z is usually a measure of spatial heterogeneity. At small scales (say, less than 0.25 m²), the changes in species richness are

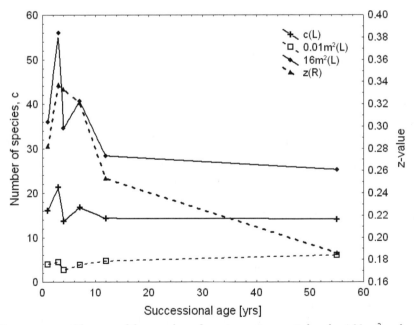

FIGURE 10.5. Changes of the number of species on two spatial scales (.01 m² and 16 m²), and c and z values of the species–area relationship ($S = cA^z$) with successional age of the field in the dry sere. Note that c is an estimate of the number of species per m². (L) and (R) mean that the corresponding scale is on the left and right y axis, respectively.

mainly affected by the ability of species to coexist, to avoid competitive exclusion. At larger spatial scales, the role of the horizontal heterogeneity of a site and dispersal limitation increases. The detailed study of changes in species richness is based on a chronosequence of three plots, located next to each other and studied for several years (Lepš and Štursa 1989). The youngest one was investigated from abandonment to its fourth year, the medium-aged plot was sampled at its seventh and twelfth years after abandonment, and these samples were compared with the stage at fifty-five-years old, the dynamics of which are very slow. The very early stages are limited by available propagules.

In central Europe, arable fields in warmer and drier regions (as are our principal study sites in the Bohemian Karst) (Osbornová et al. 1990) host higher numbers of arable weeds, and consequently, the species richness is usually higher than in colder and wetter regions. At all spatial scales, the species richness increased during the first three years of succession. The increase was more pronounced at larger spatial scales. This corresponds with colonization by scattered new colonizers that increase the number of species mostly at the larger spatial scale. Also, the z-parameter of the species area curve reached its maximum at the same time. At this time, the field hosted up to eighty species in a 64 m^2 quadrat. The dramatic drop in species richness at all spatial scales in the fourth year was caused by increased dominance of perennials, namely *Artemisia vulgaris* and *Agropyron repens*. A wind-dispersed species, *A. vulgaris* is not an important part of the soil seed bank in arable fields. It had taken four years for this species to attain dominance that suppressed the annual species. *A. repens* spreads vegetatively by underground rhizomes. Both species can be characterized as competitive ruderals. From the fourth year on, there is a continuous increase in the number of species in very small plots, and a decrease in large plots, resulting in a quite pronounced decrease of the z-value in the species-area relationship. This process is mostly driven by replacement of competitive ruderals by a steppe-like community dominated by stress tolerators. These species are often able to coexist at small spatial scales (average of six species per 0.01 m^2 plots), because of both their ecology and their smaller size.

It also seems that in younger stages the results of local competition are preemptive and consequently highly contingent on which species colonizes the patch first (the founder). In contrast, in older stages the results of competition are more predictable on the basis of species traits. Consequently, the results are similar in all the patches. This change is consistent with the shift from founder- to dominance-controlled communities predicted by models of Yodzis (1978). This theory also predicted the decrease of the z-value of the species area curve.

Stability

The hypothesis that stability increases with successional age is rather old (Odum 1969). However, various types of stability are recognized in ecology, and they might change in various directions during succession. As it has been shown many times (Lepš 1987; Prach et al. 1993), the rate of succession decreases with successional age and, according to the classical climax concept, should be zero at climax. Consequently, if the stability is defined as an opposite to the rate of directional changes, then it increases during succession. Here, we define three stability concepts, based on dynamics that are (much) faster than successional changes, and demonstrate the results of empirical tests comparing the stability of young (seven years) and old (fifty-five years) fields of the dry sere in the Bohemian Karst (Osbornová et al. 1990). The very early successional stages were not investigated, because there the successional development would interfere with dynamics that are used for the evaluation of stability. However, when appropriate data are available, we expand the discussion to other successional stages. Three stability concepts were investigated (Harrison 1979): *constancy*, an opposite of variability, is the measure of fluctuation on various temporal scales, from days to years; *resistance* is a measure of the ability of a system not to change when facing some perturbation; and *resilience* is the ability to return to the preperturbation state after the end of perturbation. The latter two concepts require the existence of some perturbation. We have used both a natural perturbation, represented by an extreme drought, and experimentally imposed perturbations. The stability is an emergent property of a community, and traditionally, there are two views on the mechanisms of community stability. Whereas one is based on the traditional "diversity begets stability" belief, the other stresses the importance of the life-history strategies of species and their interactions with the environment (Frank 1968; Lepš et al. 1982).

- Constancy: The older stage was more constant in all of the investigated parameters. In particular, the annual and interannual variation in total biomass was smaller. In particular, the dominant grasses of the older stage had a greater "permanent pool" of biomass, that is, an evergreen biomass over the winter. The constancy of biomass was probably the main cause of constancy of environmental factors. For example, both the circadian and seasonal fluctuations of soil temperature decreased with successional age (Prach 1982) (figure 10.6), not only in the two stages compared, but also in a wider range of successional ecosystems. Those factors further assured constancy of some processes such as decomposition rate (Hadincová et al. 1990).

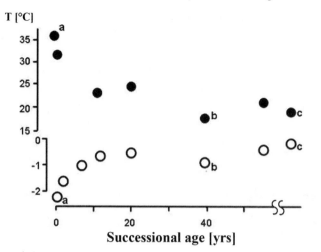

FIGURE 10.6. Maximum temperatures in the top soil layer in the first half of June 1980, and temperatures at the soil surface under the snow layer in January 1980 (five replicates each) in different stages of abandoned fields of the dry and mesic seres; (a) indicates arable land, (b) the dense shrubby stage of the mesic sere, and (c) an adjacent seminatural woodland. Adapted from Prach (1982).

- Resistance: Resistance was considered with respect to the reaction of a community to one natural factor, extreme drought, and to several experimentally introduced perturbations, including trampling, repeated mowing, application of fly ash, and application of herbicides (Lepš et al. 1982; Osbornová et al. 1990). In 1976, the precipitation during the three months preceding the vegetation peak was substantially less than normal. The resistance to drought, and also to "predictable" perturbations, that is, those that may have occured historically, such as trampling, was higher in the older stage. This was mostly caused by the prevailing species strategy. The stress-tolerant species with high root/shoot ratios of the older stage were more resistant to both drought and trampling than the competitive ruderals of the younger stage.
- Resilience: In the vast majority of cases, the resilience was higher in the younger stage. The main cause was the higher growth rate of competitive ruderals compared to stress tolerators. Our results suggest that the main determinants of community functioning are the ecological properties of constituent species, mainly dominants.

  Communities composed of ruderal and competitive-ruderal strategists are more likely to show high levels of resilience, whereas communities composed of stress tolerators are more likely to be resistant

(MacGillivray et al. 1995). Consequently, if the succession proceeds as replacement of ruderals, and competitive ruderals by stress tolerators (as is very often the case in old field succession), then we can expect an increase in constancy and resistance, but a decrease in resilience with successional age. However, the resistance is highly perturbation-specific. Moreover, the effect of each perturbation is modified by the environment. Last, but not least, biotic interactions are also very important in modifying the community response to any perturbation (Pimm 1984).

Main Patterns on Broad Geographical Scales

Reviewing published data from the geographical area considered and including our own field experience (table 10.1), the following generalizations can be tentatively suggested:

1. Unassisted succession leads in a time of approximately twenty years to seminatural stages, often valuable from the point of view of nature conservation. This is especially the case in landscapes with traditional agriculture that were not deeply altered by overfertilization or drainage, and where (semi)natural habitats still frequently occur in the vicinity of abandoned fields. Falinski (1980), Csetserits and Rédey (2001), and Ruprecht (2005, 2006) give illustrative examples. On the other hand, in largely altered landscapes, competitive ruderals or even aliens, such as *Solidago canadensis*, *Calamagrostis epigeios* or *Aster* spp., may expand, especially in nutrient-rich sites (see also Schmidt 1993, 1998). The invasion of the alien tree *Robinia pseudacacia* was reported from abandoned fields with nutrient-poor, sandy soils in Hungary. If not grazed, this may be an alternative pathway to the development of seminatural sandy vegetation (Csecserits and Rédei 2001).
2. Soil moisture and nutrient gradients are the main environmental factors responsible for vegetation patterns at regional scales.
3. The colonization by woody species is crucial in all successional seres, and seems to be restricted only under extreme site conditions. Their establishment is restricted physiologically in dry habitats, and by competition from robust, productive herbs in sufficiently wet and nutrient-rich habitats.
4. The data reviewed do not allow us to discern any latitudinal or longitudinal patterns, except the general decrease in participation of woody species toward drier and warmer regions. As reported from northern

TABLE 10.1

Late successional stages reported from central Europe under different environmental conditions

	Dry	Mesic	Wet
Nutrient poor	*Pinus sylvestris* and *Juniperus communis*[a] *Pinus sylvestris*[b] *Festuca vaginata* and *Stipa borystenica*[d] (*Robinia pseudacacia*)[d]	*Betula* sp. div. and *Pinus sylvestris*[c]	No data
Moderate in nutrients	*Festuca rupicola*[e]	*Betula pendula* and/or *Populus tremula*[f] *Crataegus* sp. div.[g] *Rosa* sp. div. and/or *Prunus spinosa*[h]	*Salix cinerea*[i]
Nutrient rich	*Festuca rupicola*[g]	*Crataegus* sp.div.[g] *Fraxinus excelsior, Betula pendula* and *Salix caprea*[i] (*Solidago canadensis*)[j]	*Phragmites australis*[g] (*Aster lanceolatus*)[k]

Note: Dominant species are listed. Alternative invasive alien dominants are given in parentheses, if reported.
[a] NE Poland, sandy soil, >30 yrs; Falinski (1980), Symonides (1985).
[b] Czech Republic, sandy soil, >10 yr; Prach, in lit.
[c] S and C Poland, >10 yr; Prach, in lit.
[d] C Hungary, sandy soils, >24 yr; Csetcserits and Rédei (2001).
[e] C Romania, >12 yr; Ruprecht (2005, 2006).
[f] Czech Republic and Slovakia, higher altitudes; >20 yr, Prach, in lit.
[g] C Czech Republic, >25 yr; Osbornová et al. (1990).
[h] Czech Republic, higher altitudes, >20 yr; Prach, in lit.
[i] S Czech Republic, >10 yr; Prach, in lit..
[j] C Germany, >12 yr; Schmidt (1983, 1988, 1998), observed and expected also in Czech Republic.
[k] SE Czech Republic, alluvial sites, >10 yr; Prach, in lit.

Europe, species diversity clearly decreased with increasing latitude, which was well evident, especially in early abandoned fields (Prach 1985). The number of species is also expected to decrease with increasing altitude, which was evident from an extensive set of vegetation data obtained in the Czech Republic.

Conclusion: Implications for Management and Restoration of (Semi)natural Vegetation

At first, we must decide what vegetation we wish to have on ex-arable land; that is, we need to define a target. Under present central European conditions, we can formulate the following targets, which then determine the next restoration strategy: (1) reserved fields; (2) cut meadows and pastures; (3)

afforestation; (4) stands of conservation value, i.e., restoration of near-natural, species-rich stands; (5) buffer zones around nature reserves, or around streams; and (6) protection against erosion.

The purpose of reserved fields is to return them to cultivation as arable land, if needed. In that case, the ultimate goal is to protect a site from the expansion of woody species and therefore arrest succession at the stage of dominant grasses. Regular cutting, at least every second year or, better yet, annually, is a reasonable measure. The grassy stage can persist for a long time and is usually dominated by the grass species mentioned above. Spontaneous succession usually runs toward those stages without any assistance. Expansion of woody species can be sufficiently controlled by regular mowing or, to a lesser degree, grazing. Goats are the most efficient grazers to suppress woody species (Bakker 1989).

If we wish to develop pastures or meadows to cut, the same strategy should be adopted, that is, to allow spontaneous succession and to start regular cutting after approximately five years from the onset of succession. In the temperate zones of the world, from which most studies originate, grasses often spontaneously start to take dominance about ten years after the onset of succession. For the support of grasses already in the early stages of succession, it is often sufficient to start mowing. Participation of fodder grasses can be enhanced by their seeding. Seeding is recommended in the initial stages, even just after abandonment, when their establishment is usually easy among the prevailing annuals, rather than later when stands of competitive broadleaf perennial forbs or undesirable *Agropyron repens* often dominate, between approximately the third and eighth years of succession. Unfortunately, various commercial seed mixtures, with ex situ sources, are usually used in central and eastern Europe to convert arable land or young, abandoned fields into grassland. However, in most seres artificial seeding is not necessary because local grasses establish spontaneously (but see Pywell et al. 2002). Moreover, the use of commercial seed mixtures is questionable because they usually contain alien species or genotypes, which may result in their local invasion or genetic "pollution." Using local seed mixtures is preferable but unfortunately not a common approach. We are aware of only two attempts in the region when a local seed mixture was successfully applied to convert arable land into seminatural grassland: dry grassland was established in several tens of hectares in the eastern part of the Czech Republic (Jongepierová, unpubl.) and an alluvial grassland in the western part of Slovakia (Šeffer and Stanová 1999).

Cutting already established shrubs or trees is another way to create a grassland if succession has proceeded further than desired. However, this is usually an expensive and labor-intensive measure that has been practiced only very locally.

Because arable soil is usually sufficiently fertile, there is no need for physical or chemical manipulation of the environment (Whisenant 1999). The only exception is the case of heavily fertilized (eutrophicated) soil, where sod cutting may be a useful measure to initiate succession toward less productive and more species-rich stages (Verhagen et al. 2001). However, this measure has not yet been practiced in the region considered.

If a woodland is the preferred target of restoration, a possible strategy is to allow spontaneous succession until woody species become established, usually between ten and twenty years, especially under mesic site conditions (see table 10.1). Unfortunately, traditional afforestation, resulting in monotonous plantations of one species, has been the dominant practice in the region. Spontaneous establishment of woody species in ex-arable land often depends on the sources of diaspores in the surroundings and the distance of a site from them (Olsson 1987; Schmidt 1993, 1998; Gardescu and Marks 2004). Edge effects are often important in this matter, contributing to the heterogeneity of seral stages (Pickett et al. 2001; Dovčiak et al. 2005).

Mowing or, alternatively, grazing, can be used to manipulate the succession toward a desired target as was well demonstrated by Csetscerits & Rédei (2001) (see table 10.1). The two quite contrasting outputs of succession, with grazing or without, demonstrate the high potential for manipulating succession (Prach et al. 2007).

It must be kept in mind that succession in abandoned fields is highly variable site by site and among various geographical areas. Recommendations for manipulation of succession are generally difficult, as well as a detailed prediction of spontaneous vegetation change. Site moisture and nutrients, history of a site (its previous cultivation), interactions with other trophic levels, character of the surrounding landscape, and various random events are among the main driving variables (Pickett et al. 1987; Tilman 1988; Walker and del Moral 2003, etc.).

The accession of the countries to the European Union in 2004 probably stimulates other changes in the agricultural use of land (Bogaerts et al. 2002). Continuation of the abandonment of arable land will depend on the subsidy policy of the European Union. At present, the subsidies are mostly invested into agricultural production. It has led, for example in Poland, to the reestablishment of agricultural activities in already abandoned agricultural land as early as the first year after the accession. Starting in 2007, the new European agricultural policy (CAP—Common Agricultural Policy) is scheduled to be focused more on agricultural activities that are oriented primarily not on productivity itself but on the other functions, such as ecofarming, agrotourism, preservation of traditional landscape structures, and so forth. We will see what will be the impact on the abandonment of arable land. As the main target for the future, we consider the protection of the traditional character of central European landscapes, especially manifested in the preservation of the mosaic of woodland and secondary treeless areas, of primary importance. Reasonably manipulated succession on ex-arable land should be the most important agent when dealing with this challenge.

Acknowledgments

We are grateful to all the colleagues who participated in the integrated research of old field succession in the Bohemian Karst in the 1970s and 1980s. We thank Viki Cramer, Wolfgang Schmidt, and an anonymous reviewer for their comments; Simona Poláková for her help with data elaboration; and Eliška Rejmánková for help with language revision. The research was supported by the grants from the Ministry of Education and the Academy of Sciences of the Czech Republic MSM6007665801, AVOZ60050516, and IAA600050702.

REFERENCES

Bakker, J. P. 1989. *Nature management by grazing and cutting.* Kluwer, Dordrecht, The Netherlands.

Bartha, S., S. J. Meiners, S. T. A. Pickett, and M. L. Cadenasso. 2003. Plant colonization windows in a mesic old field succession. *Applied Vegetation Science* 6:205–12.

Baumová, H. 1985. Influence of disturbance by moving on the vegetation of old-fields in the Bohemian Karst. *Folia Geobotanica & Phytotaxonomica* 20:245–65.

Blatt, S. E., A. Crowder, and R. Harmsen. 2005. Secondary succession in two southeastern Ontario old-fields. *Plant Ecology* 117:25–41.

Bogaerts, T., I. P. Williamson, and E. M. Fendel. 2002. The role of land administration in the accession of central European countries to the European Union. *Land Use Policy* 19:29–46.

Csecserits, A., and T. Rédei. 2001. Secondary succession on sandy old-fields in Hungary. *Applied Vegetation Science* 4:63–74.

Dovčiak, M., L. E. Frelich, and P. B. Reich. 2005. Pathways in old-field succession to white pine: Seed rain, shade, and climate effects. *Ecological Monographs* 75:363–78.

Hadincová, V., M. Kovářová, and J. Pelikánová. 1990. Decomposition and release of nutrients. In *Succession in abandoned fields: Studies in Central Bohemia, Czechoslovakia*, ed. J. Osbornová, M. Kovářová, J. Lepš, and K. Prach, 80–91. Kluwer, Dordrecht, The Netherlands.

Falinski, J. B. 1980. Vegetation dynamics and sex structure of the populations of pioneer dioecious woody plants. *Vegetatio* 43:23–38.

———. 1986. Vegetation succession on abandoned farmland (in Polish with English summary). *Wiadomosci Botaniczne* 30:25–50.

Frank, P. W. 1968. Life histories and community stability. *Ecology* 49:355–57.

Fukami, T., T. M. Bezemer, S. R. Mortimer, and W. H. van der Puten. 2005. Species divergence and trait convergence in experimental plant community assembly. *Ecology Letters* 8:1283–90.

Gardescu, S., and P. L. Marks. 2004. Colonization of old fields by trees vs. shrubs: Seed dispersal and seedling establishment. *Journal of Torrey Botanical Society* 131:53–68.

Grime, J. P. 2001. *Plant strategies and vegetation processes.* 2nd ed. Wiley, Chichester, UK.

Grime, J. P., J. G. Hodgson, and R. Hunt. 1988. *Comparative plant ecology.* Unwyn Hyman, London.

Harrison, G. W. 1979. Stability under environmental stress: Resistance, resilience, persistence, and variability. *American Naturalist* 113:659–69.

Johnstone, I. M. 1986. Plant invasion windows: A time-based classification of invasion potential. *Biological Review* 61:369–94.

Karrenberg S., P. J. Edwards, and J. Kollmann. 2002. The life history of Salicaceae living in the active zone of floodplains. *Freshwater Biology* 47:733–48.

Klaudisová, A. 1974. Succession in abandoned fields in the Bohemian Karst. MS thesis, Depon in Faculty of Natural Sciences, Charles University, Prague, CR.

Kovářová, M. 1990. Water and nutrient economy. In *Succession in abandoned fields: Studies in Central Bohemia, Czechoslovakia,* ed. J. Osbornová, M. Kovářová, J. Lepš, and K. Prach, 134–43. Kluwer, Dordrecht, The Netherlands.

Lepš, J. 1987. Vegetation dynamics in early old field succession: A quantitative approach. *Vegetatio* 72:95–102.

Lepš, J., J. Osbornová-Kosinová, and M. Rejmánek. 1982. Community stability complexity and species life-history strategies. *Vegetatio* 50:53–63.

Lepš, J., and K. Prach. 1981. A simple mathematical model of the secondary succession of shrubs. *Folia Geobotanica & Phytotaxonomica* 16:61–72.

Lepš, J., and M. Rejmánek. 1991. Convergence or divergence: What should we expect from vegetation succession? *Oikos* 62:261–64.

Lepš, J., and J. Štursa. 1989. Species-area relationship life history strategies and succession: A field test of relationships. *Vegetatio* 83:249–57.

Mellinger, M. V., and S. J. McNaughton. 1975. Structure and function of successional vascular plant communities in central New York. *Ecological Monographs* 45:161–82.

MacGillivray, C. W., J. P. Grime, S. R. Band, R. E. Booth, B. Campbell, G. A. F. Hendry, S. H. Hillier, et al. 1995. Testing predictions of the resistance and resilience of vegetation subjected to extreme events. *Functional Ecology* 9:640–49.

Odum, E. P. 1969. The strategy of ecosystem development. *Science* 164:262–70.

Olsson, E. G. 1987. Effects of dispersal mechanisms on the initial pattern of old-field forest succession. *Acta Oecologica* 8:379–90.

Osbornová, J., M. Kovářová, J. Lepš, and K. Prach, eds. 1990. *Succession in abandoned fields: Studies in Central Bohemia, Czechoslovakia.* Kluwer, Dordrecht, The Netherlands.

Pickett, S. T. A., S. L. Collins, and J. J. Armesto. 1987. Models, mechanisms, and pathways of succession. *Botanical Review* 53:335–71.

Pickett, S. T. A., M. Cadenasso, and S. Bartha. 2001. Implications from the Buell-Small Succession Study for vegetation restoration. *Applied Vegetation Science* 4:41–52.

Pimm, S. L. 1984. Stability and complexity of ecosystems. *Nature* 307:321–26.

Prach, K. 1981. Selected ecological characteristics of shrubby successional stages on abandoned fields in the Bohemian Karst (in Czech with English summary). *Preslia* 53:159–69.

———. 1982. Selected bioclimatological characteristics of differently aged successional stages of abandoned fields. *Folia Geobotanica & Phytotaxonomica* 17:349–57.

———. 1985. Succession of vegetation in abandoned fields in Finland. *Annales Botanici Fennici* 22:307–14.

Prach, K., S. Bartha, C. H. B. Joyce, P. Pyšek, R. van Diggelen, and G. Wiegleb. 2001. The role of spontaneous vegetation succession in ecosystem restoration: A perspective. *Applied Vegetation Science* 4:111–14.

Prach, K., R. Marrs, P. Pyšek, and R. van Diggelen. 2007. Manipulation of succession. In *Linking restoration and ecological succession*, ed. L. R. Walker, J. Walker, and R. J. Hobbs, 121–149. Springer, New York.

Prach, K., P. Pyšek, and P. Šmilauer. 1993. On the rate of succession. *Oikos* 66:343–46.

———. 1999. Prediction of vegetation succession in human-disturbed habitats using an expert system. *Restoration Ecology* 7:15–23.

Pywell, R. F., J. M. Bullock, A. Hopkins, K. J. Walker, T. H. Sparks, M. J. W. Burke, and S. Peel. 2002. Restoring of species-rich grassland on arable land: Assessing the limiting processes using a multi-site experiment. *Journal of Applied Ecology* 39:249–64.

Rebele, F., and C. Lehmann. 2001. Biological flora of Central Europe: *Calamagrostis epigejos* (L.) Roth. *Flora* 196:325–44.

Rejmánek, M. 1989. Invasibility of plant communities. In *Biological invasions. A global perspective*, ed. J. A. Drake et al., 369–88. Wiley, Chichester, UK.

Rejmánek, M., and J. Jeník. 1978. Biogeochemical cycles: Phosphorus problem. In *Biogeochemical cycles in countryside*, ed. B. Moldan and T. Pačes, 60–64. Geological Survey ÚÚG, Prague, CR.

Ruprecht, E. 2005. Secondary succession in old-fields in the Transylvanian Lowland (Romania). *Preslia* 77:145–57.

———. 2006. Successfully recovered grassland: A promising example from Romanian old-fields. *Restoration Ecology* 14:473–80.

Schmidt, W. 1983. Experimentelle Syndynamik—Neuere Wege zu einer exakten Sukzessionsforschung, dargestellt am Beispiel der Gehölzentwicklung auf Ackerbrachen. *Berichte der Deutschen Botanischen Gesellschaft* 96:511–33.

———. 1988. An experimental study of old-field succession in relation to different environmental factors. *Vegetatio* 77:103–14.

———. 1993. Sukzession und Sukzessionslenkung auf Brachäckern—neue Ergebnisse aus einem Dauerflächenversuch. *Scripta Geobotanica* 20:65–104.

———. 1998. Langfristige Sukzession auf brachliegenden landwirtschaftlichen Nutzflächen. Naturschutz durch Nichtstun? *Naturschutz und Landschaftsplanung* 30:254–58.

Schmiedeknecht, A. 1995. Untersuchungen zur Auswirkung von Flächenstillegungen auf die Vegetationsentwicklung von Acker—und Grünlandbrachen im Mitteldeutschen Trockengebiet. *Dissertationes Botanicae* 245:1–174.

Soukupová, L. 1990. Life histories of principal plant populations, including their allelopathic interferences. In *Succession in abandoned fields: Studies in Central Bohemia, Czechoslovakia*, ed. J. Osbornová, M. Kovářová, J. Lepš, and K. Prach, 32–38. Kluwer, Dordrecht, The Netherlands.

Symonides, E. 1985. Floristic richness, diversity, dominance and species evenness in old-field successional ecosystems. *Ekologia Polska* 33:61–79.

————. 1986. Seed bank in old-field successional ecosystems. *Ekologia Polska* 34:3–29.

Šeffer, J., and V. Stanová, eds. 1999. *Morava River floodplain meadows: Importance, restoration and management.* Daphne, Bratislava, Slovakia.

ter Braak, C. J. F., and P. Šmilauer. 2001. *CANOCO reference manual and CanoDraw for windows user's guide: Software for canonical community ordination (version 4.5).* Microcomputer Power, Ithaca, New York.

Thompson, K., J. P. Bakker, and R. M. Bekker. 1997. *The soil seed banks of northwest Europe: Methodology, density and longevity.* Cambridge University Press, Cambridge, UK.

Tilman, D. 1988. *Dynamics and structure of plant communities.* Princeton University Press, Princeton, NJ.

Verhagen, R., J. Klooker, J. P. Bakker, and R. van Diggelen. 2001. Restoration success of low-production plant communities on former agricultural soils after top-soil removal. *Applied Vegetation Science* 4:75–82.

Walker, L. R., and R. del Moral. 2003. *Primary succession and ecosystem rehabilitation.* Cambridge University Press, Cambridge, UK.

Whisenant, S. G. 1999. *Repairing damaged wildlands.* Cambridge University Press, Cambridge, UK.

Yodzis, P. 1978. *Competition for space and the structure of ecological communities.* Springer-Verlag, Berlin.

Chapter 11

Dynamics and Restoration of Abandoned Farmland and Other Old Fields in Southern France

PASCAL MARTY, JAMES ARONSON, AND JACQUES LEPART

Biogeographically, the French Mediterranean Region (FMR) is located in the northwestern quadrant of the Mediterranean Basin, where annual rainfall is relatively high: 600 to 800 mm in the lowlands. Annual rainfall is 1,200 mm or more in the mountains, that is, the southern Alps, Eastern Pyrenées, and southeastern edge of the Massif Central, which mark the northern limits of the region. On the limestone plateaus farther inland, annual rainfall drops off, and its distribution gradually changes to a continental regime (figure 11.1).

The environmental and, over the past six millennia, ecological impacts of humans and their domesticated livestock have been profound in the FMR, as has been the case throughout the Mediterranean Basin. The results have been remarkably variable, given the geological and biological heterogeneity of the region. Beginning with the westward spread of the various Neolithic innovations some 8,000–10,000 YBP (Childe 1971), and accelerating rapidly since roughly 6,000 YBP (Guilaine 1991; Price 2000), Mediterranean peoples have reworked, reshaped, and transformed vegetation and whole landscapes again and again. Biological evidence of human influence is prominent in fossil pollen strata containing cereals, chestnut, and walnut, as early as 2,600 to 6,000 YBP (Vernet 1997; Russo Ermolli and Di Pasquale 2002). During recent centuries, economic development and long-distance trade (Matvejevic 1999) also greatly contributed to the creation of anthropogenic landscapes (Vos and Stortelder 1992; Grove and Rackham 2001). As in the Mediterranean Basin generally, degradation of forest ecosystems has been a major, ongoing phenomenon (Lombard 1959; Thirgood 1981; Blondel and Aronson 1999). Since roughly 1500 AD, the shift toward a market-oriented

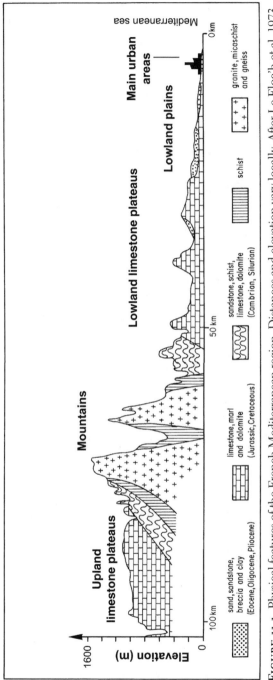

Figure 11.1. Physical features of the French Mediterranean region. Distances and elevation vary locally. After Le Floc'h et al. 1973 (*Atlas régional du Languedoc Roussillon*), modified by M. Debussche.

agriculture contributed to massive changes in agricultural and other land uses (Butlin and Dodgshon 1998).

In more recent times, three additional historical features should be considered: (1) the creation of a national railway network; (2) increasing specialization in wine production during the nineteenth century; and (3) a very active period of "forest restoration" in the FMR premontane areas, starting in the second half of the nineteenth century. Elucidating these drivers will help set the stage for the ensuing discussion of restoration and rehabilitation efforts now under way or envisioned for the future. Forest, farmland, and pasture land will be the three poles of attraction, just as they have been throughout the Neolithic and especially in the last 2,000 years. However, some of the drivers at work today are brand new. In this chapter, we compare the wide array of old field contexts present in the French Mediterranean region, the histories of their development, abandonment, and subsequent management regimes, and the various ongoing or potential interventions that are or could be applied in a spirit of restoration and management.

Transition to a New Landscape Design and Dynamic

At the end of the nineteenth century, human activity reached unprecedented levels in the FMR, which allowed, and at the same time required, intensive agricultural practices. Since this time, socioeconomic changes have progressively led to a very strong disjunction between areas of intensive agriculture (irrigated agriculture, vineyards) in the lowlands, and areas of abandoned fields in the uplands and premontane areas, where agriculture has dramatically declined. In these latter areas, natural vegetation dynamics has led to a resurgence of "preforest," or young forest land cover (Lepart and Debussche 1992; Tatoni et al. 2004).

The increasingly ubiquitous old fields of the upland regions are of two kinds. In a narrow sense, they are areas cultivated and abandoned in the last thirty years. In a larger perspective, however, they include areas that were cultivated episodically until the middle of the nineteenth century and then subsequently used primarily for grazing. Cycles of cultivation, abandonment, and clearing were a very common land use strategy in the region (Marres 1935). If we adopt the dual definition, then issues of old field restoration take on much importance in the light of the choices to be made in landscape planning and management in almost every hilly or upland area in the French Mediterranean region. But what led to this transition? What is lost and what remains?

Socioeconomic Background

The national railway system set up in the nineteenth century ushered in the first phase of economic integration of the continental French territory. Railway networks made it possible for the first time to transport wine produced in the FMR to Paris and, in the opposite direction, to import cereals from more productive regions in the northern or southwestern regions of the country (e.g., the fertile farmlands near Paris and Toulouse). From the late nineteenth century, cereal production was no longer economically viable in the FMR and, concurrently, wine production became very profitable, especially in the lowlands near the coast and in the deep river valleys (Loubere et al. 1988). Elsewhere in the FMR, widespread agricultural abandonment took place in inland rocky hill and premontane areas as a direct result of the *Phylloxera* root rot crisis that occurred at the end of the nineteenth century (Stevenson 1980). At that time, the only way available to fight against *Phylloxera* was to flood the vineyards, and this technique was suitable only in the lowlands. Consequently, a great majority of the wine producers in upland areas were financially ruined during this period, and widespread abandonment of vineyards took place, even as specialization in wine production "took root" in the lowlands.

An additional factor that led to viticultural specialization in the lowlands was that resistant American grape varieties were imported and used as rootstocks, as a complementary and more sophisticated method to limit the ravages of *Phylloxera*. As this was expensive and time consuming, it led inexorably to greater specialization in wine production. Since livestock husbandry in the lowland region had traditionally been devoted to produce manure for cereal fields, and cereals were no longer being grown, animal breeding decreased dramatically during this period.

Some decades later, however, the development of irrigation techniques and canalization systems allowed a diversification of farm activities in the lowland areas to include fruit and vegetable production. These new activities also required specialization and thus contributed to the ongoing crisis of upland agriculture. Declining agriculture and animal breeding, along with the progressive abandonment of fuelwood exploitation, so transformed the upland areas that old fields soon became ubiquitous, except where urbanization was taking place in close proximity to cities, towns, and some villages.

The third important historical factor was that in the late nineteenth century the French government initiated a public works program called *Restauration des Terrains de Montagne* (RTM) in the generally deforested mountains of the FMR. Its goal was to halt or reduce soil erosion and to restore forest cover. In modern terms (SER 2004), the aim was ecological rehabilitation rather than restoration, since a wide medley of mostly nonnative species were used, and no reference system was adopted. The results were mixed, in ecological terms (Vallauri et al. 2002), as were the socioeconomic spin-offs of these programs. Most pertinent to the discussion here is that they indirectly intensified the crisis of animal breeding systems in montane and premontane areas (Kalaora and Savoye 1986), because newly reforested areas were no longer available as grazing resources.

Agricultural Decline and Land Abandonment

In this section we examine old fields within a historical, agricultural, and ecological context. We focus on the trajectories of terraced landscapes, on the role of past cultivation in the dynamics of grazing areas, and on the balance between plantation and natural tree encroachment in old fields.

Terraces and Other Traces of Human-Built Landscapes

Some of the most striking examples of old fields resulting from former intensive management of landscapes are the abandoned terraces found throughout the FMR and indeed in most of the northern Mediterranean Basin (Ambroise et al. 1993). Several local names existed for these terraces (Blanchemanche 1995) and many local adaptations and variations arose. Terraces were devoted to vineyards (e.g., the famous Banyuls wine terraces overlooking the sea), orchards (e.g., for chestnuts in the Cévennes region and, more generally, for olive trees), or mixed-crop farming. Blanchemanche (1995) argues that, in general, building a terraced landscape took approximately one century during the eighteenth and nineteenth centuries.

In the 1950s, because of their fragmented field pattern, the impossibility of mechanizing cultivation and harvesting on narrow terraces, and the high cost of maintenance of pathways, walls, and hydraulic networks, this kind of landscape was seen by scientists and planners as being archaic as compared with the widespread modernization—that is, industrialization—of European, Australian, and North American agriculture. Small farms with limited incomes in the FMR did not survive the development of productivist agriculture that began in the 1960s. However, as we show later, the death knell has not definitively tolled for the FMR terraces.

Old Fields and Grazing/Cultivation Rotation

In addition to terraces *sensu stricto*, discontinuous stone walls, stone fences, or sparse stone piles resulting from soil cultivation are present in most of the hilly uplands of the FMR. Practices and techniques for removing stones from fields and stacking them as field and property boundary markers existed throughout Europe (Blanchemanche 1995) and the Mediterranean region, and numerous remnants and vestiges exist throughout the FMR on the calcareous plateaus called "causses," in the schist and granitic Cévennes, and elsewhere.

These stone walls and fences, occurring in what appear to have long been extensive pasture lands, indicate formerly dense patterns of old field markers (Fowler 1999) that can be used to reveal features of the past agricultural system. First of all, the agricultural system was based on a large and readily available supply of manpower and animal traction, which allowed the intensive exploitation of very small plots. Secondly, before the regional specialization that occurred after the development of a national transport network, cultivation and grazing coexisted and were highly interconnected. Grazing animals supplied manure and natural fertilizers for cultivated plots, and shifting cultivation practices allowed control of shrub and tree encroachment (Caplat et al. 2006). It is thus possible to describe the preproductivist agricultural landscape as a core of permanently cultivated land with short fallow rotations to sustain soil fertility, and a belt of rangelands periodically cultivated for short periods followed by long fallows (twenty to forty years), during which time grazing was the dominant land use (Marty et al. 2003). Clearly, this sort of labor-intensive and time-consuming land use system is no longer viable in the FMR, and those concerned with restoration should not seek to recreate it except, perhaps, on a very small scale for heuristic or experimental purposes. Some private landowners and some municipalities are actively restoring stone walls at this time, as part of personal or collective campaigns to preserve a cultural heritage. In some few cases, there are innovative agricultural activities as well, including plant nurseries, flowers, bulbs, dye plants, out-of-season vegetables, and other speciality crops that command high market prices.

Ecological Dynamics in Old Fields

In the FMR, the main features of ecological dynamics are the consequences of two historical waves of tree plantation along with a strong trend of tree and shrub encroachment.

TREE PLANTATIONS OF THE RTM PROGRAM

Old fields of the two types we have described were in some areas converted to forest in the premontane and montane areas of the FMR, following the reforestation policies adapted in the second half of the nineteenth century. The RTM programs mentioned above were justified by the forest administration on the basis of a perceived need to restore forest on lands degraded by overgrazing (Larrère et al. 1980). RTM operations were essentially located on state or rural community-owned land, especially in the southern Alps, and in the Aigoual massif of the Cévennes, at the southeastern edge of the Massif Central. Local communities offered strong resistance to forest restoration because the so-called marginal lands being taken over by foresters were critical for the agricultural systems still in use, since they provided resources for sheep grazing and were publicly managed commons for use, above all, by poor peasants. Between 1880 and 1914, in the southern Alps alone, ca. 50,000 hectares were converted to forest (Douguédroit 1980).

FOREST NATIONAL FUND PLANTATIONS

After World War II, another plantation policy allowed not only the state and municipalities but also private owners to take advantage of state funding for tree planting (Marty 2004). The Forest National Fund was designed to enhance French capacity in timber production, especially for cardboard and paper. Between 1946 and 1990, approximately 2 million ha of trees were planted in France. In the Languedoc-Roussillon region of the FMR, no fewer than 100,000 ha were planted during this period. Funding and efforts were focused on the plantation of coniferous species that were a mix of some native and many nonnative species, including Douglas fir (*Pseudotsuga menziesii*), spruce (*Picea sitchensis, P. abies*), black pine (*Pinus nigra*), fir (*Abies alba* and *A. nordmanniana*), or Atlas Mt. cedar (*Cedrus atlantica*). Plantation sites were formerly deciduous forests and former fields and meadows located in uplands heavily impacted by agricultural decline.

In spite of such massive and widespread efforts to address the regional consequences of agricultural decline by forest tree planting, natural succession was in fact the main process observed on old fields. As documented by diachronic analysis of postcards and photographs (Debussche et al. 1999) between the beginning and the end of the twentieth century (figure 11.2), natural succession on abandoned croplands (meadows and fields, terraces, vineyards and orchards) led to impressive amounts of spontaneous woodland regeneration with the native, deciduous, downy oak (*Quercus pubescens*), the evergreen holm oak (*Q. ilex*), ash (*Fraxinus* spp.), beech (*Fagus sylvatica*) and

FIGURE 11.2. Landscape changes in Mediterranean limestone uplands. Dynamics of white oak (*Quercus humilis*). Causse du Larzac, Hérault, France. (a) Postcard, early twentieth century; (b) photo taken from the same site by O. Rousset, 1998.

the introduced, but naturalized, sweet chestnut (*Castanea sativa*) dominating the emerging secondary formations.

Alternative Futures for Old Fields

In the current context of agricultural decline, land abandonment, and the massive forestation and spontaneous woodland regeneration in old fields, the main issue for policymakers is to decide whether interventions must be undertaken to (1) restore, with subsidies, cultural landscapes, with a mosaic of agricultural terraces, orchards, woodlots, vineyards, and other typical rural Mediterranean land uses, including houses; or else (2) to accept, assist, and reinforce natural successional dynamics that contribute to forest recovery in old fields and other abandoned lands. As shown in figure 11.3, both of these actions would imply action directed to slow or counter the various emerging ecosystems, which are characterized by unprecedented species assemblages and unknown trajectories of ecological functioning. Such emerging ecosystems appear increasingly in the FMR and elsewhere around the world (see Hobbs et al. 2006 for discussion). They call for serious reflection and public consultation on the part of land managers, restorationists, and conservationists.

The third path, of course, is simply to succumb to, and consider as inevitable, the relentless transformation of rural areas, formed and sculpted by successive generations of people over millennia, into suburban and rural housing developments, while preserving some hunting grounds and walking areas for recreational use. At present, all three of these options are occurring, with far too little concerted effort devoted to developing a broad, long-term overview or land use planning scheme that includes the conservation and restoration of ecosystem services. If such a broad view were to be pursued for the FMR, there are three main questions to address:

- Where and when should we use passive or active restoration?
- Is there a way to nurture and maintain a healthy, self-sustaining agricultural system that is compatible with ecosystem and environmental management plans?

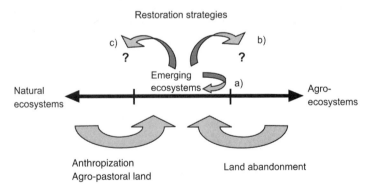

FIGURE 11.3. In the Mediterranean region, as elsewhere, natural ecosystems have been progressively transformed to agroecosystems. Nowadays, many natural and managed systems are in a state of flux as "novel" or "emerging" ecosystems. These can be self-perpetuating (a), or else, "restored." But the question arises whether to restore them to (b) intensively exploited systems in cultural landscapes, or (c) to natural ecosystems. Modified from Hobbs et al. (2006).

• Is there a future for the cultural landscapes of the past, in part or in whole?

There is a wide range of situations and settings to be considered, as will be illustrated in the following section.

Restoring Forests and Other Natural Ecosystems in the FMR: Where, Why, and How?

In this section, we examine the debates regarding forest degradation and restoration. Overgrazing was often described as the main cause for ecosystem degradation and periods of cultivation were underestimated.

FOREST ECOSYSTEMS

There are two principal questions concerning the restoration of forests over which naturalists and foresters have had heated debate—the role of the forest and grazing lands, and the means for reestablishing forest. The restoration of Mediterranean forests was a very important objective in southern France in the nineteenth and early twentieth centuries. The state agency for water and forests (*Eaux et Forêts*) launched a major program of reforestation of mountain lands that would reduce torrential floods and mudslides during heavy rains and, at the same time, establish land tenure rights for vast areas of silvicultural production. A number of approaches were proposed to restore or, more accurately, to rehabilitate these degraded lands. Over and above the physical engineering interventions that were often necessary, herbaceous, shrub, and coniferous tree cover was reestablished, via seeding or planting of seedlings. The use of deciduous trees was also tested (Vallauri et al. 2002). In the first half of the nineteenth century, methods used by French foresters, which were of largely German origin, were ill suited for mountainous regions of the FMR with its typically Mediterranean dry, hot season throughout summer (Chalvet 1998). It is not surprising that they suffered many failures as a result of ill-adapted species or provenances and near-total failure of direct seeding attempts. It was not until the large RTM projects of the nineteenth century that new silvicultural techniques better adapted to the Mediterranean climate and terrain were developed and applied. Yet deciding on objectives remained a hotly debated topic.

ROLE OF FOREST AND GRAZING LANDS

For the biogeographer Charles Flahault (1927), everything apart from cultivated lands should be reforested. No place was to be given to grazing, which he considered the primary evil leading to forest degradation. He based his arguments on the notion that primeval forest occurred almost everywhere in the FMR. Braun-Blanquet (1931) accepted this view, but argued that reforestation was inappropriate in certain situations. Kuhnholtz-Lordat (1945) took a very different position: he militated for an agro-sylvo-pastoral equilibrium, based on the complementarity of these three types of land use. He also argued that field cultivation in the past was at least as important as grazing in the degradation and transformation of Mediterranean forests. His ideas were not readily accepted by French foresters, except for a minority who were more attentive to the needs of local populations. It was only in the 1980s that the idea of reconciling agriculture, animal husbandry, and forestry became widely accepted (Etienne et al. 1998).

MEANS AND GOALS FOR REESTABLISHING FORESTS

The overriding strategy in forest restoration was often quite simple: conifers were planted with the aim of establishing productive and profitable forests. This approach was applied, for example, in numerous RTM projects on degraded mountain soils throughout southern France. Black pines (*Pinus nigra* and *Pinus laricio*), spruce (*Picea abies*), Scots pine (*Pinus sylvestris*) and, at lower altitudes, Aleppo pine (*P. halepensis*) or maritime pine (*P. pinaster*) grew rapidly to a harvestable size. Their lumber and various timber products were economically attractive. In view of their success, reforestation with these valuable species was by far the most important type of forestry intervention. However, the transition toward mature stages of succession were not as easy to achieve as expected, except in some situations in uplands where beech (*Fagus sylvatica*) showed strong development in the understory (Kunstler et al. 2005).

Ducamp, a senior forest engineer, advocated an alternative approach, since he considered fire as the main subject of concern in the Mediterranean region (Ducamp 1932, 1934). The introduction of pines in shrublands or open woodlands was known to greatly increase the risk of wildfire. However, late-successional stages of such forests should, in theory, become less prone to fire in general. Supporting evidence for this was available for beech (Joubert 1929). (This issue is of course still highly relevant and hotly debated. Being inflammable is discussed as an adaptive evolutionary strategy for plant species that mainly germinate and reestablish themselves *after* a fire [Bond and Midgley 1995]. One example is the Aleppo pine, with its serotinous cones and massive seeding following fires.)

According to Ducamp (1931, 1932), instead of using pines for rapid forest restoration, spontaneous regeneration, also known as autogenic restoration, should be attempted. He argued that promoting self-recovery in the forest environment was preferable not only in ecological terms but also for the purpose of achieving rapid economic results. Therefore, he argued, interventional actions should be based on field observations coupled with ecological knowledge. Ducamp's positions were, for his time, innovative and controversial. They were, however, implemented by the *Eaux et Forêts* Service over a short time period. But, for economic reasons, large areas were left without any intentional management, and forest stands began to spontaneously regenerate themselves (Debussche et al. 1999).

In the context of spontaneous regrowth, one group of trees of particular importance is the Mediterranean oaks, especially the deciduous ones that have been heavily impacted by past land use practices. Spontaneous recolonization by oaks is in general much slower than for conifers. Oak seedling reestablishment is greatly facilitated by the presence of shrubs, which serve as "nurse plants" and protectors from herbivores (Rousset and Lepart 2000). Young oaks can also appear under the cover of pines, but they do not grow well there due to the shade and perhaps also due to edaphic changes brought about by the pines, whose litter tends to make soils more acid (Kunstler et al. 2005).

Oak seedlings, which are very appetizing for most large animals, have great trouble establishing in the presence of domestic livestock. When ungrazed, however, oak and other tree seedlings become established during the earliest stages of succession, and a relatively dense formation of young trees can occur in as little as twenty years provided seed-bearing trees are nearby (Debussche et al. 1996). Whereas in the first half of the twentieth century most grazing lands in the FMR had an appearance of permanency due to the rare occurrence of spontaneously recurring woody plants, in recent years there has been a very notable acceleration in the rate of secondary succession, including the widespread establishment of pioneering woody plants such as the juniper (*Juniperus oxycedrus*) and growing numbers of oak seedlings as well. The establishment of dense woody vegetation takes approximately one century following abandonment (Escarre et al. 1983).

In the rare areas where seed-bearing beech trees (*Fagus sylvatica*) are still present, young beeches also appear spontaneously under cover of woody shrubs or even under pines. In lowland areas, the rate of succession changes as a function of the previous land use history and the presence or absence of grazing following agricultural abandonment. The numerous olive groves that were abandoned in the 1950s were rapidly recolonized by woody plants. Dead olive trees served as perches for fruit- and seed-eating migratory birds, and in this way a large number of tree and shrub seeds arrived and quickly created a dense and diverse thicket (Debussche et al. 1982). In addition, these olive groves were in general on rather rough terrain, and consequently were infrequently grazed by livestock. By contrast, the former cereal fields and abandoned vineyards on flatland and bottomlands were successively colonized by herbaceous annuals and biannuals, and then by low-growing, woody plants, such as thyme, lavender, and rhizomatous and perennial bunchgrasses, which come to dominate at the end of approximately fifty years.

Mediterranean Coastal Areas: From Fallow Land to Wetlands

It is no longer disputed that Mediterranean coastal areas are not natural ecosystems—without exception they are deeply modified by human activities. For the coastal marshes in and near the Camargue region, the key issue for conservation and for restoration is to define objectives combining ecosystem assessment and stakeholders' expectations. An example is the fallow fields in the lower valley of the Vistre River in the FMR, where restoration efforts were undertaken to transform fallow lands of limited ecological and naturalist interest and restore marshlands with large colonies of tree-nesting herons, particularly the Squacco heron (*Ardeola ralloides*) (Mauchamp and Grillas 2002). Hunting regulations are stricter than in other similar ecosystems in order to protect the waterbird communities.

This program also addresses the difficult issue of the choices to be made in restoration or conservation plans (Mathevet and Mauchamp 2005). Multiobjective restoration requires an assessment and prioritization of these sometimes conflicting objectives. Mauchamp and Grillas (2002) highlighted that, despite the general objective of reestablishing a floodplain, priorities among the objectives were not clearly identified, and stakeholders' views were not adequately assessed. They suggested limiting the number of objectives for such restoration plans and taking great care about the second stage of restoration: managing the restored ecosystems in the long term. After restoring an ecosystem locally, it is important to integrate it within a network of natural areas with global management guidelines while, at the same time, taking stakeholders' practices and strategies into account.

Restoring Cultural Landscapes

Past agricultural systems and rural area land use produced striking and original landscapes that are part of the cultural heritage of Europe (Antrop 2005;

Moreira et al. 2006), and especially its Mediterranean regions. Terraces, seminatural cork oak and sweet chestnut forests, and extensive polders in wetlands are among the most striking examples of human-made landscapes that survived for centuries but are now under serious threat due to recent agricultural and land use trends. Management measures aimed at restoring these landscapes are being tested and implemented in several areas.

Rebuilding Terraced Landscapes in the Cévennes Region

Since the middle of the 1990s, interest in the terraced landscapes of the northwestern Mediterranean quadrant has been rekindled (Alcaraz 2001), based on their heritage value much more than on their economic potential. Several examples of successful local restoration exist. Some restored landscapes are labeled by the French Ministry of Environment (Laurens 1997), in order to develop tourism and food product marketing.

Terraced landscapes are used for the production of sweet chestnut orchards. Widely cultivated in the western Mediterranean (Pitte 1986), chestnut ecosystems are still intensively managed in some countries (e.g., Portugal, parts of Italy) but are almost totally abandoned in France. Decline in sweet chestnut production led to a shift from chestnut orchards to chestnut forests. In 1975, only 200 ha of productive orchards were left in the Gard region. But chestnut forest is considered by local people to be a cultural landscape and a part of their heritage. Evaluation of willingness to pay for chestnut landscape maintenance by Contingent Evaluation Method (Noublanche and Chassany 1998) showed that stakeholders who own vacation homes or retirement homes in the region are willing to pay the highest contribution. Terraced landscapes can be restored by the help of labeled niche products. Following the example of the Ardèche region, producers of the Cévennes region applied to the European Union for a Controlled Appellation of Origin (AOC), also known as PDO, or Protected Designation of Origin. The objective was to reconcile agriculture production with cultural landscape restoration and not to fund management measures aiming only at maintaining aesthetic or cultural properties. Chestnut forests managed for timber production are another option but, in view of the great expense of managing the terraces, are unlikely to lead to restoration of terraced landscapes.

Restoring Cork Oak Woodlands

Cork oak (*Quercus suber*) woodlands are located between 0 and 700 m above sea level in the northwestern Mediterranean and reach somewhat higher altitudes in Morocco. In the French Mediterranean region, the three main areas where cork oak woodlands are found are the island of Corsica, French Catalonia (on the French-Spanish border), and the coastal hills of Provence, which are the only portions of the FMR where suitably sandy, acidic substrates are found (Quezel and Médail 2003). These ecosystems, when managed for cork exploitation, form a very typical Mediterranean landscape, as well as an ancient form of agroforestry (acorns collected for pigs, fuelwood gathering). Nonetheless, they are facing abandonment, and all attempts at conservation, management, and restoration must address two critical issues. First, since the 1980s, cork prices decreased dramatically and cork oak forest management in France was almost totally abandoned. In the coastal Maures massif of Provence, between Hyères and Frejus, abandonment led to shrub colonization of the understory (Amandier 2005). The highly destructive forest fires of 1989, 1990, and 2003 were all due indirectly to agricultural abandonment. Second, in the last few years, international demand for cork has increased significantly and exploitation of French cork oak has started again. But forest owners and cork producers in the past have tended to employ underqualified laborers lacking skill in the special task of removing cork bark from the trunk without damaging the tree and producing scars that allow attack from pathogenic agents. In response, populations of *Platypus cylindrus*, a tiny coleopterous insect, started to pullulate. This beetle can kill trees within two or three years and preferentially attacks recently exploited forests.

In light of these challenges, forest owners' organizations and professionals of the cork industry have tried to implement management measures in several ways to conserve and restore cork oak forest and landscape. They have sought to introduce sustainable management measures to reduce fire risk, to gather landowners together in order to decrease production costs, and to stop *Platypus* proliferation by regulating cork exploitation. A small Mediterranean Institute of Cork and Cork Oak has been created.

Scientific research and development on cork oak is very recent in France, as elsewhere. Results are scattered, and usually limited to a single discipline, that is, genetics, silviculture, or the cork industry. The few broad, interregional studies are generally out of date. However, in 1996, a five-year, European network project was organized to evaluate the genetic resources of cork oak for appropriate use in breeding and gene conservation strategies (EU-FORGEN). It was followed, in 2002, by a research and development program, funded by the European Commission, called Conservation and Restoration of European Cork Oak Woodlands (CREOAK). These efforts are evidence that interest in the conservation and restoration of cork oak woodlands exists among landowners and some policymakers. A grower's handbook was produced by the EUFORGEN group, and a broader book for general readers is in preparation by the CREOAK coalition.

Restoring Open Habitat for Biodiversity

Open landscapes of seminatural habitat were extensively developed in the FMR at the beginning of the twentieth century. French literature is full of expressions of the uniqueness and aesthetic value of open, shrub-dominated landscapes. For example, André Gide (1952) famously compared them to biblical landscapes of the Near East. Such landscapes were present from the lowest calcareous hills up to the high-altitude limestone or primary plateaus (Debussche et al. 1999). For the general public, however, until the end of the twentieth century, open Mediterranean landscapes were generally seen as being without economic interest, and therefore without interest at all, especially in the uplands where tree plantations were perceived as a serious alternative for economic development (Lepart et al. 2000). Starting in the 1980s, however, awareness of the social and ecological consequences of agricultural abandonment throughout the FMR led scientists to begin studying the effects of land use changes on biodiversity (Magnin et al. 1995; Preiss et al. 1997). Consequences of the diminution of open habitat then became a major issue for conservation and led to important discussions about how best to manage for biodiversity conservation. The Grands Causses region provides an excellent example to highlight the main points and conflicts concerning open landscapes inherited from agropastoral practices.

Located on the southeastern edge of the Massif Central, the Grand Causses region is the largest area of seminatural calcareous grasslands in France. They are undergoing rapid and massive tree and shrub encroachment, even though land abandonment and the agricultural crisis are less severe than in other circum-Mediterranean uplands. During the 1990s, several management plans aimed at preserving seminatural, open habitat were designed. Agroenvironmental schemes were implemented by government agricultural agencies and extension officers. Management measures set in place by the Cévennes National Park were designed to increase pastoral use of rangeland resources in order to limit the growth and spread of pines, oaks, and other trees, such as *Amelanchier ovalis* and *Juniperus communis*, as well as shrub species, such as boxwood (*Buxus sempervirens*) and wild rose (*Rosa canina*). A scientific research and demonstration plan, funded by the European Commission (LIFE project), was launched. The goal was to restore open habitat and reinforce adapted management practices on farms. A wide group of stakeholders was involved (farmers, hunters, naturalists, NGOs, and state administrations, among others.). Notably, in the southern part of the Causse du Larzac, restored open habitat, created by cutting trees and shrubs, followed by well-managed sheep grazing, is common and frequently used as a case study for thinking about and testing options for future management scenarios.

Interdisciplinary research on the social and physical factors in landscape dynamics (Chassany et al. 2002) led to the conclusion that ruminant grazing pressure was the major issue to be considered in any conservation and restoration plan. The assumption was made that seminatural, open habitat was the result of, and dependent on, traditional sheep breeding systems. Increasingly intensified and industrialized since the early 1960s, the emerging livestock husbandry systems no longer suppressed tree and shrub colonization (Rousset and Lepart 1999; 2000). As a result, a series of agroenvironmental schemes (AES) were designed to encourage farmers to reuse rangeland resources. The results encouraged nature managers to actively eradicate the seedlings of unwanted trees and shrubs. The Cévennes National Park administration and various professional farming bodies, backed by regional public funding, also encouraged farmers to take on contracts to "restore" open rangelands by systematically cutting both seedling and adult pine trees. In 2003, ten contracts were signed, covering a total of 456 ha. The combination of the cutting of unwanted seed-producing trees and shrubs and increased grazing pressure is necessary to preserve seminatural grasslands in the area (Etienne 2001).

Conclusion and Restoration Goals: Woodlands, Open Spaces, or Both?

Discussions about how and why to restore Mediterranean woodlands are intense and have been reviewed recently (Aronson, Le Floc'h, and Ovalle 2002; Vallejo et al. 2006; see Clewell and Aronson 2006, for a general discussion of "Why restore?"). Here the first issue we have considered is where and when extensification and intensification can be deemed appropriate and feasible, depending on the specific site and setting. As we have shown, when restoration is decided upon, it must be asked whether it can yield clear enough benefits for local stakeholders to justify the cost and lost opportunities for other land uses. The second question is, to what extent can a broad landscape and bioregional approach embrace both kinds of restoration activities?

Until the 1970s, losses in forest cover due to fuelwood gathering, overgrazing, and fires were the primary concerns for the management of Mediterranean ecosystems in the FMR. Current trends in land cover changes, however, show that decreases in woodland areas are no longer the key issue for restoration and land management (Mazzoleni et al. 2004). After a period of scientific and technical research dedicated to forest restoration, land managers and conservation bodies have to face woodland colonization as the major vegetation change related to land abandonment.

In the current context, two scenarios are conceivable. The first one is to consider that old fields, both as a cultural heritage and as habitats required by fauna and flora with high conservation value, must be conserved or restored as open habitat. The second is to let the seminatural woodlands grow back, with little or no management. For the most part, this second option is not even envisioned, even though it would constitute a form of passive restoration. Instead, the general consensus—for both tangible and intangible reasons—seems to take the relatively recent cultural landscapes, with their mosaic of open and wooded areas, as the optimal reference (see Aronson and Le Floc'h 1995; Aronson, Le Floc'h et al. 2002; and Moreira et al. 2006 for discussion of the complex issue of references in the Mediterranean context).

One element on which a consensus has emerged in France is the idea of heritage responsibility for a given area (Molina et al. 1999). According to this rationale, a species is designated as a priority species if its population in an area is significantly important compared with its population in a wider reference area, for example, a country. Following this criterion of biotic or interspecies responsibility, open, seminatural habitat, among them old fields *sensu lato*, have much higher biodiversity value than woodlands in the French Mediterranean. In addition, open-habitat species are, in general, poorly represented outside of the Mediterranean region. Thus, if the open Mediterranean habitats are lost, then numerous plant and animal species would no longer be present anywhere in France. This rationale is a justification of the need for restoring old fields on the grounds of protecting biodiversity. Combined with the cultural heritage value, this argument lends strong support for old field restoration in the direction of maintaining open spaces. In particular, insofar as it remains economically viable, specialized agriculture represents the best strategy for preserving open spaces and a particular cohort of segetal, ruderal, and meadow species.

However, these operations will have an elevated cost, and current economic conditions in the Mediterranean hills and uplands do not support the levels of human labor and capital investment required to obtain and maintain such artificially restored habitats. Therefore, in some areas it would no doubt be preferable to take advantage of spontaneous woodland dynamics and simply let the woodlands and forests grow back. New ways and means, and new legislation, may soon appear for establishing landscape and regional-scale planning criteria and models that maximize and enhance the panoply of values of Mediterranean landscapes, including forests, farmlands, and old fields maintained as open-space habitats. Such landscapes may require subsidies initially, but long term an optimal integration and use of the spontaneously growing woodlands should be found so as to realize total economic and cultural value of these landscape mosaics, including the urgent need to reduce the risk of catastrophic wildfire. In the future, research and planning must consider much longer time scales and larger spatial scales than what has been common in the past century (Merlu and Croitoru 2005).

In many cases, if nothing is done, an unknown, synanthropic, "emerging ecosystem" of disparate biological origins will result. Thus, there is a clear necessity for new kinds of evaluation, notably at the landscape scale. The criteria to be taken into account include biodiversity, ecosystem services, multiuse value for local populations, and heritage value.

Pluridisciplinary studies integrating socioeconomic and ecological data can be used to envision possible futures and assist in making decisions. In the European-funded Bioscene Program, scenarios based on the anticipation of policy changes were coupled with landscape modelling (http://www.bioscene.co.uk/). In a latter stage, sustainability assessment (Van der Vorst et al. 1999) of each scenario was conducted by adopting a participatory approach based on focus groups composed of local stakeholders. Taking into account both socioeconomic and ecological consequences, the assessment gave each scenario a score based on the points of view of stakeholders and experts. Needless to say, much more work of a transdisciplinary nature will be needed to achieve the kind of ambitious planning and action we call for at landscape and bioregional scales.

Acknowledgments

We extend our warm thanks to Viki Cramer, Christelle Fontaine, and two anonymous reviewers for helpful comments on the manuscript. JA gratefully acknowledges the European Commission for support of the CREAOK project FP5: QLRT-2001-01594. JL and PM acknowledge the European Commission for support of the BioScene project FP5: EVK2-2001-00354, and the French *Ministère de l'Ecologie et du Développement Durable* for support via the research program DIVA (*Action publique, agriculture et biodiversité*).

REFERENCES

Alcaraz, F. 2001. L'utilisation publicitaire des paysages de terrasses. *Études rurales* 157–158:195–210.

Amandier, L. 2005. La suberaie des Maures en danger. *La feuille et l'aiguille* 58:3.

Ambroise, R., P. Frapa, and S. Giorgis. 1993. *Paysages de terrasses*. Édisud, Aix-en-Provence, France.

Antrop, M. 2005. Why landscapes of the past are important for the future. *Landscape and Urban Planning* 70:21–34.

Aronson, J., and E. Le Floc'h. 1995. On the need to select an ecosystem of reference, however imperfect : A reply to Pickett and Parker. *Restoration Ecology* 3(1):1–3.

Aronson J., E. Le Floc'h, H. Gondard, and F. Romane 2002. Gestion environnementale en région méditerranéenne: Références et indicateurs liés á la biodiversité végétale. *Revue d'Ecologie (Terre et Vie)* (supplément 9):225–40.

Aronson, J., E. Le Floc'h, and C. Ovalle. 2002. Semi-arid woodlands and desert fringes. In *Handbook of ecological restoration*, Vol. 2., ed. M. Perrow and A. Davy, 466–85. Cambridge University Press, Cambridge, UK.

Blanchemanche, P. 1995. *Bâtisseurs de paysages: Terrassement, épierrement et petite hydraulique agricoles en Europe*. 17e–19e siècles. MSH, Paris.

Blondel, J., and J. Aronson. 1999. *Biology and wildlife of the Mediterranean region*. Oxford University Press, Oxford, UK.

Bond, W. J., and J. J. Midgley. 1995. Kill thy neighbour: An individualistic argument for the evolution of flammability. *Oikos* 73:79–85.

Braun-Blanquet, J. 1931. L'importance pratique de la sociologie végétale. *Bulletin de l'Association Française pour l'Avancement des Sciences*,157–64.

Butlin, R. A., and R. A. Dodgshon (eds.). 1998. An historical geography of Europe. Clarendon Press, Oxford, UK.

Caplat, P., J. Lepart, and P. Marty. 2006. Landscape patterns and agriculture: Modelling the long-term effects of human practices on *Pinus sylvestris* spatial dynamics (Causse Mejean, France). *Landscape Ecology* 21:657–70.

Chalvet, M. 1998. Connaissance sylvicole ou propagande? La transmission des savoirs sur la forêt méditerranéenne. In *Traditions Agronomiques Européennes. Elaboration et Transmission depuis l'Antiquité*, 105–16. Editions du CTHS, Paris.

Chassany, J. P., C. Crosnier, M. Cohen, S. Lardon, C. Lhuillier, and P. L. Osty. 2002. Réhabilitation et restauration de pelouses sèches en voie de fermeture sur le causse Méjan: Quels enjeux pour quelle recherche en partenariat? *Revue d'Ecologie (Terre et vie)* (supplément 9):31–49.

Childe, V. G. 1971. The neolithic revolution. In *Prehistoric agriculture*, ed. S. Struever, 15–21. Natural History Press, New York.

Clewell, A. F., and J. Aronson. 2006. Motivations for the restoration of ecosystems. *Conservation Biology* 20:420–28.

Debussche, M., J. Escarré, and J. Lepart. 1982. Ornithochory and plant succession in Mediterranean abandoned orchards. *Vegetatio* 48:255–66.

Debussche, M., J. Escarré, J. Lepart, C. Houssard, and S. Lavorel. 1996. Changes in Mediterranean plant succession: Old-fields revisited. *Journal of Vegetation Science* 7:519–26.

Debussche, M., J. Lepart, and A. Dervieux. 1999. Mediterranean landscapes changes: Evidence from old postcards. *Global Ecology and Biogeography* 8:3–15.

Douguédroit, A. 1980. Les périmètres de reboisement dans les Alpes du Sud. *Revue Forestière Française* (numéro spécial "sociétés et forêt"):37–46.

Ducamp, R. 1931. Stratégie et tactique—Science et technique forestière. *Revue des Eaux et Forêts* 69:553–65.

———. 1932. Au pays des incendies—la vérité en marche. *Revue des Eaux et Forêts* 70: 380–93.

———. 1934. Dans la sylve considérée sous ses formes naturelles permanentes, il ne saurait y avoir place pour les pins du stade régressif. *Bulletin de la Société Forestière de Franche-Comté et des Provinces de l'Est*, juin, 8.

Escarré, J., C. Houssard, M. Debussche, and J. Lepart. 1983. Evolution de la végétation et du sol après abandon cultural en région méditerranéenne: Étude de succession dans la garrigues du Montpelliérais (France). *Acta Oecologica, Oecologia Plantarum* 4:221–39.

Etienne, M. 2001. Pine trees: Invaders or forerunners in Mediterranean-type ecosystems. A controversial point of view. *Journal of Mediterranean Ecology* 2:221–31.

Etienne, M., J. Aronson, and E. Le Floc'h. 1998. Abandoned lands and land use conflicts in southern France. Piloting ecosystem trajectories and redesigning outmoded landscapes in the 21st century. In *Landscape degradation and biodiversity in Mediterranean-type ecosystems*. Ecological Studies Series No. 136, ed. P. W. Rundel, G. Montenegro, and F. Jaksic, 127–40. Springer, Berlin.

Flahault, C. 1927. Le boisement des terres incultes et des montagnes. *Bulletin de la Société Centrale Agricole de l'Aude*: 249–54, 277–88, 345–63.

Fowler, P. 1999. A limestone landscape from the air: Le Causse Mejean, Languedoc, France. *Antiquity* 73:411–19.

Gide, A. 1952. *Poésie, journal, souvenirs*. Gallimard, Paris.

Grove, A. T., and O. Rackham, eds. 2001. *The nature of Mediterranean Europe. An ecological history*. Yale University Press, New Haven.

Guilaine, J. 1991. *Pour une archéologie agraire*. Armand Colin, Paris.

Hobbs, R. J., S. Arico, J. Aronson, J. S. Baron, P. Bridgewater, V. A. Cramer, P. R. Epstein, et al. 2006. Novel ecosystems: Theoretical and management aspects of the new ecological world order. *Global Ecology and Biogeography* 15:1–7.

Joubert, A. 1929. Les quatre incendies de la forêt de Valbonne. *Revue des Eaux et Forêts* 67:534–40.

Kalaora, B., and A. Savoye. 1986. *La forêt pacifiée. Sylviculture et sociologie au XIXe siècle*. L'Harmattan, Paris.

Kuhnholtz-Lordat, G. 1945. La silva le saltus et l'ager de garrigue. *Annales de L'Ecole Nationale d'Agriculture de Montpellier* 26:1–78.

Kunstler, G., T. Curt, M. Bouchaud, and J. Lepart. 2005. Growth, mortality, and morphological response of European beech and downy oak along a light gradient in sub-Mediterranean forest. *Canadian Journal of Forest Research* 35:1657–68.

Larrère, R., A. Brun, B. Kalaora, O. Nougarède, and D. Poupardin. 1980. Reboisement des montagnes et systèmes agraires. In *Revue Forestière Française*, special issue "Société et forêts":20–36.

Laurens, L. 1997. Les labels "paysages de reconquête," la recherche d'un nouveau modèle de développement durable. *Nature Sciences Sociétés* 5:45–56.

Lepart, J., and M. Debussche. 1992. Human impact on landscape patterning: Mediterranean examples. In *Landscape boundaries. Consequences for biotic diversity and ecological flows*, ed. A. J. Hansen and F. Di Castri, 76–105. Springer-Verlag, New York.

Lepart, J., P. Marty, and O. Rousset. 2000. Les conceptions normatives du paysage. Le cas des Grands Causses. *Natures Sciences Sociétés* 8(4):16–25.

Lombard, M. 1959. Un problème cartographique: Le bois dans la méditerranée musulmane (VIIè-XIè siècle). *Annales ESC* 14(2):234–54.

Loubere, L. A., P. Adams, and R. Sandstrom. 1988. Saint-Laurent de la Salanque: From fishing village to wine town. *Agricultural History* 62(4):37–56.

Magnin, F., T. Tatoni, P. Roche, and J. Baudry. 1995. Relationship between landscape change and gastropod communities along an old field succession in Provence (Mediterranean France). *Landscape and Urban Planning* 31:249–57.

Marres, P. 1935. *Les Grands Causses, étude de géographie physique et humaine.* Arrault et Cie, Tours, France.

Marty, P. 2004. *Forêts et sociétés. Logiques d'action des propriétaires privés et production de l'espace forestier.* L'exemple du Rouergue. Publications de la Sorbonne, Paris.

Marty, P., E. Pélaquier, B. Jaudon, and J. Lepart. 2003. Spontaneous reforestation in a peri-Mediterranean landscape: History of agricultural systems and dynamics of woody species. In *Environmental dynamics and history in Mediterranean regions,* ed. E. Fouache, 179–86. Elsevier, Paris.

Mathevet, R., and A. Mauchamp. 2005. Evidence-based conservation: Dealing with social issues. *Trends in Ecology and Evolution* 20:422–23.

Matvejevic, P. 1999. *Mediterranean. A cultural landscape.* University of California Press, Berkeley, Los Angeles, CA.

Mauchamp, A., and P. Grillas. 2002. Quels objectifs de restauration pour un ancien polder de la basse vallée du Vistre? *Revue d'Ecologie (Terre et vie)* (supplément 9): 51–64.

Mazzoleni, S., G. Di Pasquale, and M. Mulligan. 2004. Reversing the consensus on Mediterranean desertification. In *Recent dynamics of the Mediterranean vegetation and landscape,* ed. S. Mazzoleni, G. Di Pasquale, M. Mulligan, P. Di Martino, and F. Rego, 281–85. John Wiley and Sons, Chichester, UK.

Merlu, M., and L. Croitoru (eds.). 2005. *Valuing Mediterranean forests: Towards total economic value.* CABI, Oxfordshire, UK.

Molina, J., J. Mathez, M. Debussche, H. Michaud, and J. P. Henry. 1999. Méthode pour établir une liste régionale d'espèces protégées. Application à la flore du Languedoc-Roussillon. In *Les plantes menacées de France,* L. J. Y. Brest. *Bulletin de la Société Botanique du Centre Ouest* (numéro spécial):399–420.

Moreira, F., A. I. Queiroz, and J. Aronson. 2006. Restoration principles applied to cultural landscapes. *Journal for Nature Conservation* 14:217–24.

Noublanche, C., and J. P. Chassany. 1998. Le rôle des acteurs locaux dans la valorisation économique du paysage. Le cas de la Vallée Française en Cévennes. *Revue de l'Economie Méridionale* 46(183):289–99.

Pitte, J. R. 1986. *Terres de Castanides, Hommes et paysages du Châtaignier de l'Antiquité à nos jours.* Fayard, Paris.

Preiss, E., J. L. Martin, and M. Debussche. 1997. Rural depopulation and recent landscape changes in a Mediterranean region: Consequences to the breeding avifauna. *Landscape Ecology* 12:51–61.

Price, T. D. 2000. *Europe's first farmers.* Cambridge University Press, Cambridge, UK.

Quezel, P., and F. Médail. 2003. *Ecologie et biogéographie des forêts du bassin méditerranéen.* Elsevier, Paris.

Rousset, O., and J. Lepart. 1999. Shrub facilitation of *Quercus humilis* regeneration in succession on calcareous grasslands. *Journal of Vegetation Science* 10:493–502.

———. 2000. Positive and negative interactions at different life stages of a colonizing species. *Journal of Ecology* 88:401–12.

Russo Ermolli, E., and G. Di Pasquale. 2002. Vegetation dynamics of south-western Italy in the last 28 kyr inferred from pollen analysis of a Tyrrhenian Sea core. *Vegetation History and Archeobotany* 11:211–19.

SER (Society for Ecological Restoration, Science and Policy Working Group). 2004. *The SER international primer on ecological restoration.* http://www.ser.org/content/ecological_restoration_primer.asp (accessed 20 January 2006)

Stevenson, I. 1980. The diffusion of disaster: The *Phylloxera* outbreak in the Département of the Hérault, 1862–80. *Journal of Historical Geography* 6:47–63.

Tatoni, T., F. Médail, P. Roche, and M. Barbero. 2004. The impact of changes in land use on ecological patterns in Provence (Mediterranean France). In *Recent dynamics of the Mediterranean vegetation and landscape,* ed. S. Mazzoleni, G. Di Pasquale, M. Mulligan, P. di Martino, and F. Rego, 107–20. Wiley and Sons, Chichester, UK.

Thirgood, J. V. 1981. *Man and the Mediterranean forest.* Academic Press, New York.

Vallauri D., J. Aronson, and M. Barbéro. 2002. An analysis of forest restoration 120 years after reforestation on badlands in the southwestern Alps. *Restoration Ecology* 10:16–26.

Vallejo, R., J. Aronson, J. G. Pausas, and J. Cortina. 2006. Restoration of Mediterranean woodlands. In *Restoration ecology: The new frontier,* ed. J. van Andel and J. Aronson, 193–209. Blackwell Science, Oxford, UK.

Van der Vorst, R., A. Grafé-Buckens, and W. R. Sheate. 1999. A systemic framework for environmental decision-making. *Journal of Environmental Assessment Policy and Management* 1:1–26.

Vernet, J. L. 1997. *L'Homme et la forêt méditerranéenne de la préhistoire à nos jours.* Errance, Paris.

Vos, W., and A. Stortelder. 1992. *Vanishing Tuscan landscapes. Landscape ecology of a submediterranean-montane area (Solano Basin, Tuscany, Italy).* Centre for Agricultural Publishing and Documentation, Wageningen, The Netherlands.

Chapter 12

Land Abandonment and Old Field Dynamics in Greece

Vasilios P. Papanastasis

Greece is a predominantly mountainous country with a total land area of 132,000 km². Most of its major mountains are quite high (up to 2,500 m above sea level) and have steep slopes resulting in a very rugged and highly dissected relief. Flat and undulating areas are restricted to valleys, mountain plateaus, and minor plains. According to the official statistics (National Statistical Service of Greece 1995), 39% of the village communities are found in the plain zone, 25% in the hilly zone, and the remaining 36% in the mountainous zone (figure 12.1).

Arable agriculture is practiced in about 30% of the whole country. The other major land use types include rangelands (i.e., grazing lands) and forests, which occupy 40% and 22%, respectively, of the total land surface, while the remaining 8% consists of settlements, water surfaces, and barren land (National Statistical Service of Greece 1995). Arable lands are found not only in plains but also in uplands (hills and mountains) where they are intermingled with seminatural communities of various types, such as grasslands, shrublands, and forests, resulting in a very diverse and heterogeneous landscape.

Historically, the area allocated to arable farmland was never fixed. On the contrary, it has been constantly changing over the years following the turbulent history of the country. At times of reduced human population, the pressure on arable farmland was low, thus restricting it to the more favorable areas in terms of topography and soil depth. In contrast, the pressure was high at periods of increased population, leading to opening up of new arable land in seminatural ecosystems, especially grasslands, located not only on lowlands but also on uplands.

Consequently, there is a dynamic equilibrium between arable lands, rangelands, and forests in the sense that quite often arable lands are opened

225

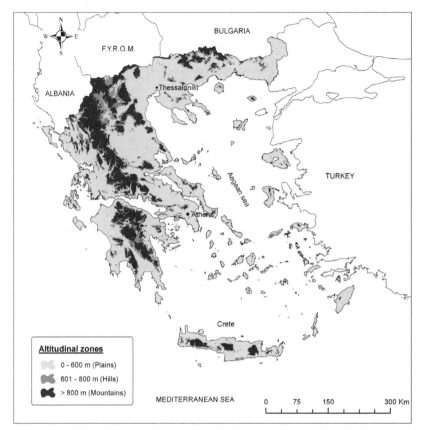

FIGURE 12.1. Map of Greece showing the distribution of main altitudinal zones. Abandoned arable lands are mostly found in the hilly and mountainous areas.

up in grasslands, shrublands, or forests for temporary cultivation, to be abandoned after a while and recolonized by herbaceous and woody species. Old fields therefore are an integral component of the Greek landscape. In this chapter, arable land abandonment and old field dynamics are reviewed and discussed and strategies for their restoration and sustainable management explored.

Brief History of Arable Farming and Land Abandonment

Arable farming in Greece dates back to the early Neolithic (seventh millennium BC) when small-scale, mixed, agropastoral use was practiced. Extensive farming was adopted much later, during the later Bronze Age and early historical period (Halstead 2000). According to van Andel et al. (1990), erosion triggered by agricultural activity was widespread during prehistoric and historic Greece. Judging from the available palynological data, however, human influence was not constant throughout history (Gerasimidis 2000).

Based on archaeological surface-survey data, Alcock (1993) concludes that the rural landscape of the classical and early Hellenistic periods was exceptionally active, characterized by the presence of numerous, dispersed, small, rural farmsteads. From the second half of the third century BC and during the whole period of Roman occupation, however, a severe reduction of such sites occurred, giving the impression of a deserted or "empty" landscape. The decline of agricultural activity during the Roman period occurred due to changes in land tenure, which encouraged either nucleated settlements or the dominance of elite landowners in the countryside. Such developments led to the abandonment of marginal land and concentration of agricultural activity on more profitable and higher quality soils, which were farmed intensively.

For the Byzantine period (fourth to the middle of the fifteenth centuries AD), the history of land use is almost unknown due to lack of literary sources and archaeological evidence. During the subsequent Ottoman period (fifteenth to nineteenth centuries), the Greek population (Christians) was forced to abandon the lowlands and move to the mountains in order to prevent mixing with the Muslims, who came from Anatolia and settled in the most fertile areas of the plains (Vakalopoulos 1964). This movement apparently resulted in opening up of arable land in the mountains. When Peloponnese and part of central Greece (roughly one-third of today's Greece) were liberated in the first half of the nineteenth century, arable land was expanded during the second half to accommodate the needs of an increasing population (Petmezas 2003). This trend continued throughout the first half of the twentieth century, when present-day Greece was created. In the meantime, new arable land was opened up by draining wetlands and marshes in order to assist in the settlement of more than one million Greeks brought from Asia Minor in the framework of the exchange of populations between Greece and Turkey. During this period and up to World War II, arable land was exposed to the highest human pressure.

One important part of ancient agriculture was the practice of terracing. There is a lot of debate on this particular issue. According to Rackham and Moody (1996), terraces in Crete go back at least to the Bronze Age (around 1500 BC) but the pressure for terracing must have been greatest in the late Venetian period (sixteenth century) when the human population was very high and fertile land in the plains was abandoned because of corsair attacks.

Extent and Nature of Old Fields

There are no official records for currently abandoned agricultural land in Greece. The only records that exist are for a category called "fallow land." The category is not described, but it should include both temporarily and permanently abandoned arable land. However, most of the fallow land category must belong to lands temporarily suspended from arable farming, specifically for less than five years, with permanently abandoned arable lands making up only a fraction of the fallow lands. This is supported by a detailed inventory of abandoned agricultural lands based on statistical data and questionnaires carried out by the Agricultural Bank of Greece for the whole country in 1983 (Tsoumas and Tasioulas 1986).

Figure 12.2 shows that the total agricultural area did not change substantially between 1961 and 2000. The land actually cultivated increased by only 6,600 ha (0.2%), but the fallow land increased by 78,200 ha (22%, or 2% of the total area under agriculture) at the same period. This means that proportionally more land was cultivated in 1961 than in 2000. The peak of fallow land occurred in 1975, but since 1980 it has started to decline, while the cultivated land increased moderately. This explains why more than 11% of the total agricultural area remained fallow in 2000.

Different proportions of abandoned land in relation to the total agricultural land were recorded in specific areas of Greece. In a project involving in-

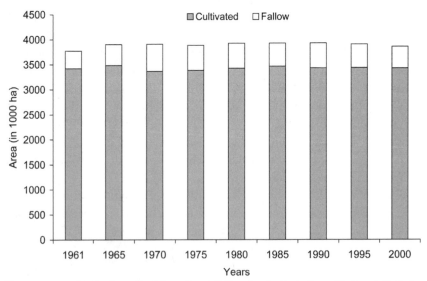

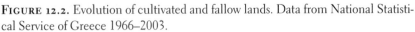

FIGURE 12.2. Evolution of cultivated and fallow lands. Data from National Statistical Service of Greece 1966–2003.

ventory of grazing lands based on aerial photo interpretation, abandoned arable lands were found to be 65,800 ha or 11.5% of the total agricultural area in the prefecture of Drama (northeastern Greece) (Papanastasis et al. 1986), but only 2.9% of the total agricultural land in western Greece (Epirus) (Platis et al. 2001). This means that the extent of old fields is not the same everywhere in the country, apparently due to regional differences in socio-economic development. In both cases, however, abandoned arable lands were suspended from cultivation (mainly with cereals) for more than five years; were predominantly covered by spontaneous, herbaceous vegetation; had a soil depth of more than 30 cm; and were exclusively used as grazing lands for domestic animals.

Altitudinal and Geographical Distribution of Old Fields

Abandoned arable lands are not evenly distributed over the various altitudinal and geographical zones in Greece. The greatest abandonment occurred in the mountainous zone where agricultural land was reduced by more than 27% between 1961 and 2000. Only a small reduction (by 2%) occurred in the hilly zone. By contrast, an increase of agricultural land by 1.45% occurred in the communities of the plain zone (figures 12.1 and 12.3).

That arable land abandonment is mainly associated with the mountain-ous areas is also confirmed by the detailed inventory of the Agricultural Bank of Greece in 1983 (Tsoumas and Tasioulas 1986). Out of the 158,820 ha

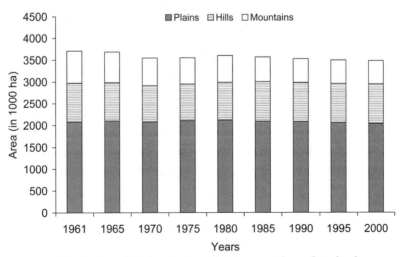

FIGURE 12.3. Evolution of total agricultural land among three altitudinal zones. Data from National Statistical Service of Greece 1966–2003.

recorded as abandoned arable lands that year, 51% were located in the mountainous zone, 35% in the hilly, and the remaining 14% in the plain zone. The largest part of the abandoned lands belonged to the rain-fed cereal crops, followed by the rain-fed orchards (e.g., almond trees), while the irrigated industrial crops (e.g., maize) and the irrigated orchards (e.g., peaches) made the lowest contribution. These differences can be attributed to socioeconomic factors.

The great reduction of arable lands in mountainous areas over the last years was documented in several parts of Greece by comparing successive series of aerial photographs. In the Portaikos-Pertouli valley of the Pindus mountain range, central Greece, for example, arable lands were reduced from 1,887 ha in 1945 to 1,000 ha in 1992 (by 47%); about 23% of these lands were abandoned, while the remaining area (685 ha) was converted to shrublands or forests either artificially through forest plantations or naturally through succession (Chouvardas 2001). In the Psilorites mountain of Crete, an area of 55,000 ha, the proportion of arable land decreased by 1,941 ha (14%) between 1961 and 1989 (Bankov 1998).

In the hills and plains, an increase rather than a decrease of areas under cultivation has been documented. Such an area is the Kolchikos watershed, covering 24,500 ha in Lagadas county, northern Greece. Here, arable lands increased by 770 ha (10%) in the years between 1960 and 1993 at the expense of grasslands (Chouvardas et al., forthcoming). Similar results were also found in the plains around the city of Chania in Crete, where the agricultural areas increased by 448 ha (13%) between 1961 and 1989, mainly due to the increase of irrigated citrus and olive groves (Papanastasis et al. 2004).

Finally, the geographic location of the abandoned arable lands varies widely. According to the Agricultural Bank of Greece inventory of 1983 (Tsioumas and Tasioulas 1986), most abandoned arable lands were found in western Greece (Epirus) and the Aegean islands (16% and 14. 5% of the total agricultural land of these regions, respectively), where mountainous and hilly areas dominate, while lower percentages were found in Crete (2.1%), Macedonia (0.6%), and Thrace (0.5%), where plains are a great proportion of the land surface. It is very likely that these percentages were increased since the year of inventory, given the steady increase of fallow land over the whole country in recent years (figure 12.2).

Abandoned Terraces

Abandoned terraces are a common agricultural feature in the mountainous and hilly zones of Greece. In the Aegean Islands, and in Crete in particular,

FIGURE 12.4. Abandoned terraces with wrecked walls in the island of Kea, eroded and spontaneously recolonized by woody species, such as *Sarcopoterium spinosum* and *Quercus ithaburensis* ssp. *macrolepis*. The latter oak species was cultivated in the past for production of tannins from acorns and for feeding domestic animals.

they are the most abundant and conspicuous cultural feature. A great part of these islands have been terraced in the past in order to create arable land on steep slopes (sometimes 45°) for the cultivation of cereals, wine, or olives. Very few of the old terraces are still cultivated, but not with cereals. Some of them are still used for vineyards and especially olive groves. However, even these cultivations are not regularly taken care of, and no repairs of the dry walls are done, resulting in their collapse and subsequent washing downslope of the terraced soil (figure 12.4). The main use of these abandoned terraces is free grazing by livestock (Margaris et al. 1998).

Socioeconomic and Policy Aspects

Agricultural land is abandoned when it ceases to generate an income for its owner and the opportunities for adjustment through changes in farming practices and farm structure are exhausted (McDonald et al. 2000). Such adjustment is particularly difficult in marginal lands, where productivity levels are close to the margin beyond which management and risks are not compensated by the profit obtained with production (Pinto Correia 1993). However, land abandonment can also take place by abrupt suspension of the farming activity, by death or retirement of a farmer without succession.

In order to understand the reasons involved in arable land abandonment in Greece, two important characteristics of its agricultural sector need to be considered. One is the predominance of small farms; almost 98% of the land holdings are smaller than 20 ha, resulting in a land fragmentation that is the highest in the Mediterranean European countries (Papadimitriou and Mairota 1998). The other reason is the family-centered social and economic organization, which is very vulnerable to socioeconomic changes.

According to Kasimis and Papadopoulos (2001), the Greek countryside was deagriculturalized during the period following World War II, due to transformations caused by two main trends: first, the rural exodus and migration of more than one million people in the period 1950–70, who left agriculture and the rural areas for the urban centers or went abroad, coupled with agricultural modernization; and second, the multiple activity of rural households who combined farm and nonfarm activities/incomes. One of the main consequences of these transformations was the increased marginalization of the mountainous and less-favored rural areas, which depend mainly on agriculture for their economic survival. The same authors claim that the European Union (EU) support of agriculture with subsidies, although it provided a satisfactory income to farmers, did not lead to any structural improvements of the family farms. This was particularly so in the marginal rural areas, where farms are not adaptive to the changing conditions of agricultural production.

Agricultural land abandonment is a complex phenomenon involving both economic and social parameters. That the majority of abandoned lands are found in the mountainous areas is attributable to their marginality, caused by reduced soil fertility, remoteness from settlements, steep slopes, high farm fragmentation, and high labor requirements, making their exploitation unprofitable. Another problem is the migration of the young people from rural to urban areas (Tsoumas and Tasioulas 1986; MacDonald et al. 2000). However, even if the arable lands in the mountains are profitable, their cultivation is suspended if their owners are too old to continue cultivation and no younger descendants are present or willing to take up farming.

Special mention should be made of the abandonment of cereal cultivation on terraces. According to Margaris et al. (1998), this abandonment can be attributed to the higher and more easily obtained cereal production in flatter areas. In these areas, production of 4 t/ha can be achieved by the use of tractors and other machines, in contrast to the terraced areas where less than 0.8 t/ha can be produced and under a greater and more painful human effort. The same authors report that terrace abandonment resulted in a decrease of per capita cultivated area from 0.43 ha in the 1960s to 0.34 ha in the 1980s.

The accession of Greece to the EU in 1981, and the Common Agricultural Policy that involved providing subsidies for farmers, should have halted the increasing trend of arable land abandonment. The increase of agricultural land in the hilly areas since the early 1980s (National Statistical Service of Greece 1966–2003) can be attributed to EU financial support, which encouraged farmers to return abandoned arable land to cultivation in order to increase the dividends received by from EU agricultural policy. A similar explanation may be provided for the intensification of arable agriculture observed mainly in the plains (Kasimis and Papadopoulos 2001; Margaris et al. 1998). The same support however did not help much in the mountainous areas where the abandonment of arable land continued to increase, suggesting that the problem of land abandonment is not only economic but also social.

The resulting transformation of landscape can be seen in several parts of Greece. In western Crete, for example, in an area of about 50,000 ha stretching from the developed north to the undeveloped south coast, the human population decreased by 36% in the hilly and 47% in the mountainous village communities from 1951 to 1991, while it changed very little in the plains. These population changes resulted in the sharp decline of cereal and vineyard cultivations in the upland areas, leading to the abandonment of terraces (Papanastasis and Kazaklis 1998).

Vegetation Development After Abandonment

The majority of arable lands in Greece have been created by clearing natural woody vegetation, particularly in the hills and mountains. Consequently, suspension of cultivation will initiate the process of secondary succession leading to some kind of woodland. The recovery of this natural vegetation, however, depends on several factors, including the dispersal and establishment capacities of plant species, prevailing environmental conditions, site history, site quality, and human interference through various management practices.

Figure 12.5 shows the development of the various plant groups in an old field of more than fifteen years since last cultivation with wheat, compared to adjacent grassland. The research was carried out in the village of Melissochori, of Lagadas county, in northern Greece, with a mean annual precipitation of 470 mm and a mean minimum temperature below zero (°C) during the coldest month. Both sites were freely grazed by sheep until 1989, when they were fenced, and the vegetation succession was studied for six years. Despite some differences from one year to the next due to fluctuation in climate, it is clear that forbs (annual and perennial) were the dominant vegetative

Forbs (annual & perennial)

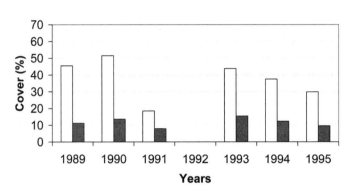

Annual legumes

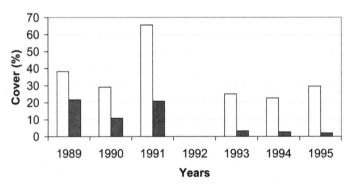

Annual grasses

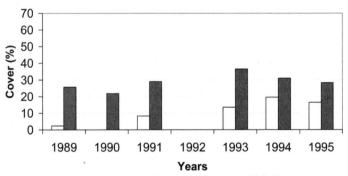

FIGURE 12.5. Succession of various plant groups in an old field (open bars) and an adjacent grassland (shaded bars) following protection from grazing. No data were obtained in 1992. After Noitsakis et al. (1992) and Karatassiou (1999).

Perennial grasses

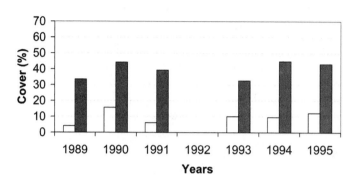

Shrubs

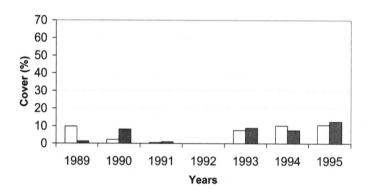

Perennial legumes

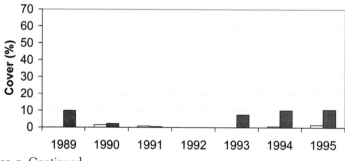

FIGURE 12.5. Continued.

cover in the old field throughout the study period, followed by annual legumes, annual grasses, and perennial grasses and shrubs; perennial legumes were the least important group. In the grassland, by contrast, the dominant vegetative group was perennial grasses followed by annual grasses. The third most important group was originally annual legumes, but these were later replaced by forbs, followed by shrubs and perennial legumes. If we disregard annual legumes, which changed widely from one year to the next, there is a clear decrease in forb cover during the last three years of the study period and an increase in annual and especially perennial grasses and shrubs in the old field. In the grassland, both forbs and annual grasses showed a decrease as opposed to the increasing trend of the perennial grasses and shrubs. By taking into account that perennial grasses are considered to be climax species in grasslands (Dyksterhuis 1949), these data show that grassland was at a more advanced successional stage than the old field. Shrubs, however, did not constitute an important component at either site.

Besides the structural differences, the two sites also differed in water-use efficiency and productivity. The species within grassland (mostly perennial) were found to have a lower water deficit at the same value of leaf water potential compared with old field species (mostly annual), resulting in higher water-use efficiency during the dry summer and, consequently, in higher productivity of grassland (Noitsakis et al. 1992; Karatassiou 1999). Indeed, herbage production was found to be significantly higher in the grassland than in the old field. Within the same growing period, biomass increased at a higher rate in grassland than in the old field when Leaf Area Index exceeded 0.6, which was attributed to the increased specific leaf weight of the species during the dry period (Noitsakis et al. 1992; Karatassiou 1999). These results suggest that vegetation succession in Mediterranean old fields is affected by the ability of each species to withstand summer drought.

In another experiment carried out also in Lagadas county but at a higher altitude (Lofiskos village), where mean annual precipitation is 570 mm, four treatments representing four stages of secondary succession following the abandonment of arable farming, namely an old field (about ten years old), a grassland, an open shrubland, and a dense shrubland were studied (Papadimitriou et al. 2004; Zarovali et al. 2004). All treatments were lightly grazed by livestock. The results of this study are summarized in table 12.1. It is clear that both annual grasses and forbs, as proportions of the herbaceous cover, decreased as the succession proceeded, while the perennial grasses and forbs increased. Herbaceous species richness, on the other hard, was significantly greater in the grassland compared with the old field, but decreased thereafter. Herbage biomass was not significantly different between the old field and

TABLE 12.1

Plant community characteristics in four successional stages after abandonment in a mesic site in northern Greece

	Vegetation type (% estimated woody cover)			
Characteristic	Old field (0)	Grassland (6)	Open shrubland (40)	Dense shrubland (80)
Species richness (species/.25 m^2)	9.0 b	13.0 a	7.5 c	5.8 d
Annual grasses (%)	81.7 a	46.6 b	18.9 c	11.3 d
Perennial grasses (%)	15.8 c	53.4 b	76.1 a	77.4 a
Annual forbs (%)	57.7 a	45.0 b	39.8 b	14.4 c
Perennial forbs (%)	42.3 c	55.0 b	57.6 b	76.9 a
Herbaceous biomass (g/m^2)	215.2 a	199.0 a	110.3 b	82.2 b

Source: Papadimitriou et al. 2004; Zarovali et al. 2004.
Note: Means within rows followed by the same letter indicate no significant differences at the 0.05 level.

grassland, but decreased thereafter. The results suggest that woody plants consisting of shrubs or small trees, such as the evergreen *Quercus coccifera* and the deciduous *Pyrus amygdaliformis* and *Quercus pubescens*, only appeared in a later stage of succession. Such shrubs at low density seem to create a favorable habitat for birds. In a special study carried out in the broader area of the previous site, higher richness and density of birds were recorded in the open than in the dense shrublands (Papoulia et al. 2003). These birds were passerines and composed of species with a great variety of feeding habits.

Different results with respect to recolonization of woody plant species after abandonment were obtained in the Aegean Islands, where the climate is typical Mediterranean and therefore drier than at the previous two sites. In such dry areas, woody plants seem to be favored due to their better adaptation to drought. In a study carried out at five sites located on four islands of the Aegean archipelago, Giourga et al. (1998) found early colonization of abandoned arable fields by woody species, particularly the phryganic species *Sarcopoterium spinosum*. The main results of their study are summarized in table 12.2. Woody plant cover and biomass were positively related with the years of abandonment, while the biomass of annual plants showed a negative relationship. The authors attributed the high presence of woody plants to the heavy grazing that resulted in the reduction of the more palatable annual species. Overgrazing of annuals leads to reduced competition with the woody seedlings and to their subsequent overgrowth (Papanastasis 1977). Protection from grazing, on the contrary, favors annual species at the expense of the unpalatable perennials, including *Sarcopoterium spinosum*, which is reduced

TABLE 12.2

Plant community characteristics in five old fields in Mediterranean islands of Greece

Experimental old field	Years since abandonment	Woody cover (%)	Standing biomass (g/m^2)		
			Woody	Annuals	Total
Lesvos I	5	5.4	167.4	201.3	403.7
Santorini	8	7.3	139.3	208.3	347.6
Lesvos II	20	16.3	264.8	65.9	362.3
Hios	30	24.3	479.8	34.2	514.0
Symi	40	21.2	496.3	17.9	526.4

Source: Giourga et al. 1998.

(Koutsidou and Margaris 1998). Nevertheless, woody species even appeared in the early-successional stages at the Santorini site, where grazing was not applied, although woody species were not as abundant as at the grazed sites. These results suggest that livestock grazing can modify secondary succession but not revert it in areas where less palatable or unpalatable woody species establish after arable land abandonment.

The establishment of phryganic species, especially the spiny ones such as *Sarcopoterium spinosum*, is a common feature of the abandoned terraces all over the Mediterranean island archipelago (figure 12.4). This particular species is adapted not only to overgrazing but also to recurrent wildfires, due to its resprouting ability and the ready germination of its seeds after burning (Papanastasis 1977; Arianoutsou-Faraggitaki 1984). A similar response is displayed by *Cistus* species, common on abandoned terraces with acid soils. Another species common on abandoned terraces is the evergreen shrub *Quercus coccifera*, resistant to both grazing and fire. These species were found to be abundant in most abandoned terraces cultivated with olive trees on the island of Lésvos. Compared with adjacent cultivated olive tree terraces, where woody plant cover was almost nonexistent and the herbaceous cover reached 78%, the abandoned terraces had a woody plant cover of more than 50%, while the herbaceous plant cover was 33% on the average for the six sites studied on the island (Veili 2005).

Restoration of Old Fields

In this section, both current and potential restoration strategies are explored, and possible ways of integrating them into broader land use within the landscape are discussed.

Current Restoration Strategies

Although agricultural lands are privately owned and used or managed exclusively by their owners, old fields become part of the grazing land of each village community, communally used by the villagers, not just by the landowner. If grazing by livestock (sheep, goats, and cattle) is excessive, the palatable herbaceous species are eliminated in favor of the less palatable or unpalatable woody species, which then dominate. This forces shepherds to set fires in order to suppress them and to promote the palatable herbaceous species. Once these latter species are overgrazed, a new fire is set. The combination of pastoral fires and overgrazing is considered the main cause of degradation and desertification of rangelands in general (Papanastasis 1977; Arianoutsou-Faraggitaki 1985), and of the abandoned terraces in particular (Margaris et al. 1998).

In several parts of Greece, however, and particularly in the mountainous areas, grazing is light (undergrazing) or has completely ceased due to rural exodus. In these areas, old fields are gradually invaded by shrubs and trees to finally become woodlands, a process that can be considered a passive restoration strategy. However, if old fields are converted to woodlands, they become forestlands under the Greek constitution and are protected by the forest law. The constitution requires a wildland to be characterized as forestland if more than 25% of the vegetative cover consists of forest species (shrubs and trees) that occupy a solid area of at least 0.3 ha. Once the old field becomes a forestland, the owner cannot cut the trees and recultivate the property. Recultivation can be done only during the early stages of abandonment when herbaceous species dominate or forest cover is below 25%. In the reforested old field, the Forest Service has the right to plant trees in order to speed up the succession to a forest stand, and the owner may lose the property unless it can be proven that it was cultivated in the past. This right has been exercised by the Forest Service in several parts of the country where old fields were planted with fast-growing species, thus implementing an active strategy to restoration of old fields to forests. The species commonly used are indigenous pines planted in areas with a potential natural vegetation of deciduous oak species.

Although these strategies serve a long-standing goal of the Forest Service to increase the forest cover of Greece, they do not necessarily promote environmental sustainability. This is because they lead to landscape homogenization by eliminating the open habitats that are crucial to several plant and animal species (Papoulia et al. 2003; Papadimitriou et al. 2004). In addition, they increase the fire risk, since the main tree species spontaneously invading the old fields or planted by the Forest Service are pines (Papanastasis and Kazaklis 1998; Chouvardas et al., forthcoming). No other efforts are made for an active restoration of the abandoned lands.

Potential Restoration Strategies

According to Naveh and Lieberman (1994), Mediterranean agricultural landscapes are specific cultural landscapes, resulting from a long history of human management adapted to restrictive environmental conditions and biological diversity. Old fields should be considered as an integral part of these landscapes as they contribute significantly to their structure and function. This is because they are open spaces that host plant and animal species unable to live in woodland ecosystems. Examples of such plants are the C_4 grasses *Dichanthium ischaemum* and *Chrysopogon gryllus* (Papadimitriou et al. 2004). Several bird species are included in the list for priority conservation (e.g., *Sylvia communis*, *Lanius collurio*, *L. minor*, *L. senator*, *Acanthis cannabina*, and *Miliaria calandra*) (Tsakiris and Stara, forthcoming). At the same time, old fields provide land uses than cannot be accommodated by more closed or forest habitats. Strategies for restoration of old fields, therefore, should be based on a conservation and landscape ecology approach combined with agricultural knowledge (Pinto Correia 1993).

Based on this framework, four alternative ways of old field restoration are proposed in order to be integrated into a functional and sustainable Greek landscape.

GROWING SPECIAL CROPS

Most old fields had been cultivated with cereals. Such crops are no longer profitable. Instead, special crops such as aromatic and medicinal plants can be used, due to their reduced environmental requirements and environmental impact, as well as the high prices their products can bring in the market. Examples of such suitable species are the Greek tea (*Sideritis scardica*) and oregano (*Origanum vulgare*). These and other crops need to be incorporated into special organic farming programs, and the necessary incentives should be given to farmers to grow them.

This strategy aims to return old fields to arable farming so that opportunities are provided to resident or new farmers to diversify their income and continue farming. Such an activity will also diversify the landscape by breaking up the continuity of natural vegetation and creating a variety of habitats (Naveh and Lieberman 1994). This strategy might work best in the mountain villages and on the islands where the climate is suitable for rain-fed crops and the landownership small. In such areas, it is not profitable for farmers to grow conventional crops because they require expensive inputs in energy and infrastructure. In order to be implemented, however, this strategy needs progressive farmers to get involved, as well as strong institutional support.

CONVERSION TO IMPROVED PASTURES

Since most old fields are grazed, their grazing capacity can be substantially increased if they are converted to improved pastures with indigenous but highly productive herbaceous and woody forage species. The species to be established depend on the specific climatic and edaphic conditions. Among potential species are the grasses *Lolium rigidum, Dactylis glomerata, Festuca arundinacea,* and *Hordeum bulbosum;* the legumes *Trifolium subterraneum, Vicia sativa, Lotus corniculatus,* and *Medicago sativa;* and the shrubs *Medicago arborea, Morus alba,* and *Atriplex halimus* (Papanastasis 1982; Papanastasis et al. 1999).

This strategy aims at arresting the succession of natural vegetation and replacing it with an artificial plant cover consisting of ecologically and economically important forage species. The leguminous species to be introduced will improve the soil fertility through atmospheric nitrogen fixation, while the perennials will improve the soil structure and stability through their proliferous root system. At the same time, this strategy will bring economic benefits to livestock farmers who will use the improved pastures for their animals. Such a strategy might work best in areas where livestock husbandry is still an important activity and there is a strong need for additional improved pastures. However, its implementation will be difficult in villages where livestock farmers are not the owners of the old fields. In this case, long-term renting of the old fields by the livestock people could be a solution.

CONVERSION TO AGROFORESTRY SYSTEMS

Old fields can be planted with timber or fodder trees so that agroforestry systems are created. These trees can produce high-quality timber or foliage and fruits to be consumed by animals during the critical periods of the year (e.g., summer). Examples of timber trees are *Fraxinus angustifolia, Acer pseudoplatanus,* and *Prunus avium;* and of fodder trees, several species of oaks, (e.g., *Quercus ithaburensis* ssp. *macrolepis*) (figure 12.4). The trees should be established with more than 10 m between in order to simultaneously grow arable crops, such as herbaceous or forage species, in the understory.

This strategy aims at replacing the final stage of succession with a kind of artificial open woodland that will combine trees and arable crops or pasture so that a variety of products are produced (e.g., timber, crops, forage, and animal products). Agroforestry systems are traditional in Greece. Silvo-arable systems in particular, which combine timber trees and crops, are very suited to small farm owners and constitute a potential restoration strategy for old fields in several parts of Greece. Farmers are willing to implement this strategy if financial support is provided by the state or the EU (Mantzanas et al. 2005).

Conversion to Conservation Spots

Old fields can be converted to seminatural grasslands with high biodiversity by using traditional management activities, such as cutting shrubs for firewood or charcoal, light to moderated grazing by livestock, and even prescribed burning. Keeping them open and free of woody plants will provide home to several plant and animals species that cannot live in woody landscapes (Papadimitriou et al. 2004; Tsakiris and Stara, forthcoming).

This strategy aims at arresting the succession of natural vegetation to the grassland stage by continuous management interventions. It might work best in national parks and protected areas where landscape and biodiversity conservation are the primary management objectives. However, the current prospects of application of such a strategy in these areas is rather limited unless the complex jurisdictional problems of the land management services involved are soon resolved.

Terrace Reconstruction

Since a large proportion of old fields are found on abandoned terraces, reconstruction of these terraces to halt soil erosion is a necessary conservation measure. It also serves historical and cultural objectives. The reconstructed terraces can then be used to apply any of the four strategy options (figure 12.6). Such a measure is also supported by the EU, and financial support is provided to farmers who are willing to repair old terraces and return them to cultivation.

Conclusion

Arable lands have been a constantly changing frontier in Greece over its long history, with abandoned fields being an almost permanent feature of the

FIGURE 12.6. Reconstructed and well-maintained terraces in the island of Kea currently used for grazing by livestock. Woody (phryganic) species have started to colonize. They will become more abundant with time, if terraces are not recultivated.

landscape. Nowadays, old fields occupy a considerable proportion of the agricultural area, particularly in the mountainous zone. They constitute the transition between arable and grazing or forest lands, and therefore they are conflict areas among different groups, such as farmers, grazers, and foresters. Their extent and nature is determined by both economic and social factors.

Vegetation development in old fields depends on several factors, both physical–ecological and anthropogenic, but it seems to follow two succession pathways. In more mesic areas, arable land abandonment results in the establishment of an herbaceous community in the first five to ten years. The first species to appear are forbs, including legumes, especially annual, coming from seeds of the soil seed bank or from adjacent noncultivated areas, followed by annual as well as perennial grasses. Woody species come in later. In drier areas, woody species appear earlier or even right after abandonment. Heavy livestock grazing, usually applied in most old fields, can modify the course of succession, either by holding back the perennial species or speeding up the encroachment of woody species, if they are unpalatable to animals. Modification is also done by fires, which suppress the woody plants in favor of herbaceous species, especially annual legumes, while their combination with heavy grazing speeds up the reestablishment of unpalatable woody species.

Old fields are important historical, cultural, and ecological assets, particularly the ones on abandoned terraces, which need to be restored and managed sustainably. The current restoration strategy aims at developing a forest cover either naturally (through succession) or artificially (through tree plantations). This strategy, however, is not environmentally sustainable, because it leads to landscape homogenization and consequently to the reduction of biodiversity and ecodiversity and as well as to an increased fire risk. An alternative strategy would be to consider old fields as parts of a diverse landscape, and compatible agricultural, grazing, and conservation activities are developed on them so that they substantially contribute to the environmental stability of rural areas.

Acknowledgments

Appreciation is expressed to Anna Sidiropoulou, Konstantinos Papanastasis, and Christakis Evangelou for their assistance in preparing this chapter. Two anonymous reviewers contributed significantly to the improvement of the manuscript with their constructive comments.

REFERENCES

Alcock, S. E. 1993. *Graecia capta: The landscape of Roman Greece*. Cambridge University Press, Cambridge, UK.

Arianoutsou-Faraggitaki, M. 1984. Post-fire successional recovery of a phryganic (East Mediterranean) ecosystem. *Acta Oecologia/Oecologia Plantarum* 5:387–94.

———. 1985. Desertification by overgrazing in Greece. *Journal of Arid Environments* 9:237–42.

Bankov, N. 1998. Dynamics of land cover/use changes in relation to socio-economic conditions in the Psilorites mountain of Crete. MS thesis. Mediterranean Agronomic Institute of Chania, Crete, Greece.

Chouvardas, D. 2001. Analysis of structure and diachronic evolution of landscapes with the use of G.I.S. MS thesis (in Greek). School of Forestry and Natural Environment, Aristotle University of Thessaloniki, Thessaloniki, Greece.

Chouvardas, D., I. Ispikoudis, and V. P. Papanastasis. Forthcoming. Evaluation of temporal changes of the landscape of Kolchikos basin of Lake Koronia with the use of Geographic Information Systems (G.I.S.). In *Rangelands of plain and hilly areas: A vehicle to rural development*, ed. P. Platis et al., 253–60 (in Greek). Ministry of Rural Development and Food and Hellenic Range and Pasture Society, Athens.

Dyksterhuis, E. 1949. Condition and management of range land based on quantitative ecology. *Journal of Range Management* 2:104–15.

Gerasimidis, A. 2000. Palynological evidence for human influence on the vegetation of mountain regions in northern Greece: The case of Lailias, Serres. In *Landscape and land use in postglacial Greece*, ed. P. Halstead and C. Frederick, 28–37. Sheffield Academic Press, Sheffield, UK.

Giourga, H., N. S. Margaris, and D. Vokou. 1998. Effects of grazing pressure in succession processes and productivity of old fields on Mediterranean islands. *Environmental Management* 22: 589–96.

Halstead, P. 2000. Land use in postglacial Greece: Cultural causes and environmental effects. In *Landscape and land use in postglacial Greece*, ed. P. Halstead and C. Frederick, 110–28. Sheffield Academic Press, Sheffield, UK.

Karatassiou, M. 1999. Ecophysiology of water use efficiency in Mediterranean grasslands. PhD thesis (in Greek). School of Forestry and Natural Environment, Aristotle University of Thessaloniki, Thessaloniki, Greece.

Kasimis, C., and A. G. Papadopoulos. 2001. The de-agriculturisation of the Greek countryside: The changing characteristics of an ongoing socio-economic transformation. In *Europe's "green ring,"* ed. L. Granberg, I. Kovách, and H. Tovey, 197–219. Ashgate, Aldershot, UK.

Koutsidou, E., and N. S. Margaris. 1998. The regeneration of Mediterranean vegetation in degraded ecosystems as a result of grazing pressure exclusion: The case of Lesvos Island. In *Ecological basis of livestock grazing in Mediterranean ecosystems*, ed. V. P. Papanastasis and D. Peter, 76–79. European Commission, EUR 18308 EN, Luxembourg.

MacDonald, D., J. R. Crabtree, G. Wiesinger, T. Dax, N. Stamou, P. Fleury, J. Gutierrez Lazpita, and A. Gibon. 2000. Agricultural abandonment in mountain areas of Europe: Environmental consequences and policy response. *Journal of Environmental Management* 59:47–69.

Mantzanas, K., E. Tsatsiadis, I. Ispikoudis, and V. P. Papanastasis. 2005. Traditional silvoarable systems and their evolution in Greece. In *Silvopastoralism and sustainable land management*, ed. M. R. Mosquera-Losada, A. Rigueiro-Rodríguez, and J. McAdam, 53–54. CABI, Wallingford, UK.

Margaris, N. S., E. J. Koutsidou, and C. E. Giourga. 1998. Agricultural transformations. In *Atlas of Mediterranean environments in Europe*, ed. P. Mairota, J. B. Thornes, and N. A. Geeson, 82–84. John Wiley and Sons, New York.

National Statistical Service of Greece. 1995. *Distribution of the country's area by basic categories of land uses* (in Greek). Athens.

———. 1966–2003. *Agricultural statistics of Greece* (in Greek). Athens.

Naveh, Z., and A. S. Lieberman. 1994. *Landscape ecology. Theory and application.* Springer-Verlag, New York.

Noitsakis, B., I. Ispikoudis, Z. Koukoura, and V. Papanastasis. 1992. Relation between successional stages and productivity in Mediterranean grasslands. In *Natural resource development and utilization*, 126–33. Commission of European Communities, Wageningen, The Netherlands.

Papadimitriou, F., and P. Mairota. 1998. Agriculture. In *Atlas of Mediterranean environments in Europe*, ed. P. Mairota, J. B. Thornes, and N. A. Geeson, 86–90. John Wiley and Sons, New York.

Papadimitriou, M., Y. Tsougrakis, I. Ispikoudis, and V. P. Papanastasis. 2004. Plant functional types in relation to land use changes in a semi-arid Mediterranean environment. In *Ecology, conservation and management of Mediterranean climate ecosystems*, ed. M. Arianoutsou and V. P. Papanastasis, 1–6. Millpress, Rotterdam, The Netherlands.

Papanastasis, V. P. 1977. Fire ecology and management of phrygana communities in Greece. In *Environmental consequences of fire and fuel management in Mediterranean ecosystems*, ed. H. A. Mooney and C. E. Conrad, 476–82. USDA Forest Service, General Technical Report. WO-3, Washington, DC.

———. 1982. Developing improved dry pastures in Greece. In *Report of the 1982 consultation on the cooperative network on fodder crop production*, 7-1-3. FAO, Rome.

Papanastasis, V. P., I. Ispikoudis, M. Arianoutsou, P. Kakouros, and A. Kazaklis. 2004. Land-use changes and landscape dynamics in western Crete. In *Recent dynamics of the Mediterranean vegetation and landscape*, ed. S. Mazzoleni, G. di Pasquale, M. Mulligan, P. di Martino, and F. Rego, 81–93. John Wiley and Sons, Chichester, UK.

Papanastasis, V. P., and A. Kazaklis. 1998. Land use changes and conflicts in the Mediterranean-type ecosystems of western Crete. In *Landscape disturbance and biodiversity in Mediterranean-type ecosystems*, ed. P. W. Rundel, G. Montenegro, and F. B. Jaksic, 141–54. Ecological Studies 136. Springer, Berlin.

Papanastasis, V. P., P. Platis, G. Halyvopoulos, and A. Tepeli-Malama. 1986. *Grazed forest lands of Drama Prefecture. A project for inventorying rangelands of northern Greece* (in Greek). Forest Research Institute of Thessaloniki, Bulletin No. 1., Thessaloniki, Greece.

Papanastasis, V. P., C. N. Tsiouvaras, O. Dini-Papanastasi, T. Vaitsis, L. Stringi, C. Cereti, C. Dupraz, D. Armand, and L. Olea. 1999. Selection and utilization of cultivated fodder trees and shrubs in the Mediterranean region, *Options Mediterraneennes*, Serie B, No. 23. Instituto Agronomico Mediterraneo de Zaragoza, Spain.

Papoulia, S., S. Kazantzidis, and G. Tsiourlis. 2003. The use of shrubland vegetation by birds in Lagadas, Greece. In *Range science development of mountainous areas*, ed. P. D. Platis and T. G. Papachristou, 117–23 (in Greek). Hellenic Rangeland and Pasture Society and Ministry of Rural Development and Food, Athens.

Petmezas, S. D. 2003. *The Greek rural economy during the 19th century: The regional dimension* (in Greek). University Publications of Crete. Heraclion, Crete, Greece.

Pinto Correia, T. 1993. Land abandonment: Changes in the land use patterns around the Mediterranean basin. In *The situation of agriculture in Mediterranean countries. Soils in the Mediterranean region: Use, management, and future trend*, 97–112. Cahiers Options Mediterraneennees, Instituto Agronomico Mediterraneo, Zaragoza, Spain.

Platis, P. D., T. G. Papachristou, and V. P. Papanastasis. 2001. Possibilities of applying the inventory project for management of rangelands in the region of Epirus. In *Range science in the threshold of the 21st century*, ed. T. Papachristou and O. Dini-Papanastasi, 43–49 (in Greek). Hellenic Range and Pasture Society, Thessaloniki, Greece.

Rackham, O., and J. Moody. 1996. *The making of the Cretan landscape*. Manchester University Press. Manchester, UK.

Tsakiris, R., and K. Stara. Forthcoming. The importance for avifauna of the habitat mosaic in the agropastoral plateaux of the Vicos-Aoos Natonal Park. In *Rangelands of plain and hilly areas: A vehicle to rural development*, ed. P. Platis et al., 423–28 (in Greek). Ministry of Rural Development and Food and Hellenic Range and Pasture Society, Athens.

Tsoumas, A., and D. Tasioulas. 1986. *Ownership status and use of agricultural land in Greece* (in Greek). Agricultural Bank of Greece, Athens.

Vakalopoulos, A. 1964. *History of modern Greece: Ottoman period (1453–1669)* (in Greek). Thessaloniki, Greece.

Van Andel, T. H., E. Zangger, and A. Demitrack. 1990. Land use and soil erosion in prehistoric and historical Greece. *Journal of Field Archaeology* 17(4):379–96.

Veili, D. 2005. Cultivation abandonment in terraces of Lesvos: Effects on structure and dynamics of ecosystems. MS thesis (in Greek). Department of Geography, University of Aegean, Mytilini, Greece .

Zarovali, M. P., Y. Tsougrakis, and V. P. Papanastasis. 2004. Herbage production in relation to land use changes in Mediterranean rangelands. *Grassland Science in Europe* 9:201–203.

Chapter 13

Old Field Dynamics on the Dry Side of the Mediterranean Basin: Patterns and Processes in Semiarid Southeast Spain

Andreu Bonet and Juli G. Pausas

Euro-Mediterranean countries have a long history of land use changes and urbanization processes (Antrop 2004). During the past century, with the advent of industrial and tourist development, these countries have experienced an important intensification of land use changes related to the abandonment of rural livelihoods and city sprawl in coastal areas.

On the northern (European) rim of the Mediterranean basin, the socioeconomic changes taking place during this period promoted a dramatic rural exodus, with the consequent abandonment of large cultivated areas (Lepart and Debussche 1992) (figure 13.1). As a result, these previously cultivated areas are now being colonized by natural vegetation, with important implications for processes such as water balance (Bellot et al. 2001), wildfire regime (Pausas 2004), and carbon sequestration (DeGryze et al. 2004). Meanwhile, landscapes on the southern (African) rim of the Mediterranean basin still suffer from overexploitation, especially overgrazing (after clearing), increasing cultivation on pronounced slopes, and associated degradation problems (Le Houérou 1993; Taiqui 1997; Redjali 2004). These contrasting patterns are related to the different socioeconomic, demographic, and political trends occurring between the northern and southern Mediterranean (Puigdefábregas and Mendizabal 1998)

Thus there is some association between land abandonment in mesic Mediterranean ecosystems and overexploitation in drier Mediterranean ecosystems. However, there are also areas in southern Europe (e.g., southeast Spain) that are considered semiarid and are undergoing the process of land abandonment (Bonet et al. 2004; Bonet et al. 2006). In this chapter we focus on such areas where studies of vegetation dynamics have been developed only recently (Bonet et al. 2001; Bonet 2004; Bonet and Pausas 2004; Pausas et al.

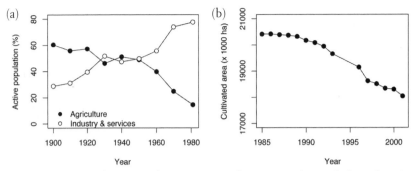

FIGURE 13.1. (a) Changes in the percentage of active population dedicated to either agriculture or industry and services, during the twentieth century in Spain. From Pausas (2004). (b) Dynamics of the overall cultivated land surface in Spain. Source: Ministerio de Agricultura, Pesca y Alimentación, 2006, and Instituto Nacional de Estadística, 2006, Spain.

2006; Pugnaire et al. 2006). Vegetation dynamics under semiarid conditions differ from that observed in more mesic areas, due to the reduced plant cover that characterizes the former, the differences in the relative importance of interspecific interactions such as facilitation and competition (Bertness and Callaway 1994), and the role that abiotic factors play in the dynamics of plant populations (Escudero et al. 1999). Many studies on secondary succession in both tropical and temperate regions have been carried out in abandoned fields after agricultural use of previously forested lands. Succession in these areas usually leads to the development of forest, although the composition may change from the original forests (Foster et al. 1998; Grau et al. 2003). In the Mediterranean Basin, some studies also indicate a forest development with succession following land abandonment (Housard et al. 1980; Tatoni and Roche 1994; Debussche et al. 1996, 2001; Debussche and Lepart 1992; Mazzoleni et al. 2004), but under semiarid conditions the forest may not be representative of late stages of succession and is usually not present in remnant vegetation, due to both low water resources and intensive human pressure since the Neolithic period (Badal et al. 1994). In semiarid Spain there is some uncertainty about which vegetation type was present prior to agricultural practices, but many authors suggest that the late-successional communities in these territories are composed of shrublands with a variable tree cover (Bolòs 1967; Rivas-Martínez 1987; Ruiz de la Torre 1990).

For practical reasons, we use a chronosequence approach (i.e., synchronic approach or space-for-time substitution) to infer successional dynamics. Although we recognize that there are some potential problems with using this approach (Pickett 1989), most predictions made with this approach have been validated when revisiting and resampling the studied communities (Debussche et al. 1996; Foster and Tilman 2000). The studies presented here were performed in Alacant Province, southeast Spain. Mean annual temperature is above 18°C and annual rainfall is about 300 mm. Soils are typically developed over marls and calcareous bedrock. Slopes over marls were terraced in the past for cultivation and then abandoned. Hand-built stone terraces permitted cultivation on slopes by preserving soil and maintaining water availability. In some cases, abandoned terraces were afforested with *Pinus halepensis*, and fires affected some of these plantations (Pausas, Bladé et al. 2004; Pausas, Ribeiro, and Vallejo 2004). Natural vegetation surrounding agricultural land is formed by a mosaic of *Stipa tenacissima* steppes, *Brachypodium retusum* grasslands with dwarf shrubs, and shrublands dominated by *Quercus coccifera* and *Erica multiflora* (Bonet et al. 2004). These formations are mixed with the *Pinus halepensis* plantations and with both active and abandoned dry woody crops like almond tree (*Prunus dulcis*), olive tree (*Olea europaea*), and carob tree (*Cerationia siliqua*). Field abandonment ranges from one to sixty years following the last cultivation, with the main set-aside process taking place during 1946–56 (Bonet et al. 2004).

Patterns in Semiarid Old Field Dynamics

Patterns in old field dynamics in the semiarid areas are determined by landscape dynamics, plant cover and composition, and species richness.

Landscape Dynamics

During the second half of the twentieth century, the vegetation changes resulting from both the abandonment of various dry crops and the plantations of Aleppo pine on other old fields transformed the landscape pathways in the Agost-Ventós catchment (Alacant). The analysis of trends in land cover categories using GIS, aerial digitized photographs (from 1946 and 1999), and fieldwork surveys (Bonet et al. 2001, 2004) allows us to understand the dynamic patterns, by counting the transition cases between land cover classes on a sequence of land cover maps in the study area (figure 13.2). These values, expressed as percentages of occupied land area, indicate the land cover changes that occurred during the last decades. Land cover categories are defined by landscape structures: urban, agricultural crops, herbaceous communities, perennial grasslands, and dwarf shrublands, including alpha grass (*Stipa tenacissima*) steppes, shrublands, and pine plantations (figure 13.2).

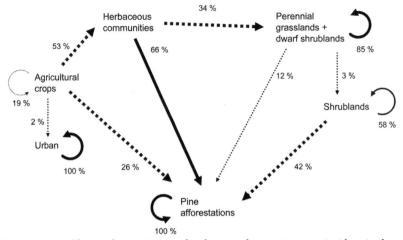

FIGURE 13.2. Observed transitions in land cover changes in a semiarid agricultural landscape (Agost-Ventós catchment, Alicante) between 1946 and 1999. Diagram shows the pathways of landscape change, estimated using aerial photographs and GIS analysis (Bonet et al., 2004). Perennial grasslands and shrublands include agricultural land outside the terrace cropping area.

Natural processes of land cover change represent 98% of the changes in recent abandoned crops (i.e., only 2% of the observed transitions were from dry crops to urban land). These land cover changes are derived from two different processes: natural vegetation recovery (as a result of autogenic vegetation dynamics dominated by old field succession) or human afforestation (pine plantations). Some afforestation has been performed in all the land cover classes (shrubland, grassland, or herbaceous). Note that natural vegetation recovery (i.e., by means of old field succession) does not include the woodland land cover category under the semiarid conditions studied.

The rate of land cover change was higher during the first period in the land abandonment sequence. The initial transitions were observed to follow a pathway from dry crops to herbaceous communities, perennial grasslands plus dwarf shrub communities, and shrublands. This landscape pattern is consistent with the one described in semiarid old field succession (Bonet 2004) and is similar to other semiarid areas (Alados, Pueyo et al. 2004). Herbaceous communities showed no permanence in the landscape between the coupled 1946–99 land cover maps, while perennial grasslands presented the higher rate of permanence (85% of observed transitions) along the abandonment pathway (from abandoned crops to shrublands). The case of permanence of perennial grasslands (such us *Stipa tenacissima* or *Brachypodium retusum* formations) is an example of community stability with slow species turnover, thus indicating that the dynamics of this vegetation is not necessarily progressive.

The main landscape changes between 1946 and 1999 were due to afforestation in old fields and affected all abandonment stages (figure 13.2). Plantations of Aleppo pine began in the study area around 1946 and increased between 1956 and 1974 over recently abandoned crops (Bonet et al. 2004). The afforestation pathway was especially relevant in herbaceous communities (66% of the observed transitions) and shrublands (42%).

Plant Cover and Composition

We studied a sixty-year chronosequence of old fields formerly planted with tree crops (fruit orchards) and that were not afforested (Bonet and Pausas 2004). In this chronosequence, a total of 222 plant species were recorded in 96 plots (100 m²). Plant cover ranged from 45.5% to 100% and was not dependent on abandonment age ($p = 0.419$); high cover values were found at any time following abandonment. However, the cover of each life-form showed a significant and different pattern along the abandonment age gradient, and there was a clear tendency in the order of dominance of the different life-forms (figure 13.3a). Annuals, forbs, and grasses showed a skewed pattern, while woody species showed a linear trend. Annuals were the first species to cover the old fields, reaching high cover values in the first five years after abandonment (mean of ca. 60%, but up to 100% in some cases). However, they were almost absent approximately twelve years after abandonment. Perennial forbs also reached a maximum during the first ten years, while perennial grasses peaked at ten to twenty-five years after abandonment. Woody species showed a significant monotonic increase throughout the time span studied (sixty years) (figure 13.3). The number of life-forms decreased with abandonment age, from four co-occurring during the first twenty years, to two life-forms, and finally (at ~ sixty years) the vegetation was practically dominated by one life-form (woody species), although one grass species (*Brachypodium retusum*) persisted (with low cover) on the sixty-year-old plots.

Plant composition, summarized as an ordination axis, clearly changed with abandonment age ($R^2 = 0.83$; $p < 0.0001$). The variability in vegetation composition along the chonosequence was mainly explained by differences in previous crop type (figure 13.4a; table 13.1). Slope was also seen to be a significant factor. Although slope is correlated with abandonment age (fields located on sites with difficult conditions for farming were the first to be abandoned), it had a significant effect on cover even when age was considered (i.e., as covariate) (table 13.1). Vegetation composition variability was higher

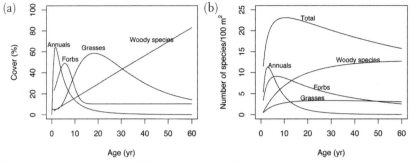

FIGURE 13.3. Changes in (a) plant cover and (b) species richness for the different life-forms along a chronosequence in Alacant, SE Spain. Fitted lines were all significant ($p < 0.0001$). Elaborated from data in Bonet and Pausas 2004.

during the first postabandonment years and tended to converge with time since abandonment (figure 13.4a).

Species Richness

Species richness per plot (100 m²) ranged from four to thirty-four, and there was a significant nonlinear relationship with abandonment age ($R^2 = 0.36$; $p < 0.0001$; figure 13.3b) (Bonet and Pausas 2004). Thus, species richness follows a hump-shaped curve as described for richness in relation to environmental variables and productivity (see Pausas and Austin 2001). Species richness also showed different patterns for the various life-forms (figure 13.3b), and the order along the age gradient was similar to cover; the main difference was that there were few species of perennial grasses with high cover. This different richness pattern for each functional group results from differences in colonization capacity and persistence. For example, most annual plants or short-lived perennials are either present in the fields as weeds during cropping or find shelter in the active field margins; thus, they can increase in number and cover quickly after abandonment. However, these species may be displaced when resources are taken over by long-lived perennial plants. The results suggest that, under semiarid conditions, total species richness shows a peak in the early stages of the chronosequence. This pattern may be attributable to both the importance of immigration processes during the early stages of succession and the dominance of extinction processes after approximately fifteen years since abandonment. Nevertheless, the species richness pattern and life-form replacement during semiarid old field succession can also vary with the precipitation gradient (Otto et al. 2006). The most outstanding differences in life-form richness and cover are shown by annual plants, whose patterns showed a strong dependence on water availability.

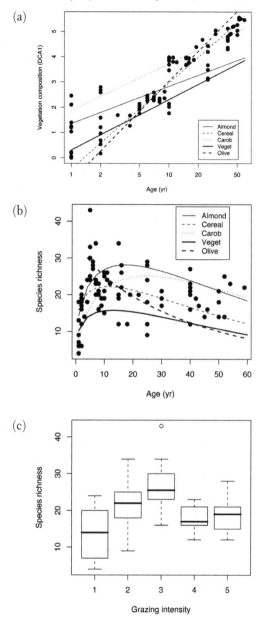

FIGURE 13.4. Variability in the relation between vegetation composition (expressed as a vegetation ordination axis, DCA1) and abandonment age (logarithmic scale) can be explained by differences in the previous crop type (a). Variability in the relationship between species richness and abandonment age can be explained both by the different, previous crop type (b) and by the grazing intensity (c). See statistics in table 13.1. Previous crop types are: almond trees, cereals, carob trees, vegetable fields, and olive trees. Explained deviance = (a) 93.7%; (b) 58.2%; (c) 31.1%; and 66%, for (b) and (c) together, i.e., the model with age, previous crop type, and grazing intensity.

TABLE 13.1

Effect of factors explaining the variability in the vegetation composition and species richness in the studied old fields

	Vegetation composition		Species richness	
	F	*p*	Chi sq.	P
Previous use	11.426	<.00001	41.302	<.00001
Slope	5.1378	.02573	10.781	.001
Altitude	1.7888	ns	.0	ns
Grazing intensity	1.7996	ns	44.269	<.00001

Note: Vegetation composition is inferred as an ordination axis (see figure 13.4), and tested with ANOVA using the F-test; Species richness is tested with a GLM (Poisson error distribution) and the significance evaluated using the Chi-squared test. Logarithm of age used as covariable in all cases.

Species richness variability throughout the chonosequence can be explained by previous crop type and grazing intensity (figures 13.4b and 13.4c; table 13.1). Old almond orchards have the highest richness, while old vegetable fields have the lowest richness. Grazing intensity explains a significant ($p = 0.001$) part of the variability in richness even when age and previous use are included in the model, suggesting that the factors are independent. Species richness in relation to grazing intensity shows a humped pattern with highest richness at intermediate levels of grazing (figure 13.4c), as has been suggested for different disturbance types (Huston 1979). Grazing promotes the coexistence of different functional groups (McIntyre et al. 1995), altering the inhibition pathway and favoring species richness (Noy-Meir 1998). Furthermore, increased dispersal by the grazers (Malo and Suárez 1995a, 1995b) could also promote richness. Other authors found differential effects on diversity in other Spanish semiarid old field communities depending on the stage of development of the vegetation and the environmental conditions (Alados, Elaich et al. 2004a; Alados et al. 2005). However, experimental trials in semiarid environments in the United States showed no relevant effects of moderate grazing on plant diversity along old field successions (Coffin et al. 1998). Thus the effect of grazing on semiarid old field richness deserves further study.

The main difference between the pattern observed in our study area and classic replacement patterns suggested for mesic Mediterranean ecosystems (Houssard et al. 1980; Escarré et al. 1983) has to do with the time lag in which these changes occur. In our case, about 45% of total woody species richness is reached ten years after abandonment, while in mesic Mediterranean conditions (southern France), woody species reached less than 20% of their total richness at this time (Escarré et al. 1983). This pattern could be partially attributed to an early colonization of some shrub species (e.g., *Rhamnus lycioides*) through the facilitation of bird-dispersed seeds by cultivated trees acting as perches.

Processes Driving Semiarid Old Field Dynamics

In areas formerly planted with tree crops (fruit orchards), it has been suggested that old field vegetation is spatially aggregated around the original crop tree (Debussche and Isenmann 1994). This process has been named "nucleation" (Yarranton and Morrison 1974) through analogy with other physical processes, or "recruitment foci" (McDonnell and Stiles 1983). This nucleation pattern can be generated by at least two different ecological processes: the perch effect and/or the facilitation effect through microenvironmental and resource improvement (Gill and Marks 1991; Verdú and García-Fayos 1996).

The perch effect refers to the process in which trees remaining from the orchards are used as perches by frugivorous birds (Debussche et al. 1982). Thus, seed rain and the resulting seedling recruitment and sapling spatial pattern should be highly patchy and largely restricted to microhabitats beneath trees (Izhaki et al. 1991; Debussche and Lepart 1992; Herrera et al. 1994; Debussche and Isenmann 1994; Alcántara et al. 2000). An alternative process for explaining the nucleation pattern is the facilitation effect (sensu Connell and Slatyer 1977). Many studies have reported improvements in soil structure, increases in soil nutrients and microbial activity, and amelioration of harsh microclimatic conditions under woody plants in semiarid environments (Jake and Coughenour 1990; Verdú and García-Fayos 1996; Moro et al. 1997; Reynolds et al. 1999). In semiarid old fields changes in soil and microclimate as described earlier are also likely to occur around the crop trees, and thus may be responsible for generating the patchy pattern of nucleation, where population dynamics could also be governed by species-specific environmental triggers (Pugnaire et al. 2006).

Nucleation patterns in old field succession have been found for several Mediterranean species, such as *Rhamnus alaternus* (Gulias et al. 2004), *R. ludovici-salvatoris* (Traveset et al. 2003), and *Pistacia lentiscus* and *Daphne gnidium* (Verdú and García-Fayos, 1996, 1998). However, to what extent the pattern is due to the perch effect or to the facilitation effect through microenvironmental and resource improvement has seldom been tested (Pausas et al. 2006). Comparing plant density beneath carob trees (*Cerationa siliqua*) and in the surrounding fields, we observe that fleshy-fruited, bird-dispersed species (e.g., *Rhanmus lycioides, Asparagus horridus*) are more abundant beneath the trees than in the open field, while nonfleshy-fruited species do not show any differential density pattern (figure 13.5). As a consequence, species density is higher beneath the trees (figure 13.5). These results cannot be explained by facilitative interactions only; increased seed rain due to the perch effect should also be invoked. This emphasizes the overwhelming role that

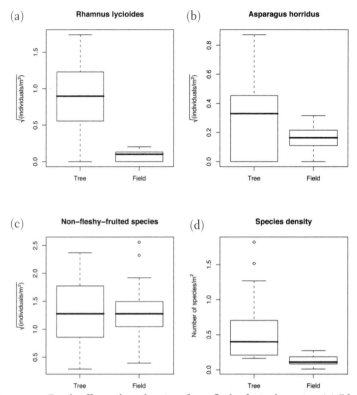

FIGURE 13.5. Perch effect: plant density of two fleshy-fruited species, (a) *Rhamnus lycioides*, and (b) *Asparagus horridus*; and of all the nonfleshy-fruited species; plus (d) the species density observed on two microsites: tree (beneath *Ceratonia siliqua*) and field. Density of both fleshy-fruited species and species density were significantly different ($p<.0001$) between the two microsites; density of nonfleshy-fruited species did not differ between microsites. Box plots indicate median (horizontal line), first and third quartiles (lower and upper sides of the box), 95% confidence intervals (vertical lines), and extreme values (dots). Elaborated from data in Pausas et al. (2006).

dispersal mode has on the dynamics of vegetation recovery in formerly cropped areas under a dry climate (Bonet and Pausas 2004; Pausas et al. 2006; Pugnaire et al. 2006). This does not mean that microenvironmental and resource improvement under trees does not play a role, but that its importance in the colonization process may be secondary.

Semiarid Old Field Restoration

We will now explore the traditional and more contemporary approaches to old field restoration in semiarid Spain.

Traditional Old Field Restoration: Pine Afforestation

Conifers, mostly pines, have long been used for reforestation in Mediterranean countries. For instance, the proportion of area reforested with conifers (compared to total area reforested) during the last decades was ca. 90%, 94%, 47%, 55%, 86%, and 71% in Spain, Turkey, Algeria, Morocco, Portugal, Greece, and Tunisia, respectively (Pausas, Bladé et al. 2004). In the case of semiarid areas of the Mediterranean Basin, including semiarid Spain, *Pinus halepensis* has been the preferred species for afforestation projects in old fields (Valle and Bocio 1996; Cortina et al. 2004). This pine species shows a high relative survivorship in stressed environments (Vilagrosa et al. 1997; Calamassi et al. 2001). Currently, these Aleppo pine plantations constitute the dominant forest land cover (more than 90% of the forested area) in Alicante Province, southeast Spain. The objectives of these plantations were mainly to increase forest productivity, and also to protect watersheds and provide employment in rural areas. The traditional strategy for reforesting in the Mediterranean was to first introduce a fast-growing pioneer species such as pine (Ceballos 1938), under the assumption that this species would facilitate the introduction (either artificial or natural) of late-successional hardwoods. Nevertheless, this latter step was seldom applied because of the costly silvicultural postplantation operations required and the current disturbance regime. Furthermore, the facilitation effect of *P. halepensis* has been questioned (Maestre et al. 2003), and, in fact, in some cases pine plantations may have negative consequences on both the natural vegetation dynamics and in different ecosystem processes (Maestre and Cortina 2004; Chirino et al. 2006). For instance, both the afforested shrublands and the afforested grasslands showed lower species richness than the original shrublands and grasslands, for total species and for most life-forms (figure 13.6). In addition, extensive pine plantations resulted in large and homogeneous areas covered with flammable, even-aged pines, interconnected through other old fields dominated by flammable shrublands. This landscape structure and composition with high fire hazard facilitated the spread of large fires during the last decades (Pausas and Vallejo 1999; Pausas 2004; Pausas, Ribeiro, and Vallejo 2004). Thus, although traditional pine plantations have partially achieved some of their initial objectives in some areas, there is a need to be critical and to revise the massive afforestation strategy.

Current Tendencies: Toward Restoring Biodiversity

Although increased forest production may still be an objective, current forest restoration actions in the Mediterranean area have new aims directed toward

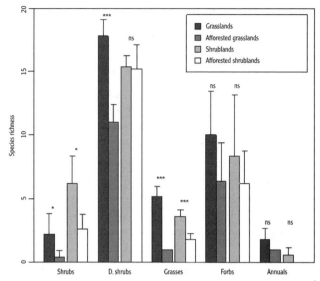

Figure 13.6. Species richness (mean and standard deviation, in 100 m²). Statistical comparisons are between natural and afforested communities. Elaborated from data in Chirino et al. (2006). ANOVA ($*p < 0.05$, $**p < 0.01$, $***p < 0.001$)

compliance with international obligations (e.g., Convention on Biodiversity, Convention to Combat Desertification, Convention on Climate Change), namely, to increase carbon fixation, to enhance biodiversity and rural development, and to reduce the fire and erosion risk. Moreover, most of the semi-arid agricultural set-aside land is privately owned. Recent European Union and national policies encourage land abandonment in less productive and marginal lands, thus providing subsidies to farmers to promote and maintain ecosystem services and natural capital in abandoned lands. Considering that the soil erosion risk on old, abandoned, terraced lands is small and shows lower rates than in other land cover types (Cerdà 1978), the restoration and conservation of biodiversity could be the main restoration management target in old field ecosystems (Bonet 2004).

The conflict between pine plantations and preexisting grassland and shrubland conservation is a common topic in Mediterranean semiarid ecosystem management (Esteve et al. 1990). It is now clear that neither plant richness nor many ecosystem functions are improved by pine afforestations and that semiarid grasslands and shrublands are the habitat for rare and protected fauna at the European scale (Tellería et al. 1988; Yanes 1994). Thus, in many cases, afforestation is unnecessary or even detrimental. Currently, efforts are directed toward developing restoration techniques for a diversity of Mediterranean shrubland species, especially resprouting species that confer high resilience to fire (Vilagrosa et al. 2003; Pausas, Bladé et al. 2004; Vallejo et al. 2006). Many of these resprouting shrubs are considered "late-successional species." However, we have shown that in many Mediterranean old fields the colonization of woody species may start very early in old woody crops, thanks both to the role of trees as perch sites for frugivorous birds (the perch effect) (Pausas et al. 2006) and to the improved microsite conditions for germination and survival (Verdú and García-Fayos 1996). Thus, the use of artificial perches (e.g., dead trees, artificial woody structures) in old fields could accelerate colonization rates (bird-mediated restoration). As attractive as this technique is for promoting succession, it has seldom been applied in Mediterranean ecosystems, and most examples come from elsewhere (e.g., McClanahan and Wolfe 1993). In this context, removal of old or dead trees after abandonment or fire (e.g., burned pine plantations) should be discouraged. Furthermore, the strong relationship between oaks and their dispersal vector (Bossena 1979) merits a deep exploration in light of Mediterranean landscapes reforestation.

Additionally, the relationship between grazing intensity and species richness (figure 13.4c) indicates that moderate grazing could promote the coexistence of different species and functional types (Lavorel et al. 1994), altering the inhibition pathway of *Brachypodium retusum* and favoring the annual species pool, and thus increasing species richness. These systems could be subjected to small-scale disturbance grazing by livestock (sheep) and wild herbivores (rabbits). Though the effect of this kind of grazing has usually been described in terms of defoliation and seeds consumption, ectozoochorous and endozoochorous dispersal would also be favored (Malo and Suárez 1995a, 1995b).

Traditional restoration plans in Mediterranean ecosystems have not considered the role of fauna. However, there is increasing evidence of the strong role of plant–animal interactions in driving succession and shaping biodiversity, and these interactions (perching, grazing, dispersal, pollination, etc.) should be encouraged for quick and cost-efficient restoration actions. Certainly, further research is needed for promoting plant–animal interactions as a restoration goal under Mediterranean conditions, but the current buildup of information suggests that there is a lot to learn in this direction.

Conclusion

Semiarid old field succession in southeast Spain is characterized by an early postabandonment peak in plant species richness. This pattern may be attributable to the high turnover rate (due to immigration and extinction processes) during the early stages of succession, which promotes the coexistence of several life-forms. Even so, the general trend toward replacing annual and biennial species with perennial forbs, grasses, and woody plants is also well apparent. This early-peak richness pattern could also be partially attributed to an early colonization of some so-called late-successional shrub species through the facilitation of bird-dispersed seeds by remnant cultivated trees acting as perches. Previous land use has been shown to be very important for understanding the variability of plant composition and richness in semiarid old fields, and the different disturbance regimes should be considered for understanding the mechanisms driving patterns through old field succession.

From our observations of old field dynamics in semiarid Spain we can suggest the following restoration priorities:

- Although soil conservation is generally assured in terraced old fields, care should be taken in some soil types where terraces are in a degradation process that promotes erosion. In such cases, soil conservation should be the priority of any restoration action.
- Ecosystem resilience should be improved with respect to human and nonhuman disturbances, that is, to ensure the regeneration capacity after fire and other disturbances (Vallejo et al. 2006).
- Biodiversity should be encouraged by (a) promoting the reintroduction of key species that have disappeared because of past land uses, and (b) promoting animal–plant interactions for quick and cost-efficient restoration actions.

REFERENCES

Alados, C. L., A. Elaich, V. P. Papanastasis, H. Ozbek, T. Navarro, H. Freitas, M. Vrahnakis, D. Larrossi, and B. Cabezudo. 2004. Change in plant spatial patterns and diversity along the successional gradient of Mediterranean grazing ecosystems. *Ecological Modelling* 180:523–35.

Alados, C. L., Y. Pueyo, O. Barrantes, J. Escos, L. Giner, and A. B. Robles. 2004. Variations in landscape patterns and vegetation cover between 1957 and 1994 in a semiarid Mediterranean ecosystem. *Landscape Ecology* 19:543–59.

Alados, C. L., Y. Pueyo, D. Navas, B. Cabezudo, A. Gonzalez, and D. C. Freeman. 2005. Fractal analysis of plant spatial patterns: A monitoring tool for vegetation transition shifts. *Biodiversity and Conservation* 14:1453–68.

Alcántara, J. M., P. J. Rey, F. Valera, A. M. Sánchez-Lafuente. 2000. Factors shaping the seedfall pattern of a bird-dispersed plant. *Ecology* 81:1937–50.

Antrop, M. 2004. Landscape change and the urbanization process in Europe. *Landscape and Urban Planning* 67:9–26.

Badal, E., J. Bernabeu, and J. L. Vernet. 1994. Vegetation changes and human action from the Neolithic to the Bronze Age (7000–4000 B.P.) in Alicante, Spain, based on charcoal analysis. *Vegetation History and Archaeobotany* 3:155–66.

Bellot J., A. Bonet, J. R. Sanchez, and E. Chirino. 2001. Likely effects of land use changes on the runoff and aquifer recharge in a semiarid landscape using a hydrological model. *Landscape and Urban Planning* 778:1–13.

Bertness, M. D., and R. M. Callaway. 1994. Positive interactions in communities. *Trends in Ecology and Evolution* 9:191–93.

Bolòs, O. 1967. Comunidades vegetales de las comarcas próximas al litoral situadas entre los ríos Llobregat y Segura. *Memorias de la Real Academia de Ciencias y Artes de Barcelona* 38(1).

Bonet, A. 2004. Secondary succession on semi-arid Mediterranean old-fields in southeastern Spain: Insights for conservation and restoration of degraded lands. *Journal of Arid Environments* 56:213–33.

Bonet, A., J. Bellot, D. Eisenhuth, J. Peña, J. R. Sánchez, and C. J. Tejada. 2006. Some evidence of landscape change, water usage and management system co-dynamics in south-eastern Spain. In *Water management in arid and semi-arid regions: Interdisciplinary perspectives*, ed. P. Koundouri, K. Karousakis, D. Assimacopoulos, P. Jeffrey, and M. A. Lange, 226–51. Edward Elgar, Aldershot, UK.

Bonet A., J. Bellot, and J. Peña. 2004. Landscape dynamics in a semiarid Mediterranean catchment (SE Spain). In *Recent dynamics of Mediterranean vegetation and landscape*, ed. S. Mazzoleni, G. di Pasquale, M. Mulligan, P. di Martino, and F. Rego, 47–56. John Wiley and Sons, London, U.K.

Bonet, A., and J. G. Pausas. 2004. Species richness and cover along a 60-year chronosequence in old-fields of southeastern Spain. *Plant Ecology* 174:257–70.

Bonet A., J. Peña, J. Bellot, M. Cremades, and J. R. Sánchez. 2001. Changing vegetation structure and landscape patterns in semi-arid Spain. In *Ecosystems and sustainable development* 3, ed. Y. Villacampa Esteve, C. A. Brebbia, and J.-L. Uso, 377–86. Wit Press, Boston.

Bossena, I. 1979. Jays and oaks: An eco-ethological study of a symbiosis. *Behaviour* 70:1–117.

Calamassi, R., M. F. Gianni Della Rocca, M. Falusi, E. Paoletti, and S. Strati. 2001. Resistance to water stress in seedlings of eight European provenances of *Pinus halepensis* Mill. *Annals of Forest Science* 58:663–72.

Ceballos, L. 1938. *Plan general para la restauración forestal de España*. ICONA, Madrid.

Cerdà, A. 1978. Soil erosion after land abandonment in a semiarid environment of southeastern Spain. *Arid Soil Research and Rehabilitation* 11:163–76.

Chirino, E., A. Bonet, J. Bellot, and J. R. Sánchez. 2006. Effects of 30-year-old Aleppo pine plantations on runoff, soil erosion, and plant diversity in a semi-arid landscape in south eastern Spain. *Catena* 65:19–29.

Coffin, D. P., W. A. Laycock, and W. K. Lauenroth. 1998. Disturbance intensity and above- and belowground herbivory effects on long-term (14 y) recovery of a semiarid grassland. *Plant Ecology* 139:221–33.

Connell, J. H., and R. O. Slatyer. 1977. Mechanisms of succession in natural communities and their role in community stability and organization. *American Naturalist* 111:1119–44.

Cortina, J., J. Bellot, A. Vilagrosa, R. Caturla, F. T. Maestre, E. Rubio, J. M. Martínez and A. Bonet. 2004. Restauración en semiárido. In *Avances en el Estudio de la Gestión del Monte Mediterráneo*, ed. R. Vallejo and J. A. Alloza, 345–406. Fundación CEAM, Valencia.

Debussche, M., G. Debussche, and J. Lepart. 2001. Changes in the vegetation of *Quercus pubescens* woodland after cessation of coppicing and grazing. *Journal of Vegetation Science* 12:81–92.

Debussche, M., J. Escarré, and J. Lepart. 1982. Ornitochory and plant succession in Mediterranean orchards. *Vegetatio* 48:255–66.

Debussche, M., J. Escarré, J. Lepart, C. Houssard, and S. Lavorel. 1996. Changes in Mediterranean plant succession: Old-fields revisited. *Journal of Vegetation Science* 7:519–26.

Debussche, M., and P. Isenmann. 1994. Bird-dispersed seed rain and seedling establishment in patchy Mediterranean vegetation. *Oikos* 69:414–26.

Debussche, M., and J. Lepart. 1992. Establishment of woody plants in Mediterranean old fields: Opportunity in space and time. *Landscape Ecology* 6:133–45.

DeGryze, S., J. Six, K. Paustian, S. J. Morris, E. A. Paul, and R. Merckx. 2004. Soil organic carbon pool changes following land-use conversions. *Global Change Biology* 10:1120–32.

Escarré, J., C. Houssard, M. Debussche, and J. Lepart. 1983. Évolution de la végétation et du sol aprés abandon cultural en région méditerranéenne: Étude de successions dans les garrigues du Montpellierais (France). *Acta Oecologica* 4:221–39.

Escudero, A., R. C. Somolinos, J. M. Olano, and A. Rubio. 1999. Factors controlling the establishment of *Helianthemum squamatum*, an endemic gypsophile of semi-arid Spain. *Journal of Ecology* 87:290–302.

Esteve, M. A., D. Ferrer, L. Ramírez-Díaz, J. F. Calvo, M. L. Suárez-Alonso, and M. R. Vidal-Abarca. 1990. Restauración de la vegetación en ecosistemas áridos y semiáridos: Algunas reflexiones ecológicas. *Ecología. Fuera de Serie* 1:497–510.

Foster, B. L., and D. Tilman. 2000. Dynamic and static views of succession: Testing the descriptive power of the chronosequence approach. *Plant Ecology* 146:1–10.

Foster, D., G. Motzkin, and B. Slater. 1998. Land-use history as long-term broad-scale disturbance: Regional forest dynamics in central New England. *Ecosystems* 1:96–119.

Gill, D. S., and P. L. Marks. 1991. Tree and shrub seedling colonization of old fields in central New York. *Ecological Monographs* 61:183–205.

Grau H. R., T. M. Aide, J. K. Zimmerman, J. R. Thomlinson, E. Helmer, and X. Zou. 2003. The ecological consequences of socioeconomic and land use changes in post agriculture Puerto Rico. *BioScience* 53:1159–68.

Gulias, J., A. Traveset, N. Riera, M. Mus. 2004. Critical stages in the recruitment process of *Rhamnus alaternus* L. *Annals of Botany* 93:723–31.

Herrera, C. M., P. Jordano, L. López-Soria, and J. A. Amat. 1994. Recruitment of a mast-fruiting, bird-dispersed tree: Bridging frugivore activity and seedling establishment. *Ecological Monographs* 64:315–44.

Houssard, C. J., J. Escarre, and F. Romane. 1980. Development of species diversity in some Mediterranean plant communities. *Vegetatio* 43:59–72.

Huston, M. A. 1979. A general hypothesis of species diversity. *American Naturalist* 113:81–101.

Izhaki, I., P. B. Walton, U. N. Safriel. 1991. Seed shadows generated by frugivorous birds in an eastern Mediterranean scrub. *Journal of Ecology* 79:575–90.

Jake, W. F., and M. B. Coughenour. 1990. Savanna tree influence on understorey vegetation and soil nutrients in north-western Kenya. *Journal of Vegetation Science* 1:325–34.

Lavorel, S., J. Lepart, M. Debussche, J. D. Lebreton, and J. L. Beffy. 1994. Small scale disturbances and the maintenance of species diversity in Mediterranean old fields. *Oikos* 70:455–73.

Le Houérou, H. N. 1993. Land degradation in Mediterranean Europe: Can agroforestry be a part of the solution? A prospective review. *Agroforestry Systems* 21:43–61.

Lepart, J., and M. Debussche. 1992. Human impact on landscape patterning: Mediterranean examples. In *Landscape boundaries: Consequences for biotic diversity and ecological flows*, ed. A. J. Handsen and F. di Castri, 76–106. Springer, New York.

Maestre, F. T., and J. Cortina. 2004. Are *Pinus halepensis* plantations useful as a restoration tool in semiarid Mediterranean areas? *Forest Ecology and Management* 198:303–17.

Maestre, F. T., J. Cortina, S. Bautista, and J. Bellot. 2003. Does *Pinus halepensis* facilitate the establishment of shrubs in Mediterranean semi-arid afforestations? *Forest Ecology and Management* 176:147–60.

Malo, J. E., and F. Suárez. 1995a. Herbivorous mammals as seed dispersers in a Mediterranean dehesa. *Oecologia* 104:246–55.

———. 1995b. Establishment of pasture species on cattle dung: The role of endozoochorous seeds. *Journal of Vegetation Science* 6:169–74.

Mazzoleni, S., G. di Pasquale, M. Mulligan, P. di Martino, and F. Rego, eds. 2004. *Recent dynamics of Mediterranean vegetation and landscape*. John Wiley and Sons, London, UK.

McClanahan, T. R., and R. W. Wolfe. 1993. Accelerating forest succession in a fragmented landscape: The role of birds and perches. *Conservation Biology* 7:279–88.

McDonnell, M. J., and E. W. Stiles. 1983. The structural complexity of old field vegetation and the recruitment of bird-dispersed plant species. *Oecologia* 56:109–16.

McIntyre, S., S. Lavorel, and R. M. Tremont. 1995. Plant life-history attributes: Their relationship to disturbance response in herbaceous vegetation. *Journal of Ecology* 83:31–44.

Moro, M. J., F. I. Pugnaire, P. Haase, and J. Puigdefábregas. 1997. Effect of the canopy of *Retama sphaerocarpa* on its understorey in a semiarid environment. *Functional Ecology* 11:425–31.

Noy-Meir, I. 1998. Effects of grazing on Mediterranean grasslands: The community level. In *Ecological basis of livestock grazing in Mediterranean ecosystems. Proceedings of the International Workshop held in Thessaloniki (Greece) on October 23–25, 1997*, ed. V. P. Papanastasis and D. Peter, 27–39. European Commission for Science, Research and Development. Brussels, Belgium.

Otto, R., B. O. Krüsi, C. A. Burga, and J. M. Fernández-Palacios. 2006. Old-field succession along a precipitation gradient in the semi-arid coastal region of Tenerife. *Journal of Arid Environments* 65:156–78.

Pausas, J. G. 2004. Changes in fire and climate in the eastern Iberian Peninsula (Mediterranean basin). *Climatic Change* 63:337–50.

Pausas, J. G., C. Bladé, A. Valdecantos, J. P. Seva, D. Fuentes, J. A. Alloza, A. Vilagrosa, S. Bautista, J. Cortina, and R. Vallejo. 2004. Pines and oaks in the restoration of Mediterranean landscapes in Spain: New perspectives for an old practice—a review. *Plant Ecology* 171:209–20.

Pausas, J. G., A. Bonet, F. T. Maestre, and A. Climent. 2006. The role of the perch effect on the nucleation process in Mediterranean semi-arid oldfields. *Acta Oecologica* 29:346–52.

Pausas, J. G., E. Ribeiro, and R. Vallejo. 2004. Post-fire regeneration variability of *Pinus halepensis* in the eastern Iberian Peninsula. *Forest Ecology and Management* 203:251–59.

Pickett, S. T. A. 1989. Space-for-time substitution as an alternative to long-term studies. In *Long-term studies in ecology: Approaches and alternative*, ed. G. E. Likens, 110–35. Springer-Verlag, New York.

Pugnaire, F. I., M. T. Luque, C. Armas, and L. Gutiérrez. 2006. Colonization processes in semi-arid Mediterranean old-fields. *Journal of Arid Environments* 65:591–603.

Puigdefábregas, J., and T. Mendizabal. 1998. Perspectives on desertification: Western Mediterranean. *Journal of Arid Environments* 39:209–24.

Redjali, M. 2004. Forest cover changes in the Maghreb countries with special reference to Morocco. In *Recent dynamics of Mediterranean vegetation and landscape*, ed. S. Mazzoleni, G. di Pasquale, M. Mulligan, P. di Martino, and F. Rego, 23–32. John Wiley and Sons, London, U.K.

Reynolds, J. F., R. A. Virginia, P. R. Kemp, A. G. De Soyza, and D. C. Tremmel. 1999. Impact of drought on desert shrubs: Effects of seasonality and degree of resource island development. *Ecological Monographs* 69:69–106.

Rivas-Martínez, S. 1987. *Memoria del mapa de series de vegetación de España*. ICONA, Madrid.

Ruiz de la Torre, J. 1990. *Mapa Forestal de España*. Memoria General. Ministerio de Agricultura y Pesca. ICONA, Madrid.

Taiqui, L., 1997. La dégradation ecologique au Rif marocain: Nécessités d'une nouvelle approche. *Mediterránea. Serie de estudios biológicos* 16:5–17.

Tatoni, T., and P. Roche. 1994. Comparison of old-field and forest revegetation dynamics in Provence. *Journal of Vegetation Science* 5:295–302.

Tellería, J. L., F. Suárez, and T. Santos. 1988. Bird communities of the Iberian shrub steppes. *Holartic Ecology* 11:171–77.

Traveset, A., J. Gulías, N. Riera, M. Mus. 2003. Transition probabilities from pollination to establishment in a rare shrub species (*Rhamnus ludovici-salvatoris*) in two habitats. *Journal of Ecology* 91:427–37.

Valle, F., and I. Bocio. 1996. Restauración de la vegetación en el sureste de la Península Ibérica. *Cuadernos de la SECF* 3:109–22.

Vallejo, R., J. Aronson, J. G. Pausas, and J. Cortina. 2006. Mediterranean woodlands. In *Restoration ecology: The new frontier*, ed. J. van Andel and J. Aronson, 193–207. Blackwell Science. Oxford, UK.

Verdú, M., and P. García-Fayos. 1996. Nucleation processes in a Mediterranean bird-dispersed plant. *Functional Ecology* 10:275–80.

———. 1998. Old-field colonization by *Daphne gnidium*: Seedling distribution and spatial dependence at different scales. *Journal of Vegetation Science* 9:713–18.

Vilagrosa, A., J. Cortina, E. Gil-Pelegrín, and J. Bellot. 2003. Suitability of drought-preconditioning techniques in Mediterranean climate. *Restoration Ecology* 11:208–16.

Vilagrosa, A., J. P. Seva, A. Valdecantos, J. Cortina, J. A. Alloza, I. Serrasolsas, V. Diego, M. Abril, A. Ferran, J. Bellot et al. 1997. Plantaciones para la restauración forestal en la Comunidad Valenciana. In *Avances en el Estudio de la Gestión del Monte Mediterráneo*, ed. R. Vallejo and J. A. Alloza, 435–546. Fundación CEAM, Valencia.

Yanes, M. 1994. The importance of land management in the conservation of birds associated with the Spanish steppes. In *Nature conservation and pastoralism in Europe*, ed. E. M. Bignal, D. I. McCracken, and D. J. Curtis, 34–40. Joint Nature Conservation Committee, Peterborough, UK.

Yarranton, G. A., and R. G. Morrison. 1974. Spatial dynamics of a primary succession: Nucleation. *Journal of Ecology* 62:417–28.

Chapter 14

Restoration of Old Fields in Renosterveld: A Case Study in a Mediterranean-type Shrubland of South Africa

Cornelia B. Krug and Rainer M. Krug

Renosterveld is part of the Cape floristic region, the smallest and most species rich of the world's floristic regions and regarded as a biodiversity hotspot (Myers et al. 2000). It can be classified as a Mediterranean-climate shrubland (sensu Di Castri 1981), being associated with the relatively fertile shale and granite soils of the Cape lowlands (Boucher and Moll 1981; Low and Rebelo 1996) and occurring under a winter or year-round rainfall regime of 250 mm to 600 mm on the southwestern tip of the African continent. The vegetation is dominated by low sclerophyllous shrubs, mostly belonging to the Asteraceae and Fabaceae, with an understory of geophytic species from a range of families (e.g., Hyacinthaceae, Iridaceae, and Orchidaceae), graminoids (Poaceae and Cyperaceae), and some Restionaceae (Moll et al. 1984; Low and Jones 1995; Low and Rebelo 1996). About one-third of the plant species occurring in renosterveld are endemic to the Cape floristic region, although few of these are endemic to renosterveld alone (Low and Rebelo 1996). The geometric tortoise (*Psammobates geometricus*), one of the rarest tortoise species in the world, only occurs in west coast renosterveld, and is severely threatened by habitat transformation and fragmentation (Baard 1993). In the past, renosterveld supported large herds of game, like mountain zebra, quagga, blue antelope (a sister species of roan and sable), red hartebeest, eland, and bontebok (Low and Rebelo 1996). Blue antelope and bontebok are renosterveld endemics, whereas the quagga also occurred in the Karoo, a semiarid grassy shrubland (Skinner and Smithers 1990). Large species like elephant, black rhinoceros, and buffalo were also common, as were large carnivores (lion, cheetah, wild dog, spotted hyena, and leopard) (Low and Rebelo 1996). Underground termitaria (termite mounds) are an interesting feature of renosterveld. These were formed about 30,000 years ago (during the

Pleistocene) when the climate was cooler and wetter than today (Midgley et al. 2002). The vegetation on those nutrient-enriched patches is related to thicket vegetation and dominated by fleshy-fruited trees and shrubs (e.g., *Olea* sp., *Rhus* sp., and *Euclea* sp.), and maintained by the deposition of bird-dispersed seeds.

Today, renosterveld has been transformed into agricultural and urban areas, the remaining natural habitat is highly fragmented, and most of the large animal species are extinct in the region. Of the original extent of west coast (Boland/Swartland) renosterveld, once covering 66,8026 ha, only 42,726 ha (6.21%) remains as natural vegetation, in 1,175 remnants (von Hase et al. 2003). More than two-thirds of these remnants are less than 5 ha in size, less than 5% are fragments larger than 100 ha, and only seven habitat remnants are larger than 1,000 ha (von Hase et al. 2003) (figure 14.1). The vegetation within the remnants itself is transformed and mostly dominated by Renosterbos, *Dicerothamnus rhinocerotis*, an asteraceous shrub.

As most of the remnants are too small and/or too isolated to carry viable plant and animal populations, it becomes essential to consider options for the restoration of this vegetation type to ensure its survival. Successful restoration, however, depends on an understanding of the ecological processes shaping an ecosystem and knowledge of past and present land uses. We cannot base our understanding solely on the processes in existing fragments, but we have to include information on historic and prehistoric changes of the vegetation, why and when they occurred, and how they can be related to anthropogenic influence.

Prehistoric and Historic Land Use

Similar to other Mediterranean-climate ecosystems, renosterveld has been influenced by human activity over thousands of years, although the impacts were not as severe as in the Mediterranean Basin (Newton 2007). Here we consider three phases in which considerable changes due to human activity took place: Khoi herders (2,000 YBP until the 1700s), European settlement (1652–early 1900s), and modern agriculture (early 1900s until today).

Prehistory: Khoi-San Hunter-Gatherers, Indigenous Large Herbivores, and Fire

From around 10,000 YBP, the climate in the western and southwestern Cape became more similar to today's climate (Newton 2007), favoring a vegetation dominated by shrubs. The presence of shrub species was likely controlled by

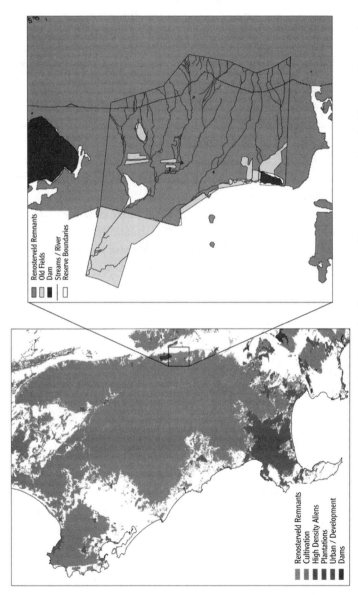

Figure 14.1. Remaining west coast (Boland/Swartland) renosterveld remnants as identified by von Hase et al. (2003), and location of old fields in the natural vegetation remnant on Elandsberg private nature reserve. Data for transformed habitats from Cape Conservation Unit (2003); geographic information for Elandsberg from CapeNature (2004).

a range of factors, such as water and nutrient availability, soil type and fertility, palatability of the species, and grazing pressure. Composition of the large herbivore fauna, as described by Skead (1980), Klein (1974), and du Plessis (1969), suggests a mixed shrub–grass vegetation. Most species were either mixed feeders like eland and red hartebeest, or pure browsers (e.g., black rhinoceros), with mountain zebra being the only pure grazer occurring in the area (Skinner and Smithers 1990). Shrub patches were likely interspersed with more or less extensive grassy areas, consisting mainly of bunch grasses like *Ehrharta, Pentaschistis, Merxmuellera, Tribolium, Cymbopogon*, and *Eragrostis* species, and locally abundant *Themeda triandra* and *Cymbopogon marginatus* (Low and Rebelo 1996). Geophytes formed an abundant understory beneath shrubs and among the grasses.

Anthropogenic influence at that time was relatively low, as only nomadic Khoi-San hunter-gatherers inhabited the region. Geophytes formed an important part of their diet (Deacon 1992; Parkington 1977) and the Khoi used small-scale burns to increase the growth and abundance of these plants, but also to attract and snare antelope (Deacon 1992). In addition to these small-scale, anthropogenic fires, larger-scale fires occurred naturally in renosterveld. These fires most likely originated in the fire-prone fynbos vegetation bordering on renosterveld. Burning favors grasses (Cowling et al. 1986), and some geophyte species germinate only after fire (Manning et al. 2002). This attracts game species to freshly burned areas (Luyt 2005; Nicola Farley, pers.comm.).

Prehistory: Khoi Herders, the Introduction of Livestock, and Fire Management

At around 2,000 YPB, the lifestyle of the Khoi-San changed from hunter-gatherer to herder of domestic livestock, first with goats and sheep, and later with cattle (Schweitzer and Scott 1973; Schweitzer 1979; Klein 1986). This lifestyle change impacted directly on the vegetation, as the Khoi used fire to promote the growth of grasses for grazing their livestock (Thom 1952). After a period of intense browsing and grazing, the Khoi burned the vegetation and moved on, returning after one to four years to complete a cycle (Thom 1952, 1954). Through short burning intervals and short but intense grazing pressure, shrub cover was reduced, leading to a substantial increase in grass cover.

The presence of the indigenous, large, herbivore fauna was not impacted by these practices, as hunting pressure by the Khoi was rather low (Klein 1974), and game species profited from the freshly burned areas. The activities of the Khoi (burning and intensive grazing) coupled with the activity of the large herbivores (browsing and grazing) led to large tracts of grassland (Thom 1952), though parts of renosterveld remained relatively undisturbed.

European Settlement: Introduction of Cropping and Habitat Transformation

In 1652, Jan van Riebeck established a provisioning station for the Dutch East India Company in Table Bay, Cape Town, with fruit and vegetable gardens as well as livestock compounds. Twenty years later, the Cape was purchased from the Khoi (Newton 2007), and the expansion inland began a few years later, setting the scene for the transformation of renosterveld into agricultural lands for crop and livestock farming. Cereal crops and vines were planted on the nutrient-rich renosterveld soils to the north and east, cattle were grazed on natural vegetation and fallow fields. One hundred and fifty years after van Riebeck's arrival, grapes were the main crop of the Boland (Paarl/Stellenbosch/Somerset West areas), while cereals, mostly wheat, were planted extensively in the Swartland (Wellington/Malmsbury areas) (Newton 2007). By this time, the land was already severely degraded, and erosion of the topsoil was common (Newton 2007), as even steep slopes were plowed and planted.

The settlers farmed mainly cattle and sheep traded from the Khoi (Leibbrandt 1902, 1901; Thom 1958, 1954, 1952) or brought in from Europe. In contrast to the nomadic Khoi, these livestock herds were not moved, but remained on the same pastures, usually on less fertile, poorer soils of the farms, throughout the year. To encourage the growth of grasses, the vegetation was burnt, and immediately grazed again. However, veld burning was prohibited by law as early as 1687 to prevent excessive burning and runaway fires and to protect crops and livestock (Newton 2007); thus, naturally occurring fires were suppressed.

The Europeans also hunted the abundant game birds and large game mammals of the renosterveld, often on a large scale, for domestic consumption (Thom 1958). The large herds of game were quickly diminished, and two species became extinct: bluebock (*Hippotragus leucophaeus*), around 1800 (Skinner and Smithers 1990), and quagga (*Equus quagga*), by 1875 (Skinner and Smithers 1990). The other indigenous large herbivores disappeared from the Western Cape before the turn of the twentieth century.

Modern Agriculture: Transformation and Fragmentation

The discovery of gold and diamonds in the north of the country at the beginning of the twentieth century, coupled with the development of an efficient transport system, led to the mechanization and intensification of agriculture in the fertile regions of the Western Cape, mostly within renosterveld (Talbot 1947). This led to the extreme fragmentation present today. Larger areas were planted, including marginal lands, and rest periods were shortened, leading to severe soil exposure, compaction, and erosion by the 1950s (Newton 2007). Only thirty years later was farming regulated to control erosion. Plowing of the marginal areas and steep slopes was abandoned, but no special care was taken with small remnants, and they tended to received the same treatment as the fields; for example, they were sprayed and fertilized (Donaldson et al. 2002). This had negative effects on species composition and diversity in the remnants and encouraged the invasion by alien species, especially grasses (Musil et al. 2005).

Change in Ecological Processes

We have used a state and transition model (figure 14.2) to illustrate the pattern and dynamics of renosterveld, and how this has changed with human impact. We consider renosterveld as a mosaic of grass and shrub patches, with herbivory and fire as the main processes driving the shift between states. We do not include thicket patches in the model. The changes occurring during European settlement, through the removal of large herbivores and the suppression of fire, had a considerable impact on ecosystem processes in renos-

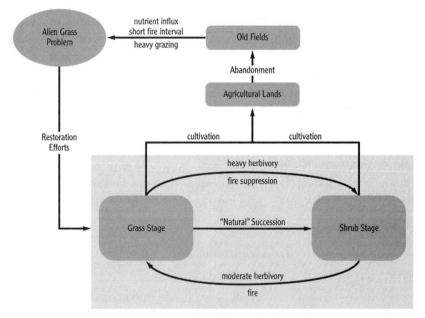

Natural Vegetation Dynamics

FIGURE 14.2. Conceptual state and transition model for renosterveld vegetation dynamics.

terveld and led to the dominance of a single species, *Dicerothamnus rhinocerotis*, in many west coast renosterveld remnants.

The role of large herbivores in renosterveld is threefold: they disperse seeds of grasses and some geophytes over long distances (Shiponeni 2003; Shiponeni and Milton 2006); create gaps in shrubby patches through browsing and trampling, thus enabling germination and establishment of geophytes and grasses; and reduce grass cover through grazing, facilitating shrub establishment (Midoko-Iponga 2004; Midoko-Iponga et al. 2005). The removal of large herbivores restricted the long-distance spread of dung-dispersed species. Without large browsers, fewer gaps were created in the shrub patches, restricting the establishment of shade-intolerant species and favoring longer-lived shrub species (Boucher 1983). Overgrazing of the natural vegetation by cattle, coupled with the suppression of fires, resulted in monospecific stands of renosterbos (*Dicerothamnus rhinocerotis*), a long-lived, wind-dispersed, unpalatable asteraceous shrub.

Fire reduces shrub cover, leads to the establishment of grasses (Cowling et al. 1986), and encourages the germination of geophytes. Some geophyte species occurring in renosterveld, for example, Amaryllids, are only able to germinate after fire (Manning et al. 2002). Increased fire occurrence and frequency led to the development of vast grasslands in the Western Cape. Remaining shrubby patches allowed for the regeneration of longer-lived shrub species. Heavy grazing by livestock immediately after burning led to an increase in unpalatable species, especially *Dicerothamnus rhinocerotis* (Levyns 1956), a reduction of the grass component, and limited germination of geophyte species dependent on fire.

Fragmentation also affected ecological processes in the renosterveld remnants. Many very small fragments were (and are still) treated as part of the surrounding fields, and were sprayed with fertilizers and pesticides (Donaldson et al. 2002), and burnt with the harvested wheat fields. In addition, many of these small fragments were used for grazing livestock and were often oversown with legumes or grasses to increase the fodder value. Nutrient enrichment, especially of nitrogen and potassium, favors introduced grasses over indigenous species (Orlander et al. 1996). Overgrazing and frequent burning increases the proportion of introduced grass species occurring in a remnant and exacerbates the invasion problem (Van Rooyen 2004). To a lesser extent, woody, introduced species are also a threat to vegetation remnants. Rouget et al. (2003) estimate that about 5% of west coast renosterveld remnants are invaded to a level of 75% cover or more.

Survival of populations of indigenous plant species within the vegetation remnants is also threatened by the limited reproductive potential of many indigenous species. The movement of pollinators across a fragmented landscape may be restricted, and some taxa that are important pollinators in renosterveld (e.g., monkey beetles) are sensitive to changes in habitat characteristics (Donaldson et al. 2002). The abundance of pollinators may be affected by fragment size (Donaldson et al. 2002). Some of the orchid species occurring in renosterveld have been documented to be pollinator-limited (Donaldson et al. 2002) as their pollinators, such as oil-collecting bees, have very specific habitat requirements (Pauw 2004). However, Kemper et al. (1999) have shown that pollinator resources (i.e., geophyte density) do not decline with decreasing fragment size, and a diversity of insect pollinators can be maintained (Donaldson et al. 2002). Species composition changes with fragment size (Donaldson et al. 2002) or degree of isolation (Vrdoljak and Samways 2005), and fragment size and isolation of a remnant influences seed or fruit set (Donaldson et al. 2002). Transformation also has an effect on seed dispersal (Kemper et al. 1999), and species with short dispersal ranges are not able to colonize distant fragments (Shiponeni 2003).

Incentives for Restoration

Due to the highly transformed state of renosterveld, all remaining remnants have been declared 100% irreplaceable in relation to conservation targets (Cowling and Heijnis 2001; von Hase et al. 2003). In theory, this should prevent any further transformation of the vegetation for agricultural or development purposes. However, to successfully implement the fine-scale conservation plans for the Cape floristic region, and to maintain ecological processes across the landscape, the restoration of abandoned fields and degraded habitat remnants is required. Restoration of old fields can contribute to increasing the area under conservation, providing habitat for rare or endangered species (e.g., the geometric tortoise), creating corridors or stepping stones linking fragments and buffer zones between transformed and natural areas. Restored sites can also play an important role by serving as propagule sources for future restoration or natural expansion of habitat (Musil et al. 2005).

Wheat cropping was heavily subsidized during the twentieth century in South Africa (Newton 2007). With the removal of these subsidies (Donaldson 2002), an increase in labor costs due to improved labor laws in post-apartheid South Africa, and taxation of all land, the feasibility of dryland agriculture is more difficult. Dryland cropping of cereals is being abandoned, and farmers are searching for alternative crops, for example, olives, wine grapes, or canola (Fairbanks 2004). However, increased interest in the conservation of native vegetation (Winter 2003; Winter and Esler 2005), and the provision of financial incentives to set aside land for conservation (Botha

2001), have led to a search for alternative use and income strategies from natural renosterveld vegetation, based on activities such as game farming or ecotourism. For example, CapeNature, the provincial nature conservation agency, has created the Stewardship Program, where land is set aside for conservation under contract. The Biodiversity and Wine Initiative, supported by South African Wine and Brandy, markets products whose production is coupled with the conservation of biodiversity. These programs assist landowners to conserve natural vegetation, and farmers contribute in-kind by clearing aliens, restoring abandoned land, and conducting erosion control.

As the incentives for restoration differ, so do the purposes of the sites to be restored. At the onset of restoration, a goal needs to be established on what is to be achieved with the restoration of an abandoned field, and what the land will be used for. Is it to increase the area of natural habitat, for example, for game farming; to increase the proportion of suitable habitat for a species; or to return bulb species to an area where they occurred before, for example, to attract tourists to visit a site? Or will the restored sites act as corridors or stepping stones, linking larger fragments and thus allowing for dispersal and gene flow between populations? Each of these goals needs a different approach. As renosterveld is a shrubland with a bulb and grass component, the primary aim for the restoration of a site should be to establish a mix of shrubs, grasses, and bulb species. When the site is to be used for game farming, the grass and shrub component should be rather high, to provide graze and browse for the antelope species. Bulb species provide an important food source for grazers. On the other hand, when bulb species are the main focus of restoration, their proportion needs to be larger. Most bulb and grass species need open areas between shrubs, so disturbances, for example, small-scale fires, are necessary to create open patches within the vegetation.

Old Field Restoration in Renosterveld: Underlying Vegetation Dynamics and Ecological Processes

The focus of the Renosterveld Restoration Project were abandoned agricultural fields at the Elandsberg Private Nature Reserve on the Farm Bartholomeus Klip (19°03' E, 33°27' S), near Hermon in the Western Cape, South Africa (figure 14.1). About half of the property (approx. 3,500 ha) is a working sheep and wheat farm, while the other half was proclaimed a private nature reserve in 1973 to protect the geometric tortoise, *Psammobates geometricus*. Since then, a number of indigenous game species have been reintroduced. The vegetation in the low-lying areas of the reserve is classified as Alluvium Sand Fynbos and Swartland Shale Renosterveld (Mucina and

Rutherford 2004), with Mountain Fynbos on the mountain slopes. The lowland portion of the reserve is about 1,000 ha in size. Elandsberg is part of the largest remaining renosterveld vegetation complex in the Western Cape (von Hase et al. 2003; Newton 2006).

A number of abandoned agricultural fields of different ages are incorporated into the reserve. All of these fields were plowed and planted with oats (*Avena sativa*) in the 1960s and 1970s. The fields were classified into different successional stages according to age class: last plowed before 1997 (five years of recovery), before 1987 (fifteen years), and before 1967 (thirty-five years). The youngest abandoned fields, which were oversown with European grasses after oat planting was discontinued, serve as grazing grounds for the reintroduced game species and act as buffers between the natural vegetation and the agricultural fields (figure 14.1).

The old field used for experimental studies was abandoned and incorporated into the reserve in 1987. Until 1985, oats were the main crop on the field; the field was oversown with European grasses (*Briza* sp., *Lolium* sp., and *Vulpia* sp.) and used for sheep grazing (Mike Gregor, pers. comm.). Although the field is directly adjacent to an extensive patch of natural vegetation, introduced grasses still dominate the vegetation. Little indigenous vegetation is present, with the exception of *Dicerothamnus rhinocerotis*, *Athanasia* sp., *Relhania* sp., and *Asphalatus* sp., which have established in the drainage lines closest to the edge of the field that borders the natural vegetation. Some geophytes, mostly weedy *Oxalis* species that are adapted to disturbances, have also established on the old field. Game species regularly grazing the sites are bontebok, burchell's zebra, black wildebeest, eland, and springbok.

Underlying Vegetation Dynamics and Processes

A study on secondary succession in old fields at Elandsberg has shown that although the structure of vegetation within the remnants and the oldest old fields is similar, species richness and species diversity are lower in the old fields (C. B. Krug et al. 2004; Walton 2005). This is due to there being fewer understory species (forbs, geophytes, and grasses) in the old fields. While structural recovery appears complete, a number of species either are not able to recolonize after disturbance or have been lost from the seed bank.

The recolonization ability of a species depends on its seed dispersal properties and the successful establishment of its seedlings, which is determined by the competitive ability of the species and resource availability (light, nutrients, moisture) at the site. In a fragmented landscape, seed dispersal is limited (Bakker and Berendse 1999), and the chance of seeds being dispersed to abandoned sites is very low. The main seed dispersal agents in renosterveld are wind and large herbivores, with about 41% of the species reaching the old fields dispersed by wind, and the same proportion dispersed in the dung of game species. The remaining species reached the site through other methods, mostly through tumbling (Shiponeni 2003; R. M. Krug et al. 2004; Shiponeni and Milton 2006). The seed density of indigenous species was closely related to distance from natural vegetation (Shiponeni 2003; Krug 2007), indicating that very few seeds of most species are dispersed over large distances. Seed rain was dominated by introduced grasses, and only two indigenous species, the shrub *Dicerothamnus rhinocerotis* and the grass *Tribolium hispidum*, contributed significantly to the seed rain (Shiponeni 2003). No seeds of geophyte species present in the natural vegetation were found in the old field. Corms of geophytes are dispersed by porcupines; however, these distances are usually rather short. Large herbivores transport mostly seeds of grasses and annuals. These were, however, mainly species occurring on the old fields, and introduced grass species dominated the samples (Shiponeni 2003; Krug 2007). Although the large herbivores dispersed a large number of seeds and species, they contributed very little to movement of species from the natural vegetation to the old field. The results indicate that the recolonization of old fields by native species in renosterveld is partly dispersal limited.

Even though the seeds of indigenous species are dispersed, very few seedlings are observed in the old fields. Germination of seeds is inhibited by the existing vegetation (Thompson et al. 1977; Grace 1999). Grasses, particularly introduced grasses, are known to interfere with the survival and growth of woody species and forbs (Davies 1985). Herbivory also influences species establishment, either through seedling removal or by trampling, and the maintenance of grazing lawns by game species inhibits the recolonization of indigenous shrubs. Competition and herbivory are known to negatively affect species establishment (Bonser and Reader 1995). In recent experiments, grass competition had a greater effect on seedling growth and survival than herbivory (Midoko-Iponga 2004; Midoko-Iponga et al. 2005). The negative influence of competition from alien grasses on the establishment of indigenous shrub species, combined with limited seed dispersal, impacts severely on the capacity of indigenous renosterveld species to recolonize abandoned agricultural areas.

Implications for Successful Restoration Interventions

To facilitate the return of indigenous species to these old fields, introduced grass species must be removed and seeds of indigenous species broadcast on the cleared sites. Ideally, large herbivores (game and livestock) should be prevented from entering the area to be restored, to prevent seedling losses from grazing or trampling, and to prevent dispersal of introduced grasses onto the site.

Abandoned fields set aside for restoration are often adjacent to fields that are still being worked and are thus subjected to an influx of nutrients through fertilizers. In addition, agricultural fields are plowed along contour lines, and drainage lines are established, changing the surface water flow and soil hydrology. High nutrient levels in the soil favor alien grasses (Orlander et al. 1996; Milton 2004), while the changes in soil hydrology might prevent the establishment of indigenous species (Memiaghe, unpubl. data; Andrej Rozanov, pers. comm.). The drift of herbicides and pesticides may also have a negative impact on restored plant and animal (particularly pollinator) communities on the site. In a highly fragmented landscape, local species pools are often diminished. Due to geographical and topographical variability of the landscape, suitable ecotypes of species might not be available. As very few large fragments still exist, sourcing of seeds in the quantities required for restoration may pose a problem.

The highly fragmented nature of the ecosystem may have considerable negative impacts on the persistence of indigenous species and the survival of the populations after restoration. The restoration of plant communities does not necessarily imply that other taxa will follow. For example, the specific habitat requirements of specialized pollinators might not be met within restored sites or adjacent small fragments, due to subtle differences in microclimate, soil type, or soil structure, or the availability of host plants for larvae. The dispersal ranges of animals are limited by the connectivity of their habitat, and the availability of corridors or stepping stones between fragments (Beier and Noss 1998). The extent to which ecosystem processes can be re-created and maintained on restored sites, and the landscape within a mosaic of different vegetation fragments, needs to be considered. In the case of renosterveld, where a high degree of transformation took place over a relatively short time period, it is nearly impossible to restore the original vegetation. Therefore, restoration efforts should focus on a specific goal, as discussed earlier, rather than on aiming to recreate a vegetation type that has been largely transformed.

Old Field Restoration of Renosterveld: Application and Methodology

Our recommendations for the restoration of renosterveld are largely based on our current understanding of ecological processes. Although preliminary field trials have been conducted (Midoko-Iponga 2004; Midoko-Iponga, unpubl. data; Memiaghe, unpubl. data), the methods suggested need to be tested in large-scale field trials, some of which are currently under way.

Methods for Renosterveld Restoration

Current restoration efforts in renosterveld aim to reduce the cover of introduced grasses while at the same time maintaining or even increasing species richness and diversity of indigenous target species (geophytes, indigenous grasses, and shrubs). Below we detail some of these restoration experiments conducted in the renosterveld. Midoko-Iponga (2004) compared the effectiveness of brush cutting, burning, and herbicide (grass-specific, postemergence) application on the removal of introduced grasses. The plots were treated once, in autumn, and were then monitored over a twelve-month period (Midoko-Iponga 2004; Midoko-Iponga, unpubl. data) to determine the effects of the treatments on species richness and diversity and cover abundance of target species. The plots were monitored again twenty-four and thirty months after treatment to determine longer-term vegetation changes after a once-off treatment. In addition, seeds of an indigenous grass, *Ehrharta calycina* (Poaceae), and an indigenous shrub, *Eriocephalus africanus* (Asteraceae), were sown onto the treatment plots, and the survival and growth of the emerging seedlings was monitored.

Herbicide application and burning proved equally effective in reducing the cover of introduced grasses (figure 14.3). However, indigenous species richness and diversity was lowest after burning (tables 14.1 and 14.2). A year after treatment, introduced grass cover was similar in all treatments and gradually declined over the next year and a half (figure 14.3). Thirty months after the treatment, richness and diversity of indigenous species across all treatments were very similar (table 14.2). Musil et al. (2005) found that herbicide application, and a combination of a light burn with a subsequent herbicide application, were most effective methods to control introduced grasses, while at the same time maintaining a high diversity of indigenous species, especially geophytes.

Based on our understanding of the ecological processes, and on the results of the preliminary field trials, we recommend for successful restoration of abandoned old fields in renosterveld the application of a grass-specific herbicide at the onset of winter, preceded by an autumn burn to achieve a flush of the target alien grasses. We further recommend a repeat application of a grass-specific herbicide to control the recruitment of alien grasses. However, herbicides need to be applied with caution, as they can damage the roots of

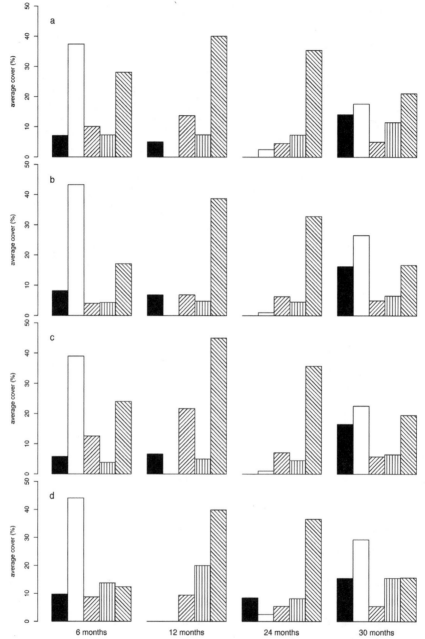

FIGURE 14.3. Average cover of forbs (shaded bars), geophytes (open bars), shrubs (upward diagonal hatches), indigenous grasses (vertical hatches), and introduced grasses (downward diagonal hatches) after treatment: (a) control, (b) autumn burn, (c) autumn brush cut, (d) autumn herbicide application.

TABLE 14.1

Mean species richness (± Standard Deviation) of indigenous species posttreatment

Treatment	Control		Burn		Brush cut		Herbicide			
Months posttreatment	Mean	SD	Mean	SD	Mean	SD	Mean	SD	F-value	p-value
6	11.00	1.00	7.67	1.53	10.00	1.00	10.00	1.73	3.27	.080	ns
12	4.67	1.53	3.33	2.08	2.67	1.53	3.67	1.53	.74	.560	ns
24	7.33	1.53	7.67	.58	7.33	1.15	7.67	.58	.10	.956	ns
30	8.67	1.53	10.00	1.00	8.33	2.52	9.00	1.00	.58	.642	ns

TABLE 14.2

Simpson diversity index D (Krebs 1999) and Simpson equity index E (Krebs 1999) of indigenous species (forbs, geophytes, shrubs, and grasses) posttreatment

Treatment

Months posttreatment		Control	Burn	Brush cut	Herbicide
6	Diversity	4.93	2.69	4.55	4.74
	Equity	.45	.35	.46	.47
12	Diversity	2.47	2.91	2.32	2.69
	Equity	.60	.90	.91	.77
24	Diversity	4.67	4.33	3.81	4.27
	Equity	.65	.57	.52	.55
30	Diversity	6.13	5.51	4.69	5.23
	Equity	.71	.55	.58	.58

indigenous vegetation even at low concentrations (Cornish and Burgin 2005). Oversowing with a mix of indigenous species should be considered, especially in those areas with limited seed sources. Sowing times of the species used are dependent on their time of flowering and seed set, usually before, during, and after the winter rains, or between March and October. Indigenous grasses should not be sown too early after herbicide application, as the seedlings are destroyed and resources are wasted. However, both Musil et al. (2005) and Midoko-Iponga (2004) found that herbicide application was the most expensive treatment method, costing about R13,380/ha (US$2,230/ha) and R36,100/ha (US$6,000/ha), respectively. Therefore, Musil et al. (2005) recommend mowing grass before seed set and removing the cut grass for fodder to offset costs, as the most feasible method for clearing introduced grasses in renosterveld. This strategy, however, as with the repeat application of grass-specific herbicide, affects the recruitment of indigenous grasses, which are an integral part of renosterveld vegetation. As we have conducted research only on old fields covered with introduced grasses, no recommendations can be made on how to restore sites dominated by introduced annual dicots.

Problems and Potential Solutions

Although suitable restoration strategies for west coast renosterveld could be identified within this project, a number of problems and pitfalls still remain. Seeds of indigenous species are not easily available and need to be collected by hand, adding to the total cost of the restoration. Natural vegetation remnants that can be used as seed sources are often small and have a diminished species pool. As topography, microclimate, and soil are highly variable, particular ecotypes might not be suitable for the area to be restored. Phenology of species also differs; species flower and set seed before, during, or after the winter rainfall period, which determines the time of seed collection and seed broadcasting. It might therefore be necessary to oversow the restoration site repeatedly after treatment. Many of the introduced grasses are now resistant to the herbicides most commonly available (Mike Gregor, pers. comm.). The state of the land surrounding a restored site will also have an impact on the success of restoration. Restored sites that are adjacent to, or surrounded by, active agricultural fields are more likely to have a persistent alien cover of introduced grasses than sites adjacent to natural remnants. Heavy grazing and frequent burning also leads to an increase in introduced grasses, and the timing and severity of these two disturbances need to be carefully considered.

Conclusion

Renosterveld is a highly transformed and fragmented ecosystem. Very little is known about the original ecosystem processes, and we can only speculate about the role large herbivores and fire played in the system before transformation by European settlers. Renosterveld was most likely a mixed shrub grassland, with a large geophyte component. Very few large renosterveld fragments remain, and all vegetation remnants are considered irreplaceable. Vegetation on abandoned fields, and in small remnants, is dominated by introduced grasses. Restoration of renosterveld can serve a number of needs, be it to create buffer zones between natural and agricultural areas, to create "nuclei" for further restoration of large tracts of lands, or to create stepping stones and corridors connecting fragments; at the same time, restoration can be used as an alternative to traditional land use of cropping.

Currently the most suitable method for the restoration of renosterveld is the repeat application of a grass-specific herbicide, coupled with a light burn, to reduce the cover of introduced grasses, while at the same time maintaining a high richness and diversity of indigenous species. We are currently investigating soil hydrology and soil nutrient status in abandoned agricultural fields, and how these affect the persistence of introduced grasses and the establishment of indigenous vegetation, with the aim to better understand the drivers of vegetation dynamics on old fields, and to improve our restoration guidelines.

Acknowledgments

Funding for the Renosterveld Restoration Project was provided by a generous grant of the WWF SA / Table Mountain Fund (ZA 5035). Additional funding was provided through the National Research Foundation of South Africa, grant GUN 2053674, to S.J. Milton, the WWF/Life-project in Namibia, the Africa America Institute, and the Government of Gabon. We are very grateful to Sue Milton for her support, guidance, and advice throughout.

We thank the late Dale Parker and his wife, Elizabeth, at Farm Bartholomeus Klip for all their support and willingness to accommodate students and researchers on the private nature reserve. Very special thanks to Mike Gregor, Nicola Farley, and Bernard Wooding at Elandsberg for their enthusiasm for the project and generously provided logistic support.

The following students conducted research within the Renosterveld Restoration Project: Ndafuda N. Shiponeni (MSc 2003), Donald Midoko-Iponga (MSc 2004), Benjamin Walton (MSc 2005), Nicola Farley (MTech 2007), and Rainer M. Krug (PhD 2007). Their enthusiasm and dedication contributed greatly to the success of the project. We thank Ian Newton for access to his literature collection for the duration of the project. Last but not least, we thank Raphael Y. Kongor and Herve R. Memiaghe for the continued monitoring of the restoration plots.

We are very grateful for the comments and suggestions of Viki Cramer and two anonymous reviewers, which improved the manuscript considerably.

References

Baard, E. H. W. 1993. Distribution and status of the geometric tortoise, *Psammobates geometricus*, in South Africa. *Biological Conservation* 63:235–39.
Bakker, J. P., and F. Berendse. 1999. Constraints in the restoration of ecological diversity in grassland and heathland communities. *Trends in Ecology and Evolution* 14:63–68.

Beier, P., and R. F. Noss. 1998. Do habitat corridors provide connectivity? *Conservation Biology* 12:1241–52.

Bonser, S. P., and R. J. Reader. 1995. Plant competition and herbivory in relation to vegetation biomass. *Ecology* 76:2176–83.

Botha, M. 2001. *Incentives for conservation on private land: Options and opportunities.* Cape Conservation Unit, Botanical Society of South Africa, Kirstenbosch, Cape Town.

Boucher, C. 1983. Floristic and structural features of the coastal foreland vegetation south of the Berg River, Western Cape Province, South Africa. *Bothalia* 14:669–74.

Boucher, C. B., and E. J. Moll. 1981. South African Mediterranean shrublands. In *Vegetation types of the world: Mediterranean-type shrublands,* ed. F. Di Castri, 233–48. Elsevier Scientific Publishing, New York.

Cape Conservation Unit. 2003. Transformed areas: ArcView 3.2 shapefiles. Botanical Society of South Africa, Kirstenbosch, Cape Town.

CapeNature. 2004. Elandsberg nature reserve including old fields: ArcView 3.2 shapefiles. Scientific Services, Jonkershoek.

Cornish, P. S., and Burgin, S. 2005. Residual effects of glyphosate herbicide in ecological restoration. *Restoration Ecology* 13:695–702.

Cowling, R. M., and C. E. Heijnis. 2001. Broad habitat units as biodiversity entities for conservation planning in the Cape floristic region. *South African Journal of Botany* 67:15–38.

Cowling, R. M., S. M. Pierce, and E. J. Moll. 1986. Conservation and the utilization of south coast renosterveld, and endangered South African vegetation type. *Biological Conservation* 37:363–77.

Davies, R. J. 1985. The importance of weed control and the use of tree shelters for establishing broadleaved trees on grass-dominated sites in England. *Forestry* 58:167–80.

Deacon, H. J. 1992. Human settlement. In *The ecology of fynbos: Nutrients, fire and diversity,* ed. R .M. Cowling, 260–70. Oxford University Press, Cape Town, SA.

Di Castri, F. 1981. *Vegetation types of the world: Mediterranean-type shrublands.* Elsevier Scientific Publishing, New York.

Donaldson, J. S. 2002. Biodiversity and conservation farming in the agricultural sector. In *Mainstreaming biodiversity in development: Case studies from South Africa,* ed. S. M. Pierce, R. M. Cowling, T. Sandwith, and K. MacKinnon, 43–55. World Bank, Washington, DC.

Donaldson, J. S., I. Nänni, C. Zachariades, and J. Kemper. 2002. Effects of habitat fragmentation on pollinator diversity and plant reproductive success in renosterveld shrublands of South Africa. *Conservation Biology* 16:1267–76.

Du Plessis, S. F. 1969. The past and present geographical distribution of *Perissodactyla* and *Artiodactyla* in Southern Africa. MS thesis, University of Pretoria, Pretoria, SA.

Fairbanks, D. H. K., C. J. Hughes, and J. K. Turpie. 2004. Potential impact of viticulture expansion on habitat types in the Cape floristic region, South Africa. *Biodiversity and Conservation* 13:1075–1100.

Grace, J. B. 1999. The factors controlling species density in herbaceous plant communities: An assessment. *Perspectives in Plant Ecology, Evolution and Systematics* 2:1–28.

Kemper, J., R. M. Cowling, and D. M. Richardson. 1999. Fragmentation of South African renosterveld shrublands: Effects on plant community structure and conservation implications. *Biological Conservation* 90:103–11.

Klein, R. G. 1974. A provisional statement on terminal Pleistocene mammalian extinctions in the Cape biotic zone (Southern Cape Province, South Africa). *South African Archaeological Society*. Goodwin Series 2:39–45.

——. 1986. The pre-history of Stone Age herders in the Cape Province of South Africa. In *Prehistoric pastoralism in southern Africa*, ed. M. Hall and A. B. Smith. South African Archaeological Society. Goodwin Series 5:5–12.

Krebs, C. J. 1999. *Ecological methodology*. Addison Wesley Longman, Menlo Park, CA.

Krug, C. B., R. M. Krug, D. Midoko-Iponga, N. N. Shiponeni, B. A. Walton, and S. J. Milton. 2004. *Restoration of west coast renosterveld: Facilitating the return of a highly threatened vegetation type*. Proceedings of the 10th MEDECOS Conference, 2004, Rhodes, Greece.

Krug, R. M. 2007. Modelling seed dispersal in restoration and invasion. PhD thesis, University of Stellenbosch, Matieland, SA.

Krug, R. M., C. B. Krug, N. Farley, D. Midoko-Iponga, I. P. Newton, N. N. Shiponeni, B. A. Walton, and S. J. Milton. 2004. *Reconstructing west coast renosterveld: Past and present ecological processes in a Mediterranean shrubland of South Africa*. Proceedings of the 10th MEDECOS Conference, 2004, Rhodes, Greece.

Leibbrandt, H. C. V. 1901. *Precis of the archives of the Cape of Good Hope. Journal, 1662–1670*. W. A. Richards and Sons, Cape Town, SA.

——. 1902. *Precis of the archives of the Cape of Good Hope. Journal, 1671–1674 & 1676*. W. A. Richards and Sons, Cape Town, SA.

Levyns, M. 1956. Notes on the biology and distribution of the rhenoster bush. *South African Journal of Science* 56:141–43.

Low, A. B., and F. E. Jones. 1995. *The sustainable use and management of renosterveld remnants in the Cape floristic region*. Flora Conservation Committee, Botanical Society of South Africa, Kirstenbosch, SA.

Low, A. B., and A. G. Rebelo. 1996. *Vegetation of South Africa, Lesotho and Swaziland*. Department of Environmental Affairs and Tourism, Pretoria, SA.

Luyt, E. du C. 2005. Models of Bontebok (*Damaliscus pygargus pygargus*, Pallas 1766) habitat preferences in the Bontebok National Park and sustainable stocking rates. MSc thesis, University of Stellenbosch, Matieland, SA.

Manning, J., P. Goldblatt, and D. Snyman. 2002. *The color encyclopedia of Cape bulbs*. Timber Press, Cambridge, UK.

Milton, S. J. 2004. Grasses as invasive alien plants in South Africa. *South African Journal of Science* 100:69–75.

Midgley, J. J., C. Harris, H. Hesse, and A. Swift. 2002. Heuweltjie age and vegetation change based on $\delta^{13}C$ and ^{14}C analyses. *South African Journal of Science* 98:202–204.

Midoko-Iponga, D. 2004. Renosterveld restoration: The role of competition, herbivory and other disturbances. MSc thesis, University of Stellenbosch, Matieland, SA.

Midoko-Iponga, D., C. B. Krug, and S. J. Milton. 2005. Competition and herbivory influence growth and survival of shrubs on old fields: Implications for restoration of threatened renosterveld shrubland, South Africa. *Journal of Vegetation Science* 16:685–92.

Moll, E. J., B. M. Campbell, R. M. Cowling, L. Bossi, M. L. Jarman, and C. B. Boucher. 1984. A description of major vegetation categories in and adjacent to the fynbos biome. *Council for Scientific and Industrial Research* 83:11–14.

Mucina, L., and M. C. Rutherford. 2004. *Vegetation map of South Africa, Lesotho and Swaziland: Shapefiles of basic mapping units*. Beta Version 4.0, February 2004. National Botanical Institute, Cape Town, SA.

Musil, C. F., S. J. Milton, and D. W. Davies. 2005. The threat of alien invasive grasses to lowland Cape floral diversity: An empirical appraisal of the effectiveness of practical control strategies. *South African Journal of Science* 101:337–44.

Myers, N., R. A. Mittermaier, C. G. Mittermaier, G. A. B. da Fonseca, and J. Kent. 2000. Biodiversity hotspots for conservation priorities. *Nature* 403:853–58.

Newton, I. P. 2007. Recent transformations in west-coast renosterveld: Patterns, processes and ecological significance. PhD thesis, University of the Western Cape, Bellville, SA.

Orlander, G., U. Nilsson, J. E. Hällgren, and J. A. Griffith. 1996. Competition for water and nutrients between ground vegetation and planted *Picea abies*. *New Zealand Journal of Forest Science* 26:99–117.

Parkington, J. 1977. Soaqua: Hunter-fisher-gatherers of the Olifants river valley, South Africa. *South African Archaeological Bulletin* 32:150–57.

Pauw, A. 2004. Effect of habitat fragmentation on pollination syndromes in renosterveld. PhD thesis, University of Cape Town, Cape Town, SA.

Rouget, M., D. M. Richardson, R. M. Cowling, J. W. Lloyd, and A. T. Lombard. 2003. Current patterns of habitat transformation and future threats to biodiversity in terrestrial ecosystems of the Cape floristic region, South Africa. *Biological Conservation* 112:63–85.

Schweitzer, F. R. 1979. Excavations at Die Kelders, Cape Province, South Africa: The Holocene deposits. *Annals of the South African Museum* 78:101–233.

Schweitzer, F. R., and K. J. Scott. 1973. Early occurrence of domestic sheep in sub-Saharan Africa. *Nature* 241:547.

Shiponeni, N. N. 2003. Dispersal of seeds as a constraint in revegetation of old fields in renosterveld vegetation in the Western Cape, South Africa. MSc thesis, University of Stellenbosch, Matieland, SA.

Shiponeni, N. N., and S. J. Milton. 2006. Seed dispersal in the dung of large herbivores: Implications for restoration of renosterveld shrubland old fields. *Biological Conservation* 15:3161–75.

Skead, C. J. 1980. *Historical mammal incidence in the Cape Province*. Vol. 1. *The Western and Northern Cape*. Department of Nature and Environmental Conservation of the Provincial Administration of the Cape of Good Hope, Cape Town, SA.

Skinner, J. D., and R. H. N. Smithers. 1990. The mammals of the southern African subregion. University of Pretoria, Pretoria, SA.

Talbot, W. J. 1947. *Swartland and sandveld*. Oxford University Press, Cape Town, SA.

Thom, H. B. 1952. *Journal of Jan van Riebeck, Volume 1: 1651–1655*. Van Riebeck Society, Cape Town, SA.

———. 1954. *Journal of Jan van Riebeck, Volume 2: 1656–1658*. Van Riebeck Society, Cape Town, SA.

———. 1958. *Journal of Jan van Riebeck, Volume 3: 1659–1662*. Van Riebeck Society, Cape Town, SA.

Thompson, K., J. P. Grime, and G. Mason. 1977. Seed germination in response to diurnal fluctuations of temperature. *Nature* 267:147–49.

Van Rooyen, S. 2004. Factors affecting alien grass invasion into west coast renosterveld fragments. MSc thesis, University of Stellenbosch, Matieland, SA.

Von Hase, A., M. Rouget, K. Maze, and N. Helme. 2003. *A fine-scale conservation plan for the Cape lowlands renosterveld technical report: Main report*. Cape Conservation Unit, Botanical Society, Cape Town, SA.

Vrdoljak, S. M., and M. J. Samways. 2005. A special case for the generalists? Flower visitors in the lowlands of the CFR. Entomological Society of Southern Africa Congress, 2005, Rhodes University, Grahamstown, SA.

Walton, B. A. 2005. Vegetation patterns and dynamics of renosterveld at Agter-Groeneberg Conservancy, Western Cape, South Africa. MSc thesis, University of Stellenbosch, Matieland, SA.

Winter, S. J. 2003. Attitudes and behavior of landholders towards the conservation of Overberg coastal renosterveld, a threatened vegetation type in the Cape floral kingdom. MSc thesis, University of Stellenbosch, Matieland, SA.

Winter, S. J., and K. J. Esler. 2005. An index to measure the conservation attitudes of landowners towards Overberg coastal renosterveld, a critically endangered vegetation type in the Cape floral kingdom, South Africa. *Biological Conservation* 126:383–94.

Chapter 15

Prospects for the Recovery of Native Vegetation in Western Australian Old Fields

VIKI A. CRAMER, RACHEL J. STANDISH, AND RICHARD J. HOBBS

The wheat and sheep farming area of southwest Western Australia (figure 15.1), known locally as the wheatbelt, is an apparent ecological and economic contradiction: despite extensive clearing and widespread land degradation, it remains one of the most ecologically and agriculturally important areas in Australia. In many catchments (watersheds), up to 90% of native vegetation has been cleared for broad-scale agriculture (Hobbs et al. 1995), yet the area retains an extraordinarily rich endemic flora (Hopper and Gioia 2004). Although rainfall is highly variable and land degradation due to soil acidity, sodicity, and salinity is increasing, the wheatbelt continues to produce around 50% of Australia's wheat and other grain exports (Department of Agriculture 2005).

Although some economic and environmental indicators would suggest that land abandonment in the region is likely to be widespread, this has not been the case so far. Most farms remain profitable, although the bottom quartile of farm businesses (based on rate of return to capital) are under sustained financial pressure and many may eventually leave the industry (Kingwell and Pannell 2005). Declining commodity prices since the 1960s have been offset by increases in productivity, particularly during the 1990s when advances in agronomy led to rapid increases in wheat yields (Passioura 2002; Allison and Hobbs 2004). While rainfall in early winter has decreased by 15%–20% in the past thirty years, and autumn and winter temperatures have increased (Indian Ocean Climate Initiative 2002), water availability has not been the primary limitation of wheat yields, except on the drier margins of the wheatbelt (McFarlane et al. 1993; Passioura 2002). The lower water use of wheat crops compared with native vegetation has been a primary driver of secondary

FIGURE 15.1. Map of Western Australia showing the agricultural area in the southwest.

salinity, one of the most significant forms of land degradation in the area (Cramer and Hobbs 2002).

Australian agriculture has traditionally depended on large overseas markets for products that can be produced cheaply using large tracts of land (National Land and Water Resources Audit 2001b). This has led to low rural population densities that will further decline as labor-saving technologies and increases in farm size reduce on-farm employment opportunities and the number of farm families (Kingwell and Pannell 2005). A decline in the cultural relevance of farming as a lifestyle identity (National Land and Water Resources Audit 2002) and the lack of identification of urban Australians with agricultural landscapes means that there is little social or political will to sustain rural communities (Abensperg-Traun et al. 2004). Hence a combination of economic and social factors—lower average earnings in inland regions, fewer educational opportunities, few employment opportunities outside of agriculture, and the changed life aspirations of women and youth—may further prompt some farmers to leave the industry in coming decades (National Land and Water Resources Audit 2002; Kingwell and Pannell 2005).

Yet, even with significant social change and population decrease, the abandonment of productive land is unlikely. Kingwell and Pannell (2005) predict continued production growth from yield improvements and an increased proportion of the landscape sown to crops and fodder for the next twenty-five years. The trend toward increasing farm size means that farms "abandoned" by one owner are likely to be amalgamated into neighboring properties. If increases in productivity and improvements in efficiency continue to offset declining terms of trade, then land abandonment in southwest Western Australia may be confined only to land considered "hard" (i.e., hard-setting and rocky soils, difficult to cultivate) and to areas where production on degraded land (through soil acidity, sodicity, or salinity) cannot be maintained profitably through either improved cultivars, agronomy, or broader natural resource management. Current producers in the central wheatbelt estimate that 7.5%–10% of their farm is unproductive (National Land and Water Resources Audit 2001b), and this may give some indication of the area that may be abandoned if current economic and agronomic trends continue.

"A Million Acres a Year": Development of Agriculture in Southwest Western Australia

By the 1880s, after the first sixty years of agriculture in Western Australia, only 69,000 acres of land had been cleared for cropping (Allison and Hobbs 2004). By the 1940s, another sixty years later, 14 million acres of land had been cleared and over 70% of native vegetation in the central wheatbelt had been removed (Arnold and Weeldenberg 1991, cited in Hobbs et al. 1993). The dramatic increase in clearing was driven by the need for the state to develop an export income, and wheat production—actively assisted by government—allowed the maximum possible returns in the shortest period of time

(Beresford et al. 2001). Yet the 1940s marked only the beginning of what is probably the most rapid and indiscriminate clearing of land for agriculture in Australia's history. Between 1945 and 1961, the area of land cleared increased from 14 million acres to 25 million acres, followed by the pinnacle of clearing in the early 1960s, when more than one million acres a year were "opened up" under various land settlement schemes (Beresford et al. 2001). Despite early evidence of significant land degradation and two periods when economic circumstances threatened (1914–19) or caused widespread land abandonment (1930s), the Western Australian government aggressively pursued expansionist agricultural policies throughout most of the twentieth century. A combination of significant government incentives, the high cultural value attached to the farming lifestyle, and technological improvements in agriculture has led to a land use history where rapid phases of agricultural development and consequent land degradation have been the norm.

Today, most catchments in the wheatbelt retain <10% of their native vegetation cover (Shepherd et al. 2002). The pattern of land clearing, based on preferential clearing of particular vegetation types, has meant that very few large areas of eucalypt woodland remain intact, and there remains a significantly greater proportion of shrub communities than woodland communities. Native vegetation is closely linked to landform and soil type, with eucalypt woodlands occurring on heavier or loamy soils that are relatively more fertile, and shrublands and heath occurring on loamy sands and deep sands of less fertility. Woodlands are poorly represented in conservation reserves, and many of the woodland remnants on private property or along road verges have been degraded by stock grazing and the invasion of nonnative herbaceous species (Hobbs et al. 1993).

Historical Agricultural Practices and Land Degradation

The sandy soils of the wheatbelt region are among the most ancient and highly weathered in the world and contain very low amounts of phosphorus. As a consequence, most soils in the wheatbelt were largely unprofitable for agriculture until superphosphate became available in Western Australia in the 1920s (Bolland et al. 2003). Farmers were encouraged to apply large amounts of the fertilizer, aided by a bounty paid by the Commonwealth Government (Bolland et al. 2003). A typical cropping rotation at this time was a cereal crop, grown with the addition of 70–100 kg ha^{-1} superphosphate, followed by a volunteer pasture phase that was grazed, and then bare-fallowed in its second year in preparation for the following cereal crop (Nulsen 1993). A decline in wheat prices and the onset of the Depression in the 1930s led to many farms being abandoned and absorbed into neighboring farms (York Main 1993). Low rainfall during 1935–38, fallowing practices, and the destruction of both crops and vegetation by plagues of rabbits left topsoil exposed to strong winds. The shallow and fragile layer of topsoil was lost in severe dust storms (York Main 1993). The link between secondary salinity and land clearing was widely recognized during this period, yet government commissioners endeavored to discredit expert opinion, and land release and clearing continued (Beresford et al. 2001).

The establishment of the ley-farming system in the 1950s, where a cereal crop was followed by two years of grazing on a legume-based pasture, was a response to the high wool prices following World War II. Despite appearing to be a stable farming system, the continued multiple cultivation of the heavier soils and the burning of stubble to control plant diseases destroyed the structure of the soil (Nulsen 1993). The identification of micronutrient deficiencies in the 1950s, followed by the introduction of compound (NPK) fertilizers and new varieties of wheat in the 1960s, led to the clearing of the light-textured, yellow sand and loamy sand soils. These areas had largely remained intact because their scrubby vegetation was difficult to clear and the application of superphosphate did not improve cereal yields (Nulsen 1993; York Main 1993; Passioura 2002). These light-textured soils are especially prone to wind erosion.

Another major change in farming practices occurred in the 1970s with the introduction of lupins (*Lupinis angustifolius* and *L. alba*), which were ideally suited to the yellow sands and loamy sands of the wheatbelt. The benefits obtained from increased soil nitrogen and organic matter from a lupin rotation were outweighed by increases in soil acidity in these sandy soils that are poorly buffered against changes in pH (Nulsen 1993). Soil acidity remains the most widespread soil degradation problem in the wheatbelt, followed by secondary salinity (National Land and Water Resources Audit 2001b). In the late 1980s, canola became an important "break" crop, due to the remarkable yields of the wheat crops that followed it (Passioura 2002). Canola does not host the major root diseases of cereals, and with root diseases better under control, farmers were prepared to apply more nitrogen fertilizer to their wheat crops. The 1980s also saw an increase in the use of herbicides for weed control, resulting in reduced tillage (Nulsen 1993). More recently, Western Australia has had the greatest adoption of no-tillage cultivation in Australia, with over 80% of farm managers using no-till practices for a proportion of cropping (D'Emden and Llewellyn 2004). The sustainability of no-till cultivation is threatened, however, by increases in herbicide-resistant weed populations.

Barriers to Natural Regeneration on Abandoned Land in Southwest Western Australia at Landscape and Local Scales

The clearing and land use history of the wheatbelt suggests the likely ecological state of an abandoned piece of farmland: isolated from remnant vegetation, with degraded soil structure but increased levels of nutrients from residual fertilizer, and prone to invasion from weed species. The clearing history, particularly the extent of clearing and the focus on clearing particular vegetation types, has determined the landscape-scale barriers to old field regeneration, while land use history plays the dominant role in determining local barriers to regeneration. Superimposed on these barriers created by human actions are barriers related to climate and to intrinsic qualities of the vegetation itself.

Landscape-scale Barriers in a Heavily Fragmented System

The most obvious landscape-scale barrier to the establishment of native vegetation on abandoned farmland in the wheatbelt is the distance from an intact remnant and, thus, a source of seed. The native vegetation left in the region is present as a large number of mostly small (<10 ha) remnants; this is particularly so for the woodlands on heavier soils that have been cleared most extensively (Hobbs 1998). In the central wheatbelt, the distance from abandoned farmland to the nearest native vegetation ranges from <50 m to almost 1.5 km (Arnold et al. 1999). Although evidence for long-distance dispersal of seeds exists (He et al. 2004), in general, seed dispersal distances are likely to be only tens of meters, except for small, wind-dispersed seeds (e.g., members of the Asteraceae). Further, fragmentation may affect pollinator abundance, pollination and seed set, and seed predation and dispersal (Hobbs and Yates 2003). Many remnants on farms have been grazed by sheep, further disturbing understory vegetation and restricting plant regeneration (Hobbs 1998).

The broad-scale clearing of vegetation has altered landscape-scale processes and exacerbated the hot and dry nature of the Mediterranean climate. While global changes in atmospheric circulation patterns are the primary cause of the 15%–20% reduction in winter rainfall in the region (Indian Ocean Climate Initiative 2002), changes in land surface characteristics such as albedo, surface roughness, and canopy resistance after land clearing have led to reduced incidence of cloud formation over cultivated areas compared with that over native vegetation (Lyons 2002). Decreasing rainfall has been accompanied by increased daytime and nighttime temperatures, particularly in winter and autumn. Seedling germination and establishment occurs early or later during the winter rainfall period when both soil temperature and soil moisture conditions are optimal (Yates et al. 1996). The onset of the summer drought period provides a severe barrier to seedling survival. Reduced rainfall and warmer temperatures have implications for soil water stores during spring and early summer, essentially lengthening the summer drought period. Intact vegetation also significantly reduces wind speeds near the ground (Linacre and Hobbs 1977), and wind speeds even a few meters inside remnant vegetation are considerably less than in adjacent abandoned paddocks (V. Cramer, personal observation). Higher wind speeds over cleared land will affect the microclimate around seedlings (discussed later), and increase evapotranspiration rates in larger plants, particularly during the summer drought period.

The rapid and broad-scale clearing of land in the wheatbelt has altered the hydrological cycle, leading to the development of shallow water tables and soil salinity (Peck and Williamson 1987; George 1992; McFarlane et al. 1993; Cramer and Hobbs 2002). One million hectares of farmland in Western Australia are currently salt-affected, with a further three million hectares considered to be "at risk" (National Land and Water Resources Audit 2001a; Pannell 2004). The salinization of the soil profile and the waterlogging often associated with shallow water tables have serious implications for plant health and ecosystem processes in these systems (Cramer and Hobbs 2002). The successful production of salt-tolerant wheat cultivars has been slow (Munns 2005), and the productivity of salt-affected areas is likely to remain low. Hence, salt-affected land is the most likely to be abandoned. Yet little regeneration of native species is likely in these areas. Studies of native remnant vegetation affected by secondary salinity show that species diversity, species richness, and structural complexity is lost, with little colonization by salt-tolerant native species (Lymbery et al. 2003; Cramer, Hobbs, and Atkins 2004; Cramer, Hobbs, et al. 2004).

Local Barriers to the Regeneration of Native Vegetation

Microclimate, soil modification caused by cultivation, and competition with introduced annual species are the most important local barriers to the regeneration of native species on abandoned farmland in the wheatbelt. These barriers are likely to be further strengthened by positive feedback between them (see chapter 3). Summer temperatures in the wheatbelt regularly exceed 40 °C, and daytime relative humidity may be less than 10%. While no data are available for soil temperatures in abandoned paddocks in the wheatbelt, significant differences in soil temperature between grazed and ungrazed remnants have been recorded in both summer and winter, with soil temperatures in grazed remnants approaching 50°C in summer (Yates, Norton, and Hobbs 2000). The greater cover of exotic annuals than native perennials in grazed remnants, and the lower litter cover, are considered to be a major contributing factor to the differences in soil temperatures (Yates, Norton, and Hobbs 2000). Differences in soil temperature are likely to be even greater between abandoned paddocks, where there will be little or no perennial cover, and intact remnants. Under these conditions of high evaporative demand, the successful establishment of perennial vegetation is dependent on access to water stored in the soil profile.

Cultivation practices in the wheatbelt have caused a decline in soil structure and led to the homogenization of the soil surface, conditions which reduce the availability of soil water to plants both directly and indirectly. Soil organic matter plays a vital role in stabilizing soil structure by binding aggregates (Dalal and Chan 2001). Losses of organic matter from the top 0–0.1 m of soils may exceed 60% after fifty years of cereal cropping in soils of the Australian cereal belt (Dalal and Chan 2001). The low organic matter of soils in the wheatbelt in their natural state means that they are structurally fragile and even small losses of organic matter will lead to a deterioration in soil structure (Nulsen 1993). Even fourteen years after abandonment, soil organic C in wheatbelt soil cultivated for sixty years was significantly lower than in an adjacent remnant (Standish et al., 2006). The compaction of cultivated soils is widely reported for the wheatbelt (Hamza and Anderson 2003). Compaction increases bulk density, penetration resistance, and water repellency through a reduction in porosity and a change in pore size structure (Nulsen 1993; Pettit et al. 1998; Yates, Norton, and Hobbs 2000), further exacerbating changes in soil structure caused by the loss of organic matter.

Cultivation of soil also reduces soil surface heterogeneity and the presence of microsites or microcatchments that may be important for seedling establishment after germination. Sandy soils in the southern part of Australia are commonly water repellent, with a thin, intensely repellent, surface layer acting as a barrier to an underlying wettable soil (Garkaklis et al. 1998). The diggings of small mammals that feed on hypogeous fungi have lower water repellency than surrounding undisturbed soils (Garkaklis et al. 1998), and were likely to have played an important role in providing microsites with greater water infiltration for seedling establishment. These mammals have all but disappeared from the wheatbelt. The pores created by the activity of ants and termites are important conduits for root growth and water movement in soils, and soil modified by termites contains greater levels of organic carbon (Lobry de Bruyn and Conacher 1990; Hobbs et al. 1993; Passioura 2002). While ant activity (albeit of different species) will continue in the presence of cultivation, termite activity will not (Hobbs et al. 1993). Termite diversity on abandoned land in the wheatbelt remains low for decades (Abensperg-Traun and de Boer 1990).

The retention of nitrogen, phosphorus, and potassium in old fields soils after the application of NPK fertilizer is variable and may largely depend on soil type. Arnold et al. (1999) found no significant differences in total soil nitrogen or available phosphorus between twenty-seven sites with different land use histories (undisturbed, cleared, cleared and cultivated). The lack of significant difference in soil N and P concentrations between the sites is likely a reflection of the short clearing or cultivation times for the majority of sites (one year) and the sandy loams and deep sand soils present. Standish et al. (2006) found no significant differences in inorganic N in soils in three old fields and adjacent remnants, although total N was significantly greater in one old field. Available (Colwell) P was significantly greater in old field soil than the adjacent remnants at two sites, despite these sites having different land use histories with regard to years of cultivation and time since abandonment (figure 15.2). Deep leaching of N in sandy soils, denitrification, reduced soil organic matter, and nutrient removal in product and plant residue may explain the general absence of increased soil N in these old fields (Dalal and Chan 2001). While decreased organic matter would also lead to decreases in organic P, the majority (50%–75%) of the water-soluble P applied as fertilizer in any growing season is absorbed by reacting with iron, aluminium, and calcium at the surface of soil constituents (Bolland et al. 2003). This residual soil P becomes available to plants in the years following its application and may persist in the soil for years.

Indigenous plant species are adapted to low levels of soil P, and high concentrations of available P may have a toxic effect during plant establishment (Handreck 1997; Shane et al. 2004). Elevated levels of soil P also favor the establishment of introduced annual species over native perennials (Hobbs and Atkins 1988; Hobbs and Atkins 1991; Hester and Hobbs 1992) (figures 15.2 and 15.3). Soil disturbance further exacerbates this effect. Although native species respond to nutrient additions, introduced annuals such as *Avena* spp. (wild oats), *Arctotheca calendula* (cape weed), *Ursinia anthemoides* (South African marigold) and *Hypochaeris glabra* (smooth catsear) germinate earlier and grow faster than both annual and perennial native species (Hobbs and Atkins 1988; Standish et al., 2007) (figure 15.4a). These species are common, and often dominant, pasture weeds in the wheatbelt and form persistent soil seed banks (Ellery and Chapman 2000). Such species are likely to dominate the soil seed bank and vegetation following abandonment (figures 15.3 and 15.4b). The peak of germination in *Ursinia anthemoides* occurs at lower tem-

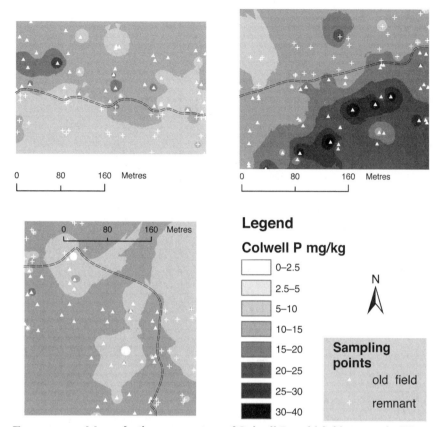

Figure 15.2. Maps of soil concentrations of Colwell P at old field sites in the Western Australian wheatbelt: Letchford (upper left), Pullen (upper right), Woolering (lower left). Maps were produced by interpolation to grid from point data using an inverse distance weighted algorithm (variable search radius and power of two). Concentrations of Colwell P were significantly greater in the old fields at Letchford and Pullen than in their adjoining remnants. Reproduced from Standish et al. 2006, with kind permission of Springer Science and Business Media.

peratures than for a number of native Asteraceae (Schütz et al. 2002) and *Eucalyptus* spp. (Yates et al. 1996). While native species typically germinate early or later in the winter period when soil temperatures are warmer, *Ursinia anthemoides* can continue to germinate throughout the entire winter rainfall period. As rainfall decreases and soil temperatures rise in spring, introduced annual species are already well established and have a competitive advantage for increasingly limited soil water resources. Further, there is evidence that invasive species may alter nutrient-cycling processes and create positive feedback loops that enhance their invasiveness (Ehrenfeld 2003).

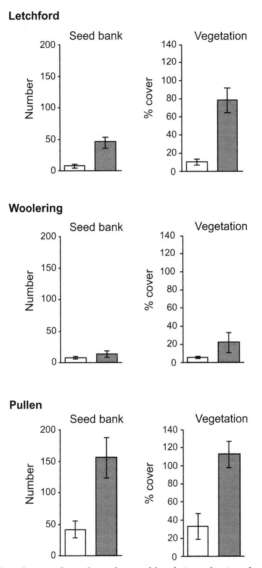

FIGURE 15.3. Abundance of weeds in the seed bank (number) and vegetation (% cover) in the remnants (open bars) and old fields (shaded bars) of the Letchford, Woolering, and Pullen sites (data adapted from Standish et al. 2007). Sites with significant differences in soil concentrations of Colwell P between remnant and old field (Letchford and Pullen; figure 15.2) had the greatest abundance of weeds in old field vegetation, although the difference in the abundance of weeds in remnants and old fields is significant at all sites (see Standish et al. 2007). This suggests that elevated soil P concentration is an important (but not the only) factor in promoting the establishment and maintaining the dominance of weeds on abandoned farmland in the wheatbelt.

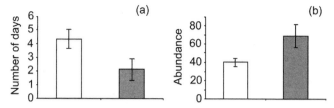

FIGURE 15.4. (a) Number of days to the appearance of the first germinant of each native species (open bars) and weed species (shaded bars) in the soil seed bank of old field sites in the wheatbelt of Western Australia (mean ± SE). (b) Abundance of germinants of native (open bars) and weed (shaded bars) species within the same seed bank trial (control group; $n = 75$). Data adapted from Standish et al. (2007).

Letting Nature Take Its Course: Resilient Alternative States or Long-term Transient States in Wheatbelt Old Fields?

Only two studies have directly assessed the return of native vegetation to abandoned farmland in southwest Western Australia. The first of these studies (Arnold et al. 1999) focused on the recovery of shrubland communities at sites that were cleared only or were cleared and cultivated. All of the sites that were cleared only were abandoned after four years. Only one of the cultivated sites was farmed for more than six years before being abandoned. Time since abandonment ranged from twenty-two to sixty years (average thirty-one years). The cultivated sites had a lower total cover of woody vegetation than cleared and undisturbed sites, but substantially greater cover of grass species. Arnold et al. (1999) speculated that dispersal limitation and competition with grass species were responsible for the lower establishment of woody native perennials at the cultivated sites. Although high richness and diversity of plant species were found on the cultivated sites, the structure and composition of these communities were different from undisturbed communities.

Our recent study in the wheatbelt has focused on a detailed spatial analysis of soil properties, vegetation cover and type, seed dispersal, and the soil seed bank in three old fields adjacent to intact remnant vegetation (Standish et al. 2006; Standish et al. 2007). The three fields have different land use histories and the colonization of native vegetation differs greatly between the sites (figure 15.5). All three sites were abandoned because of the difficulty in plowing their stony soils. One site (Letchford), cleared in the 1950s, was abandoned after only one year in cultivation. Vegetation in this field is dominated by native (*Aristida holathera, Austrostipa eremophila*) and introduced grasses (*Avena* spp.), although some native shrubs (mostly *Hakea recurva* and *Acacia* spp.) have colonized the area in the past ten to fifteen years

(M. Barnes, pers. comm., 2003). Herbaceous native annuals are also present, particularly at the boundary between the old field and the remnant (figure 15.5a). The second site (Woolering), cleared in the late 1950s, was cultivated for four years before being abandoned. Native vegetation dominates all strata, apart from one small area of *Avena* spp. The ground cover in this old field is composed of native annual species and is largely dominated by members of the Asteraceae (figure 15.5b). The high spatial heterogeneity in plant floristics in the wheatbelt means that direct comparison between the old field and remnant is not necessarily valid, but based on the dominant species within each strata the recovery of native vegetation in this old field has been excellent. It is important to note the presence of native vegetation on three edges of this old field (compared with one edge for the other two remnants). Two major disturbance events, a fire in 1963 and a severe storm in 1978, may also have influenced the establishment of native species. The third site (Pullen) has had a more typical land use history: cleared in 1930 then cropped until 1990, when it was no longer economical to cultivate. The ground cover is dominated by introduced annual grass and herb species, most notably *Avena* spp., *Pentaschistus airoides*, *Arctotheca calendula*, *Hypochaeris glabra*, *Erodium* spp., and *Trifolium* spp. Native grasses (particularly *Aristida holathera*) and native herbs (mostly Asteraceae) have colonized the area within 50 m of the boundary between remnant and old field. Only one native perennial species (*Acacia acuminata*) has colonized this old field (figure 15.5c).

Didham et al. (2005) have suggested that differentiating between resilient alternative states (i.e., equilibrium states) and long-term transient states may be difficult in abiotically controlled systems such as the wheatbelt. They consider that pervasive landscape alteration, constraints on native propagule access, high levels of species invasions, and the resultant stochastic priority effects on community assembly generate transient alternative states where the rate of "succession" may grossly exceed the timeframe of human observation. With observational data available from few field sites and limited experimental data available, it is difficult to draw generalizations on whether the ecosystem states reported above represent resilient alternative states, or whether the state dominated by introduced grasses and herbs is simply a long-term transient state within a successional trajectory modified (and largely stalled) by the presence of alien species. The landscape context in which the old field exists is clearly important in overcoming propagule limitation in native species (see Woolering example) and influencing the course of ecosystem assembly. Further work is required to determine the relative importance of land use history and disturbance events after abandonment in determining ecosystem assembly on abandoned farmland in the wheatbelt.

FIGURE 15.5. Vegetation communities on old fields in the wheatbelt of Western Australia. (a) Letchford, vegetation (foreground) dominated by native (*Aristida holathera*) and alien (*Avena* spp.) grass species, with some native annuals (*Podolepis canescens*); (b) Woolering, vegetation dominated by native perennial and annual species (*Acacia acuminata*, *Waitzia nitida* and *Schoenia cassiniana*); and (c) Pullen, vegetation (foreground) dominated by alien annual grasses and herbs (*Avena* spp., *Trifolium* spp.) with little colonization by native species. See text for details of land use history at each site.

Overcoming Barriers to Vegetation Regeneration and Setting Realistic Restoration Goals

The barriers to the regeneration of native vegetation in old fields are similar to those in woodlands degraded by heavy grazing (Yates, Norton, and Hobbs 2000; Yates, Hobbs, and Atkins 2000) and management of some of these barriers is relatively straightforward, if not labor intensive. Studies of old fields and degraded remnant vegetation in the wheatbelt consistently find that the presence of introduced annual species limits the recruitment and survival of native species (Hobbs and Atkins 1991; Yates, Hobbs, and Atkins 2000; Yates and Broadhurst 2003). The manual creation of microcatchments that locally increase water availability (by deep soil ripping or divot creation with a spade, for example) can increase the survivorship of planted seedlings in areas with a dense cover of annual weed species (Yates, Hobbs, and Atkins 2000; Standish, unpublished data) (figure 15.6). Soil disturbance will also improve water infiltration in compacted soils (Nulsen 1993). Yet the effective seed dispersal and staggered dormancy mechanisms of introduced annuals, com-

FIGURE 15.6. Creation of microcatchments (here accompanied by the removal of weed species) may improve the survivorship of planted seedlings by improving soil moisture availability during the summer drought period. In this experiment, planting *Eucalyptus* seedlings in microsites in areas with a dense cover of annual grasses (such as *Avena* spp.) improved their survival: all seedlings not planted in microsites died. Removing competition from weed species, however, had a much greater influence on seedling survival than creating microsites (Standish, unpublished data).

bined with their sheer abundance and ability to persist in the seed bank (Paterson et al. 1976; Hobbs and Atkins 1988; Ellery and Chapman 2000; Standish et al. 2007), means that controlling the establishment of these species within an old field is essential until a canopy of native vegetation is able to limit the establishment of weed species (Hobbs and Atkins 1991).

The dispersal of seeds of native species into the old field may be limited, even when the old field is close to remnant native vegetation (Standish et al. 2007). Limited propagule dispersal into old fields can be overcome by direct seeding. Direct seeding of native species has successfully been used in bushland regeneration and in establishing diverse plantations as hosts for sandalwood (*Santalum spicatum*) farm forestry (Woodall and Robinson 2002). Yet the question remains whether these "plantations" of native species become self-sustaining plant populations in the absence of continued management. The recruitment of some *Eucalyptus, Acacia,* and *Melaleuca* species in the wheatbelt occurs only after large-scale disturbances such as fires, severe storms, droughts, floods, and above-average rainfall in the first summer following germination (Burrows et al. 1990; Yates et al. 1996; Yates and Hobbs 1997; Gaol and Fox 2002; Yates and Broadhurst 2003; Jasper 2004). Stochastic disturbances related to weather events and climate cannot be managed, while large-scale fires seldom occur in the wheatbelt due to low fuel loads in agricultural land and active fire suppression (Yates and Broadhurst 2003). Much of the success in the continued regeneration of native vegetation in old fields may be dependent on disturbance events outside of the control of restorationists.

Clearly, in a highly fragmented and abiotically harsh landscape such as the wheatbelt, the restoration of vegetation to a state approaching that which existed prior to clearing will require significant management of the restoration trajectory over the long term. Landscape context and land use history will be important factors in guiding the development of realistic restoration goals that are achievable with the resources available. In relatively intact parts of the landscape, where land that is abandoned has been cleared only or under cultivation for a short time, restoration of an ecosystem similar to that which existed prior to clearing may be possible with limited management. In heavily cleared areas with a long-term, high-input history of farming, the restoration of the vegetation structure present prior to clearing, but containing only a subset of species, may be a more suitable goal. Moderate cover of weed species may be tolerated in these old fields. Although floristically different to intact vegetation, these old fields would help maintain ecosystem processes at the local and landscape scale and may play a role in meta-population dynamics (Hobbs et al. 1993).

Conclusion: Ecological Restoration in a Linked Social–Ecological System

Barriers to ecological restoration of abandoned farmland in the wheatbelt exist not only in our technical ability to overcome seed limitation or improve soil conditions, but also in our financial and social ability to fund and manage restoration projects. A predicted consequence of declining rural populations is greater difficulty in maintaining or improving environmental outcomes that are not closely linked to production benefits (Kingwell and Pannell 2005). There is little political will to sustain rural communities in Western Australia in the manner common in Europe (for example, through subsidies), and the capacity of farmers to fund restoration work depends on global terms of trade, climate, and a bottom-up, people-driven landcare system (Abensperg-Traun et al. 2004). The traditional reliance on voluntary conservation and restoration efforts in agricultural landscapes is less tenable as the resident population falls (Kingwell and Pannell 2005). Yet local communities, in collaboration with the nongovernment sector, have instigated innovative, landscape-scale restoration projects, such as Gondwana Link (www.gondwanalink.org), which promotes species-diverse sandalwood (*Santalum spicatum*) plantations and farm forestry in the southern, higher-rainfall areas of the wheatbelt as part of a strategy to reconnect large tracts of vegetation and prevent further land degradation. Such integrated restoration options, however, are not currently possible in lower rainfall areas, as profitable perennial options are not available (Pannell 2001). In the absence of new perennial farming systems for these areas, ecological restoration of abandoned farmland may be reliant on continued community engagement and funding through programs such as the Living Landscapes process developed by Greening Australia (Western Australia). While the priority for such programs is to restore intact remnant vegetation, the restoration of some of the structural and functional properties of native vegetation in old fields may provide important elements of connectivity in this highly fragmented landscape.

Acknowledgments

Thanks to Helen Allison and Alan Rossow for supplying and adapting, respectively, figure 15.1.

REFERENCES

Abensperg-Traun, M. A., and E. S. de Boer. 1990. Species similarity and habitat differences in biomass of subterranean termites (Isoptera) in the wheatbelt of Western Australia. *Australian Journal of Ecology* 15:219–26.

Abensperg-Traun, M., T. Wrbka, G. Bieringer, R. Hobbs, F. Deininger, B. York Main, N. Milasowszky, N. Sauberer, and K. P. Zulka. 2004. Ecological restoration in the slipstream of agricultural policy in the old and new world. *Agriculture, Ecosystems and Environment* 103:601–11.

Allison, H. E., and R. J. Hobbs. 2004. Resilience, adaptive capacity, and the "lock-in trap" of the Western Australian agricultural region. *Ecology and Society* 9(1):3 [online]. http://www.ecologyandsociety.org/vol9/iss1/art3.

Arnold, G. W., M. Abensperg-Traun, R. J. Hobbs, D. E. Steven, L. Atkins, J. J. Viveen, and D. M. Gutter. 1999. Recovery of shrubland communities on abandoned farmland in southwestern Australia: Soils, plants, birds and arthropods. *Pacific Conservation Biology* 5:163–78.

Beresford, Q., H. Bekle, H. Phillips, and J. Mulcock. 2001. *The salinity crisis: Landscapes, communities, politics.* University of Western Australia Press, Perth.

Bolland, M. D. A., D. G. Allen, and N. J. Barrow. 2003. *Sorption of phosphorus by soils: How it is measured in Western Australia.* Agriculture Western Australia, Perth.

Burrows, N., G. Gardiner, B. Ward, and A. Robinson. 1990. Regeneration of *Eucalyptus wandoo* following fire. *Australian Forestry* 53:248–58.

Cramer, V. A., and R. J. Hobbs. 2002. Ecological consequences of altered hydrological regimes in fragmented ecosystems in southern Australia: Impacts and possible management responses. *Austral Ecology* 27:546–64.

Cramer, V. A., R. J. Hobbs, and L. Atkins. 2004. The influence of local elevation on the effects of secondary salinity in remnant eucalypt woodlands: Changes in understorey communities. *Plant and Soil* 265:253–66.

Cramer, V. A., R. J. Hobbs, G. Hodgson, and L. Atkins. 2004. The influence of local elevation on soil properties and tree health in remnant eucalypt woodlands affected by secondary salinity. *Plant and Soil* 265:175–88.

D'Emden, F. H., and R. Llewellyn. 2004. No-till adoption and cropping issues for Australian grain growers. In *New directions for a diverse planet: Proceedings of the 4th International Crop Science Congress, Brisbane, Australia, 26 September–1 October 2004.* Regional Institute, Brisbane.

Dalal, R. C., and K. Y. Chan. 2001. Soil organic matter in rainfed cropping systems of the Australian cereal belt. *Australian Journal of Soil Research* 39:435–64.

Department of Agriculture. 2005. *Western Australia's agri-food, fibre and fisheries industries 2003–2004.* Bulletin 4627. Department of Agriculture, Perth.

Didham, R. K., C. H. Watts, and D. A. Norton. 2005. Are systems with strong underlying abiotic regimes more likely to exhibit alternative stable states? *Oikos* 110:409–16.

Ehrenfield, J. G. 2003. Effects of exotic plant invasions on soil nutrient cycling processes. *Ecosystems* 6:503–23.

Ellery, A. J., and R. Chapman. 2000. Embryo and seed coat factors produce seed dormancy in capeweed (*Arctotheca calendula*). *Australian Journal of Agricultural Research* 51:849–54.

Gaol, M. L., and J. E. D. Fox. 2002. Reproductive potential of *Acacia* species in the central wheatbelt: Variation between years. *Conservation Science Western Australia* 4:147–57.

Garkaklis, M. J., J. S. Bradley, and R. D. Wooller. 1998. The effects of Woylie (*Bettongia penicillata*) foraging on soil water repellency and water infiltration in heavy textured soils in southwestern Australia. *Australian Journal of Ecology* 23:492–96.

George, R. J. 1992. Estimating and modifying the effects of agricultural development on the groundwater balance of large wheatbelt catchments. *Journal of Applied Hydrogeology* 1:41–54.

Hamza, M. A., and W. K. Anderson. 2003. Responses of soil properties and grain yields to deep ripping and gypsum application in a compacted loamy sand soil contrasted with a sandy clay loam soil in Western Australia. *Australian Journal of Agricultural Research* 54:273–82.

Handreck, K. A. 1997. Phosphorus requirements of Australian native plants. *Australian Journal of Soil Research* 35:241–89.

He, T., S. L. Krauss, B. B. Lamont, B. P. Miller, and N. J. Enright. 2004. Long-distance seed dispersal in a metapopulation of *Banksia hookeriana* inferred from a population allocation analysis of amplified fragment length polymorphism data. *Molecular Ecology* 13:1099–1109.

Hester, A. J., and R. J. Hobbs. 1992. Influence of fire and soil nutrients on native and non-native annuals at remnant edges in the Western Australian wheatbelt. *Journal of Vegetation Science* 3:101–8.

Hobbs, R. J. 1998. Impacts of land use on biodiversity in southwestern Australia. In *Landscape disturbance and biodiversity in Mediterranean-type ecosystems*, ed. P. W. Rundel, G. Montenegro, and F. M. Jaksic, 81–106. Springer-Verlag, Berlin.

Hobbs, R. J., and L. Atkins. 1988. Effect of disturbance and nutrient addition on native and introduced annuals in plant communities in the Western Australian wheatbelt. *Australian Journal of Ecology* 13:171–79.

———. 1991. Interactions between annuals and woody perennials in a Western Australian nature reserve. *Journal of Vegetation Science* 2:643–54.

Hobbs, R. J., R. H. Groves, S. D. Hopper, R. J. Lambeck, B. B. Lamont, S. Lavorel, A. R. Main, J. D. Majer, and D. A. Saunders. 1995. Function of biodiversity in the Mediterranean-type ecosystems of southwestern Australia. In *Mediterranean-type ecosystems. The function of biodiversity*, ed. G. W. Davis and D. M. Richardson, 233–84. Springer-Verlag, Berlin.

Hobbs, R. J., D. A. Saunders, and G. W. Arnold. 1993. Integrated landscape ecology: A Western Australian perspective. *Biological Conservation* 64:231–38.

Hobbs, R. J., D. A. Saunders, L. A. Lobry de Bruyn, and A. R. Main. 1993. Changes in biota. In *Reintegrating fragmented landscapes: Towards sustainable production and nature conservation*, ed. R. J. Hobbs and D. A. Saunders, 65–106. Springer-Verlag, New York.

Hobbs, R. J., and C. J. Yates. 2003. Impacts of ecosystem fragmentation on plant populations: Generalising the idiosyncratic. *Australian Journal of Botany* 51:471–88.

Hopper, S. D., and P. Gioia. 2004. The southwest Australian floristic region: Conservation of a global hotspot of biodiversity. *Annual Review of Ecology and Systematics* 35:623–50.

Indian Ocean Climate Initiative. 2002. *Climate variability and change in south west Western Australia*. Indian Ocean Climate Initiative, Perth.

Jasper, R. 2004. Flood as an agent of renewal: Lessons for revegetation. *Ecological Management and Restoration* 5:210–11.

Kingwell, R., and D. Pannell. 2005. Economic trends and drivers affecting the wheatbelt of Western Australia to 2030. *Australian Journal of Agricultural Research* 56:553–61.

Linacre, E., and J. Hobbs. 1977. *The Australian climatic environment*. John Wiley and Sons, Brisbane, Australia.

Lobry de Bruyn, L. A., and A. J. Conacher. 1990. The role of termites and ants in soil modification: A review. *Australian Journal of Soil Research* 28:55–93.

Lymbery, A. J., R. G. Doupé, and N. E. Pettit. 2003. Effects of salinisation on riparian plant communities in experimental catchments on the Collie River, Western Australia. *Australian Journal of Botany* 51:667–72.

Lyons, T. J. 2002. Clouds prefer native vegetation. *Meteorology and Atmospheric Physics* 80:131–40.

McFarlane, D. J., R. J. George, and P. Farrington. 1993. Changes in the hydrologic cycle. In *Reintegrating fragmented landscapes: Towards sustainable production and nature conservation*, ed. R. J. Hobbs and D. A. Saunders, 146–89. Springer-Verlag, New York.

Munns, R. 2005. Genes and salt tolerance: Bringing them together. *New Phytologist* 167:645–63.

National Land and Water Resources Audit. 2001a. *Australian dryland salinity assessment 2000. Extent, impacts, processes, monitoring and management options.* Land and Water Australia, Canberra, Australia.

National Land and Water Resources Audit. 2001b. *Land use change, productivity and diversification.* Bureau of Rural Sciences, Canberra, Australia.

National Land and Water Resources Audit. 2002. *Australians and natural resource management* 2002. Land and Water Australia, Canberra, Australia.

Nulsen, R. A. 1993. Changes in soil properties. In *Reintegrating fragmented landscapes: Towards sustainable production and nature conservation*, ed. R. J. Hobbs and D. A. Saunders, 107–45. Springer-Verlag, New York.

Pannell, D. J. 2001. Dryland salinity: Economic, scientific, social and policy dimensions. *Australian Journal of Agricultural and Resource Economics* 45:517–46.

———. 2004. *Politics and dryland salinity: History, tensions and prospects* (working paper). School of Agriculture and Resource Economics, University of Western Australia, Perth.

Passioura, J. B. 2002. Environmental biology and crop improvement. *Functional Plant Biology* 29:537–46.

Paterson, J. G., N. A. Goodchild, and W. J. R. Boyd. 1976. Effect of storage temperature, storage duration and germination temperature on the dormancy of seed of *Avena fatua* L. and *Avena barbata* Pott ex Link. *Australian Journal of Agricultural Research* 27:373–79.

Peck, A. J., and D. R. Williamson, eds. 1987. Hydrology and salinity in the Collie River Basin, Western Australia. *Journal of Hydrology*, special edition 94:1–198.

Pettit, N. E., P. G. Ladd, and R. H. Froend. 1998. Passive clearing of native vegetation: Livestock damage to remnant jarrah (*Eucalyptus marginata*) woodlands in Western Australia. *Journal of the Royal Society of Western Australia* 81:95–106.

Schütz, W., P. Milberg, and B. B. Lamont. 2002. Seed dormancy, after-ripening and light requirements of four annual Asteraceae in south-western Australia. *Annals of Botany* 90:707–14.

Shane, M. W., M. E. McCully, and H. Lambers. 2004. Tissue and cellular phosphorus storage during development of phosphorus toxicity in *Hakea prostrata* (Proteaceae). *Journal of Experimental Botany* 55:1033–44.

Shepherd, D. P., G. R. Beeston, and A. J. M. Hopkins. 2002. *Native vegetation in Western Australia: Extent, type and status.* Department of Agriculture, Western Australia, Perth.

Standish, R. J., V. A. Cramer, R. J. Hobbs, and H. T. Kobryn. 2006. Legacy of land-use evident in soils of Western Australia's wheatbelt. *Plant and Soil* 280:189–207.

Standish, R. J., V. A. Cramer, S. L. Wild, and R. J. Hobbs. 2007. Seed dispersal and recruitment limitation are barriers to native recolonisation of old-fields in Western Australia. *Journal of Applied Ecology* 44:435–45.

Woodall, G. S., and C. J. Robinson. 2002. Direct seeding Acacias of different form and function as hosts for Sandalwood (*Santalum spicatum*). *Conservation Science Western Australia* 4:130–34.

Yates, C. J., and L. M. Broadhurst. 2003. Assessing limitations on population growth in two critically endangered *Acacia* taxa. *Biological Conservation* 108:13–26.

Yates, C. J., and R. J. Hobbs. 1997. Woodland restoration in the Western Australian wheatbelt: A conceptual framework using a state and transition model. *Restoration Ecology* 5:28–35.

Yates, C. J., R. J. Hobbs, and L. Atkins. 2000. Establishment of perennial shrub and tree species in degraded *Eucalyptus salmonophloia* (salmon gum) remnant woodlands: Effects of restoration treatments. *Restoration Ecology* 8:135–43.

Yates, C. J., R. J. Hobbs, and R. W. Bell. 1996. Factors limiting the recruitment of *Eucalyptus salmonophloia* in remnant woodlands. 3. Conditions necessary for seed germination. *Australian Journal of Botany* 44:283–96.

Yates, C. J., D. A. Norton, and R. J. Hobbs. 2000. Grazing effects on plant cover, soil and microclimate in fragmented woodlands in south-western Australia: Implications for restoration. *Austral Ecology* 25:36–47.

York Main, B. 1993. Social history and impact on landscape. In *Reintegrating fragmented landscapes: Towards sustainable production and nature conservation*, ed. R. J. Hobbs and D. A. Saunders, 23–58. Springer-Verlag, New York.

PART 3

Synthesis: Old Field Dynamics and Restoration

In the following chapter we aim to draw together a number of aspects on the dynamics of abandoned farmland that have emerged throughout the book. We first of all revisit the reasons why we think the abandonment of land and what happens to it afterward are becoming increasingly important in today's world. We then examine the suite of case studies presented in part 2 and draw out commonalities and differences among them with a view to seeking generalities, if possible, and highlighting why the differences might be important. We then consider what lessons might be learned from the case studies presented, together with old field studies in general, in terms both of concepts and ideas in ecology and in the more practical application of restoration in different parts of the world.

Chapter 16

Old Field Dynamics: Regional and Local Differences, and Lessons for Ecology and Restoration

RICHARD J. HOBBS AND VIKI A. CRAMER

The history of land abandonment is long; fields have been abandoned since ancient times, and different parts of the world have different chronologies. In chapter 1, and in several of the case studies, the increasing importance of land abandonment in many parts of the world was highlighted. We know considerably more about some parts of the world than others—for instance, the history and results of land abandonment are relatively well documented in the eastern United States (chapters 8 and 9), but less well known elsewhere. In this book we have gathered case studies from a range of countries and regions but have by no means produced a comprehensive picture. There are, for instance, recent studies from Mexico (Castellanos et al. 2005), China (Zhang 2005), Egypt (El-Sheikh 2005), the Austrian and Italian Alps (Dirnböck et al. 2003; Laiolo et al. 2004), and eastern Canada (Benjamin et al. 2005; Blatt et al. 2005), all of which focus on aspects of the ecology of abandoned land in a variety of ecosystem types ranging from desert to alpine.

Why Should We Care About Old Fields?

The range of ecosystems and regions covered in this book, in the recent studies listed, and in numerous other studies available on the Web and elsewhere indicates that old fields are important ecosystems across the globe. Data presented in chapter 1 also indicate that the extent of abandoned land is increasing rapidly. Thus, while we frequently see a focus on deforestation and land transformation, there is a growing trend for transformed land to be abandoned. What happens to that land is important from many perspectives. The ecosystem development following abandonment may provide opportunities for the restoration of aspects of the native ecosystem and contribute to the achievement of conservation outcomes in the region. Alternatively, it may pose a threat to conservation values if the ecosystem that develops replaces a system that is valuable or threatened regionally. Development of woody vegetation on abandoned fields in the tropics and elsewhere also has the potential to contribute in a positive way to carbon accounting, providing fast-growing carbon sinks. This links to the socioeconomic consequences of land abandonment, which may be positive in some areas, but negative in others, if the abandonment is accompanied by, or leads to, rural depopulation, loss of traditional industries, and reduced income from tourism.

For these reasons, it seems that improving our understanding of why land is abandoned and how the ecosystem responds to that abandonment is likely to be increasingly important. It is thus essential that we examine this across the range of histories of abandonment, ecosystem types, and development dynamics apparent in the case studies in this book and elsewhere.

What Do the Case Studies Tell Us?

The extent to which the environment favors rapid recolonization and regrowth following abandonment strongly influences the outcome (table 16.1). As discussed by Hobbs et al. (2006), environmental harshness will vary across life zones, depending on temperature, fertility, and moisture availability. Ewel (1999) suggested that abiotic stress was likely to display a nonlinear relationship with environmental harshness (figure 16.1a); similarly, as environmental harshness declines, the opportunity increases for either more species to grow and thrive or for particular species to become dominant, leading to increased competition and predation, which Ewel aggregated into "biotic stress." If abiotic and biotic stresses are combined, total stress is greatest at either end of the gradient: in harsh environments the constraints to establishment and/or growth are primarily abiotic, while in more benign environments the constraints are mostly biotic, arising from the preexisting mix of species present.

Hobbs et al. (2006) went on to use the inverse image of this graph (figure 16.1b) to describe the ease with which an ecosystem will redevelop following disturbance or human modification. Since aggregate stress is lowest in the midrange of the x-axis, ecosystem redevelopment can be hypothesized to occur with most ease in this region. Redevelopment to a preexisting composition can be expected to be limited by abiotic conditions at one end of the graph and biotic conditions at the other. This appears to describe what happens after abandonment of fields in different parts of the world. The tropics could be considered as having favorable environmental conditions and high

TABLE 16.1

Analysis of how environmental, biological, and human factors affect ecosystem recovery in old fields

				Michigan and New Jersey	Northern Europe	Puerto Rico	Brazil	Queensland	France	Greece	Spain	South Africa	Western Australia
Environment	Climate 1	Wet ✓	Dry ✗	✓	✓	✓	✓	✓	✗	✗	✗	✗	✗
	Climate 2	Mild/cool ✓	Warm/hot ✗	✓	✓	✗	✗	✗	✗	✗	✗	✗	✗
	Soils nutrients	High ✓	Low ✗	✓	✓	✓	✗	✓/✗	✓	✓	✓	✓/✗	✗
Modification	Land use intensity	Traditional ✓	Industrial ✗	✗	✓/✗	✓/✗	✓/✗	✗	✓	✓	✓/✗	✗	✗
	Fragmentation	Low ✓	High ✗	✗	✗	✓/✗	✓/✗	✗	✓/✗	✓/✗	✓/✗	✗	✗
	Adaptation to agricultural disturbance	High ✓	Low ✗	✓	✓	✗	✗	✗	✓	✓	✓	✗	✗
Biota	Dispersal ability	Good ✓	Poor ✗	✓/✗	✓	✓/✗	✗	✓	✓	✓	✓	✗	✗
	Aggressive invasive species	Low abundance ✓	High abundance ✗	✓/✗	✓	✓/✗	✓/✗	✗	✓	✓	✓	✗	✗

TABLE 16.1

Continued

		Michigan and New Jersey	Northern Europe	Puerto Rico	Brazil	Queensland	France	Greece	Spain	South Africa	Western Australia
Rate of ecosystem development	Fast / Slow	✓ / ✗	✓	✓/✗	✓/✗	✓/✗	✓	✓	✓	✗	✗
Likelihood of arrested succession	Low / High	✓ / ✗	✓	✓/✗	✓/✗	✓/✗	✓	✓	✓	✗	✗
Dominant mode of restoration required	Passive / Active	✓ / ✗	✓	✓/✗	✓	✓/✗	✓	✓	✓	✗	✗
Old fields requiring long-term management		No	No	Site specific	Site specific	Yes	No	No	No	Yes	Yes

Note: Upper part of the table outlines general characteristics of old fields within each geographical area; lower part summarizes how these characteristics will affect ecosystem development and inform the type of restoration strategy required. Ticks indicate the more favorable alternative for each characteristic, that is, the one most likely to favor autogenic development on abandoned land; crosses represent the condition more likely to impede development. Boxes that contain both a tick and a cross indicate that a particular characteristic of old fields in that area are variable. Total number of ticks for each region indicates the overall favorableness of the abiotic and biotic environment for rapid ecosystem redevelopment following abandonment. The analysis recognizes general trends as outlined in the case studies. However, as with all natural systems, there are likely to be exceptions to the generalizations.

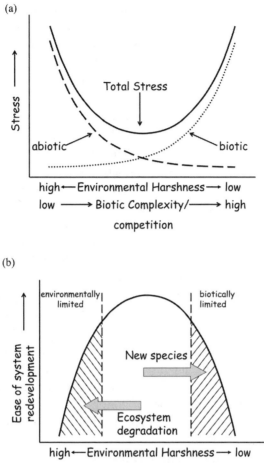

FIGURE 16.1. (a) Stress on an ecosystem is related to environmental harshness and biotic complexity: in harsh environments the constraints to establishment and/or growth are primarily abiotic, while in more benign environments the constraints are mostly biotic, arising from the preexisting mix of species. Total stress is greatest at either end of the gradient. The inverse image of this graph (b) portrays the ease with which an ecosystem will redevelop following disturbance or human modification. Modified from Hobbs et al. 2006.

biotic stress (via competition, etc.) and the Mediterranean areas as having more stressful environmental conditions and less biotic stress, with the temperate systems somewhere in the middle.

Overlaid on this continuum of abiotic and biotic stress is the level of adaptation of the native community to the type and degree of disturbance caused by clearing and cropping practices (table 16.1). In many of the ecosystems discussed in the case studies, clearing and cultivation are novel mechanical disturbances to which native communities are not adapted, despite their adaptation to often severe and widespread disturbances such as fire (Stylinski and Allen 1999). Where a suite of introduced species adapted to both the environmental conditions and to agricultural disturbance are present, ecosystem development after abandonment is unlikely to follow a trajectory toward a predisturbance state, particularly when agriculture has degraded abiotic conditions. This is particularly so in the Mediterranean-climate ecosystems outside of the Mediterranean basin. Mediterranean-basin annual herbs and grasses are highly adapted to frequent urban and agricultural disturbance and are able to germinate before native species and grow more quickly, thus becoming superior competitors for soil resources (Stylinski and Allen 1999; Standish et al. 2007). Invasive pasture grasses may also arrest succession in tropical ecosystems, but over ecologically short time periods of ten to fifteen years (chapter 6). The dichotomy between the life strategies of introduced and native species is most extreme in the case study of Western Australia (chapter 15). Low nutrient environments favor perennial plants with long-lived leaves and root systems, while strategies consisting of short leaf life spans, rapid growth, and prolific seed output are not well developed (M. Westoby, pers. comm., 2006). Thus, the majority of the flora in these ecosystems simply do not have the life-history strategies to compete with introduced annual herbs and shrubs adapted to agricultural disturbance. Conversely, in the "young" postglacial landscapes of northeastern America, native species have life strategies that generally make them competitive with introduced species after agricultural disturbance. Thus the presence of exotic species in these ecosystems may not greatly alter the trajectory of community development after abandonment, although some species are of concern (chapter 8).

Within this context, are there generalizable principles that are evident despite the observed differences? Within the limitations imposed by climate and soil characteristics, it does appear that the combination of past land use type and land use intensity explains much of the difference in patterns within similar ecosystem types. It also appears that seed dispersal of native species is limiting everywhere, but that the importance of this depends on a variety of factors, including size of the field and/or proximity to seed sources, whether bird dispersal is an important process in the region, and the prevalence of invasive species with prolific seed output.

Lessons for Ecology

The foundation of any branch of science is the attempt to find generalizable concepts and theories that apply across a wide range of situations. In ecology, concepts such as succession and community assembly have all mostly developed in the temperate areas of the Northern Hemisphere. The collection of case studies in this book allows a consideration of how well concepts and generalities translate from their "home" systems to others. As we saw above, patterns observed in temperate regions would not necessarily be repeated in other regions. A clear message from this is that the ideas developed in the temperate systems relating to patterns and processes in old field dynamics are less applicable in the tropical and Mediterranean-type systems studied here (table 16.1). Hence generalizations need to be bounded. While existing successional models explain community development in some cases studies, and other case studies were able to make use of and adapt existing succession models (e.g., Ganade, in chapter 5, used the models of Connell and Slatyer [1977] to explain the dynamics of old fields in Brazil), it is clear that existing successional models do not adequately explain community development throughout all of the case studies. In these cases, particularly shown in the South African and Western Australian examples, where the likelihood of slow ecosystem development and "arrested succession" is high, theories of ecosystem assembly and alternative stable states may provide more useful conceptual models for understanding ecosystem development.

Nevertheless, the finding that overall differences in dynamics could be explained on the basis of past land use, climate, soil characteristics, community adaptation to particular disturbance regimes, and the type and degree of invasive species present provides a framework on which to build generalized models of ecosystem development on abandoned farmland. Similarly, the generality of dispersal limitation across most systems provides interesting evidence from which to improve ideas on community assembly.

Lessons for Restoration

We can discuss how all of this can inform restoration in two ways. First, there is the question of whether active restoration activities can or should be undertaken in areas of abandoned agricultural land. Second, there are the broader implications of the findings from old fields for restoration ecology in general.

The first topic clearly relates to determining when and where active restoration is needed or is appropriate on abandoned agricultural lands. Where ecosystem redevelopment proceeds along a path that leads to a desirable outcome, then there is little need for active intervention (table 16.1). The notion that "autogenic" restoration is both desirable and cost-effective has been discussed by Whisenant (2002) and Whisenant et al. (1995), among others, and is analogous to the idea that successional processes can be effectively used as restoration tools (Hobbs et al. 2007). Active restoration is, however, required where the developmental pathway after abandonment is inappropriate from either a conservation or land use perspective or where the system remains "stuck" in an inappropriate state. Development after abandonment may be deemed inappropriate if, for instance, the system becomes dominated by invasive or undesirable species, if species that detract from the conservation value of the area invade (for instance, if woody plants invade areas that were previously valuable grassland habitat), or if the development adversely affects the value of the area for tourism or other uses.

In cases where the system becomes stuck in a particular state, this is a problem only when that state is deemed inappropriate and then measures need to be taken to nudge the development in an appropriate direction. This involves determining the key factors preventing ecosystem development from progressing and finding means of dealing with these factors. This may simply be dispersal limitation, as discussed earlier, or it may relate to other factors such as the presence of competing species or adverse soil conditions. Where more than one of these factors is operating at the same time, the level of intervention involved may be large, and whether this is feasible or not will depend on the level of resources and societal will available. Several of the case studies identified the need to move beyond considering multiple factors independently to considering the interactions between these factors, particularly where such interactions create feedback loops that reinforce degrading processes and maintain the ecosystem in an inappropriate state. The worst-case scenario, as exemplified by the Western Australian case study, may be that broad-scale restoration of abandoned farmland presents an impossibly difficult and expensive problem. On the other hand, changing economic settings and novel opportunities, such as those presented by carbon trading, could provide other options for dealing with such situations.

This highlights an important consideration that arises from a comparison between the case studies in "natural" versus "cultural" landscapes. Where agriculture is relatively recent and has led to the reduction and fragmentation of native vegetation, abandonment of agricultural land is viewed primarily as a potential means to return areas to something approaching the natural vegetation of the area. The success or otherwise of restoration is thus measured in these terms. Where landscapes are viewed as cultural, such as in the European case studies, this reference back to "natural ecosystems" is less prevalent, and therefore there is more consideration of a range of options after abandonment (see also Aronson and Vallejo 2006; Vallejo et al. 2006).

These considerations arising in the context of abandoned land readily feed into broader issues relating to restoration. Restoration projects need to determine the nature and extent of intervention necessary, based on an assessment of what has led to ecosystem degradation or is preventing system recovery. Similarly, a clear enunciation of goals is essential and needs to be related to the ecological realities and socioeconomic context within which the restoration project is being undertaken. Studies on abandoned land seem certain to continue to play an important role both in the development of ecological ideas and understanding of ecosystem dynamics and also in the development of the science and practice of ecological restoration.

REFERENCES

Aronson, J., and R. Vallejo. 2006. Challenges for the practice of ecological restoration. In *Restoration ecology: The new frontier*, ed. J. van Andel and J. Aronson, 234–47. Blackwell, Oxford, UK.

Benjamin, K., G. Domon, and A. Bouchard. 2005. Vegetation composition and succession on abandoned farmland: Effects of ecological, historical and spatial factors. *Landscape Ecology* 20:627–47.

Blatt, S. E., A. Crowder, and R. Harmsen. 2005. Secondary succession in two southeastern Ontario old-fields. *Plant Ecology* 177:25–41.

Castellanos, A. E., M. J. Martinez, J. M. Llano, W. L. Halvorson, M. Espiricueta, and I. Espejel. 2005. Successional trends in Sonoran Desert abandoned agricultural fields in northern Mexico. *Journal of Arid Environments* 60:437–55.

Connell, J. H., and R. O. Slatyer. 1977. Mechanisms of succession in natural communities and their role in community stability and organization. *American Naturalist* 111:1119–44.

Dirnböck, T., S. Dullinger, and G. Grabherr. 2003. A regional impact assessment of climate and land-use change on alpine vegetation. *Journal of Biogeography* 30:401–17.

El-Sheikh, M. A. 2005. Plant succession on abandoned fields after 25 years of shifting cultivation in Assuit, Egypt. *Journal of Arid Environments* 61:461–81.

Ewel, J. J. 1999. Natural systems as models for the design of sustainable systems of land use. *Agroforestry Systems* 45:1–21.

Hobbs, R. J., S. Arico, J. Aronson, J. S. Baron, P. Bridgewater, V. A. Cramer, P. R. Epstein et al. 2006. Novel ecosystems: Theoretical and management aspects of the new ecological world order. *Global Ecology and Biogeography* 15:1–7.

Hobbs, R. J., L. R. Walker, and J. Walker. 2007. Integrating restoration and succession. In *Linking restoration and succession in theory and in practice*, ed. L. R. Walker, J. Walker, and R. J. Hobbs, 168–179. Springer, New York.

Laiolo, P., F. Dondero, E. Ciliento, and A. Rolando. 2004. Consequences of pastoral abandonment for the structure and diversity of alpine avifauna. *Journal of Applied Ecology* 41:294–304.

Standish, R. J., V. A. Cramer, S. L. Wild, and R. J. Hobbs. 2007. Seed dispersal and recruitment limitation are barriers to native recolonisation of old-fields in Western Australia. *Journal of Applied Ecology* 44:435–45.

Stylinski, C. D., and E. B. Allen. 1999. Lack of native species recovery following severe exotic disturbance in southern California shrublands. *Journal of Applied Ecology* 36:544–54.

Vallejo, R., J. Aronson, J. G. Pausas, and J. Cortina. 2006. Restoration of Mediterranean woodlands. In *Restoration ecology: The new frontier*, ed. J. van Andel and J. Aronson, 193–207. Blackwell, Oxford, UK.

Whisenant, S. G. 2002. Terrestrial systems. In *Handbook of ecological Restoration*. Vol. 1. *Principles of restoration*, ed. M. R. Perrow and A. J. Davy, 83–105. Cambridge University Press, Cambridge, UK.

Whisenant, S. G., T. L. Thurow, and S. J. Maranz. 1995. Initiating autogenic restoration on shallow semiarid sites. *Restoration Ecology* 3:61–67.

Zhang, J.-T. 2005. Succession analysis of plant communities in abandoned croplands in the eastern Loess Plateau of China. *Journal of Arid Environments* 63:458–74.

ABOUT THE EDITORS

Viki A. Cramer teaches environmental restoration at Murdoch University in Perth, Western Australia. Her research interests are in ecosystem restoration, landscape ecology, plant ecology, and nature conservation and natural resource management in social–ecological systems. Much of her research has been conducted in the agricultural landscapes of southeast Queensland and southwest Western Australia. Her current research focuses on the conservation and restoration of plant communities in urban areas, from both an ecological and a social perspective.

Richard J. Hobbs is a professor of Environmental Science at Murdoch University, and holds an Australian Professorial Fellowship. His particular interests are in vegetation dynamics and management, fragmentation, invasive species, ecosystem restoration, conservation biology, and landscape ecology, and he has research experience in Australia, the United Kingdom, Europe, and North America. He serves or has served in executive positions in a number of learned societies and on numerous editorial boards. He was president of the Ecological Society of Australia, 1998–1999; president of the International Association for Landscape Ecology, 1999–2003; he is currently editor in chief of the journal *Restoration Ecology*. He is listed by ISI as among the most highly cited researchers in ecology and environmental science, and is a Fellow of the Australian Academy of Science.

T. Mitchell Aide is a professor in the Department of Biology at the University of Puerto Rico, Rio Piedras. His research interests include tropical community ecology, restoration ecology, and conservation biology. Dr. Aide has worked throughout Latin America and recently has begun research on land use practices in tropical China.

James Aronson is head of the Restoration Ecology group at the Center of Functional and Evolutionary Ecology, Centre National de la Recherche Scientifique (CNRS), in Montpellier, France. He is also Curator of Restoration Ecology at the Missouri Botanical Garden, USA. He has worked on projects and programs related to the restoration and rehabilitation of degraded ecosystems in many parts of the world. With an international coalition he is now concentrating on the transdisciplinary approach called *Restoring Natural Capital*, especially as it applies in developing countries.

Andreu Bonet is a professor in the Department of Ecology and coordinator of the Font Roja Natura Research Station (Font Roja Natural Park), University of Alicante, Spain. His research interests are in plant community ecology and land use and land cover changes. Much of his work has focused on vegetation dynamics in Spanish old fields and landscapes, but he is now also working in ecosystem conservation and restoration in Spain and Cuba.

Mary L. Cadenasso is an assistant professor and ecologist in the Department of Plant Sciences at the University of California, Davis. Her research investigates the reciprocal link between spatial heterogeneity in landscapes and plant community dynamics and ecosystem function. She works across disciplines and scales, and her current research is in forest, savanna, and urban systems.

Carla P. Catterall is an ecologist in the School of Environment at Griffith University, Brisbane, Australia. Her research interests are centered on the responses of wildlife to variation in the quality and quantity of their habitats, especially patterns of change in forest cover, and related conservation issues. Working with colleagues and students, she has been active in discovering and monitoring the effects on biodiversity of human-induced land-cover changes, including those due to urbanization, pastoralism, and vegetation restoration.

Sarah M. Emery is a postdoctoral Research Fellow in the Ecology and Evolutionary Biology Department at Rice University and a recipient of a National Parks Ecological Research Fellowship. Her postdoctoral work focuses on understanding the role of fungal mutualists in regulating invasion in Great Lakes sand dune plant communities. Her related interests include the role of dominant species in communities, relationships between diversity and compositional stability in grasslands, prairie and dune restoration, and the management and population dynamics of invasive species.

Peter D. Erskine is a Research Fellow in the Centre for Mined Land Rehabilitation at the University of Queensland. His research interests include the restoration and rehabilitation of plant communities on degraded lands and the functional benefits of biodiversity. Much of his research has been conducted in tropical rain-forest landscapes of Australia and Southeast Asia, but he has recently begun to work on mined land rehabilitation in a variety of climatic zones.

Gislene Ganade is a professor of ecology at Unisinos University in southern Brazil, and part of the Projeto Dinâmica Biológica de Fragmentos Florestais in the Instituto Nacional de Pesquisas da Amazônia and Smithsonian Tropical Research Institute in Manaus, Brazil. Her research interests are in plant succession, community ecology, and ecosystem restoration. Much of her work has been carried out in tropical and subtropical forests and old fields of Brazil, but she is also conducting research on the relationship between multitaxa diversity and ecosystem functioning in exotic and native tree plantations.

Katherine L. Gross is a University Distinguished Professor in the Department of Plant Biology and director of the W. K. Kellogg Biological Station, Michigan State University. Her research interests are in plant community ecology, particularly the determinants of variation in species diversity across community types. Her current work focuses on native grasslands and successional old fields, and she is continuing long-term research on weed communities and diversity in row-crop agriculture.

Karen Holl is a professor of environmental studies at the University of California, Santa Cruz. She studies the effect of local- and large-scale processes on the restoration of tropical forests in Costa Rica and grassland and riparian forests in California.

John Kanowski is a Research Fellow in the Centre for Innovative Conservation Strategies, Griffith University, Brisbane, Australia. His research interests include the ecology and conservation of rain-forest mammals and the restoration of cleared rain-forest landscapes in eastern Australia.

Cornelia B. Krug was a postdoctoral researcher at the Department of Conservation Ecology and Entomology at the University of Stellenbosch, in Stellenbosch, South Africa, and has recently joined the Department of Zoology at the University of Cape Town as a Research Associate. Her research interests combine basic ecology, landscape, and restoration ecology, with current research focusing on the effects of habitat fragmentation and transformation on biological diversity. Previous work investigated suitable restoration techniques for west coast renosterveld, how species adapt to an arid environment, and zoological communities on set-aside land with different land use history and management.

Rainer M. Krug conducted his PhD studies at the Department of Conservation Ecology and Entomology at the University of Stellenbosch in Stellenbosch, South Africa, focusing on the role of seed dispersal in restoration and biological invasion. Originally trained as a physicist specializing in ecological modelling, he obtained an MSc in Conservation Biology from the University of Cape Town. His research uses theoretical ecology, particularly ecological modelling, and its integration and interaction with field experiments and data, to investigate the role of population and ecosystem processes in the conservation of species and ecosystems.

David Lamb teaches ecology at the School of Integrative Biology at the University of Queensland, Australia. He is interested in tropical forest restoration, landscape ecology, and conservation biology. In recent years, much of his research has concerned ways of integrating restoration within the social and economic systems of people living in degraded landscapes.

Jacques Lepart is a researcher with the Centre National de la Recherche Scientifique (CNRS) at the Centre for Functional and Evolutionary Ecology in Montpellier, France. He is an expert in plant population ecology and landscape ecology. His main topics of research are landscape dynamics, plant succession, and plant demography. In recent years, most of his work has addressed the issue of tree and shrub encroachment and its consequences on landscape and biodiversity.

Jan Lepš is a professor in the Department of Botany, University of South Bohemia, České Budějovice, Czech Republic. His research interests are in community ecology, particularly the mechanisms controlling the diversity of communities, consequences of species diversity variation for ecosystem functioning, interactions of plants with higher trophic levels, and multivariate analysis of ecological data. Although most of his work has been done in central Europe, he is also engaged in tropical rain-forest research.

Ariel E. Lugo is director of the USDA Forest Service International Institute of Tropical Forestry in Rio Piedras, Puerto Rico, where he is an active scientist in tropical forest ecology with research focused on the effects of disturbances on tropical forests species composition and functioning. Dr. Lugo is a highly quoted scientist and has to his credit over 400 scientific publications. He is also a member of numerous editorial boards of science journals and routinely offers his time for national and international consultations in tropical forestry.

Pascal Marty is a researcher with the Centre National de la Recherche Scientifique (CNRS) at the Centre for Functional and Evolutionary Ecology in Montpellier, France. He is a geographer specializing in landscape and land use changes. His main area of research is in the understanding of interactions between socioeconomic systems and ecological systems at the landscape scale. He is particularly interested in the role of past and current agricultural practices in seminatural landscapes changes.

Scott J. Meiners is an associate professor in the Department of Biological Sciences, Eastern Illinois University–Charleston, and is currently leader of the Buell–Small Succession Study, a long-term study of postagricultural succession. His research interests revolve around factors that influence the dynamics and regeneration of plant communities, typically focusing on old fields or other communities challenged by anthropogenic disturbances.

Vasilios P. Papanastasis is a professor of forestry and natural environment at the Aristotle University of Thessaloniki in Thessaloniki, Greece. His research interests are in rangeland ecology, particularly the role of livestock grazing in Mediterranean ecosystems. He is now working on agroforestry, desertification, and restoration ecology of Mediterranean communities and landscapes.

Juli G. Pausas is a Research Scientist at CEAM (Centro de Estudios Ambientales del Mediterráneo, Mediterranean Centre for Environmental Studies, Valencia, Spain) and an Assistant Professor at the Department of Ecology, University of Alicante (Spain). His research focuses on regeneration ecology and vegetation dynamics in Mediterranean and fire-prone ecosystems, and specifically on the role of fire in shaping Mediterranean species, communities, and landscapes.

Steward T. A. Pickett is Distinguished Senior Scientist at the Institute of Ecosystem Studies in Millbrook, New York. His interests include natural disturbance and vegetation dynamics, the role of landscape and community heterogeneity in temperate forest and African savanna, and the structure and function of urban ecosystems. He directs the Baltimore Ecosystem Study, Long-Term Ecological Research program, and conducts work at Kruger National Park in South Africa.

Karel Prach is professor of botany in the Faculty of Biological Sciences, University of České Budějovice, and researcher at the Institute of Botany, Academy of Sciences of the Czech Republic, Třeboň. His main research interests concern vegetation succession, plant invasions, and ecology of river floodplains, with emphasis on restoration aspects.

Marcel Rejmánek is a professor of ecology at the University of California, Davis. His research interests are in plant invasions, dynamics of tropical forests, and classification of plant communities.

Rachel J. Standish is a postdoctoral Research Fellow in the School of Environmental Science at Murdoch University, Perth, Western Australia. She has a broad interest in ecology and its application to the management and restoration of native ecosystems. Much of her work has focused on the response of plant communities to change within the fragmented agricultural landscapes of New Zealand and Western Australia.

Lawrence R. Walker is a professor in the School of Life Sciences at the University of Nevada, Las Vegas. His research focuses on mechanisms and theories of plant succession and their application to restoration ecology. His work spans many types of habitats and disturbances around the globe and emphasizes cross-site comparisons.

Jess K. Zimmerman is a professor in the Institute for Tropical Ecosystem Studies, University of Puerto Rico, San Juan. His research interests are focused on tropical forest community dynamics, particularly with respect to patterns and dynamics as influenced by hurricane and human disturbance. He also studies restoration ecology, climate controls on tropical forest phenology, orchid reproductive ecology, and ant ecology.

Island Press Board of Directors

Victor M. Sher, Esq. *(Chair)*
Sher & Leff
San Francisco, CA

Dane A. Nichols *(Vice-Chair)*
Washington, DC

Carolyn Peachey *(Secretary)*
Campbell, Peachey & Associates
Washington, DC

Drummond Pike *(Treasurer)*
President
The Tides Foundation
San Francisco, CA

Robert Baensch
Director, Center for Publishing
New York University
New York, NY

William H. Meadows
President
The Wilderness Society
Washington, DC

Merloyd Ludington Lawrence
Merloyd Lawrence Inc.
Boston, MA

Henry Reath
Princeton, NJ

Will Rogers
President
The Trust for Public Land
San Francisco, CA

Alexis G. Sant
Trustee and Treasurer
Summit Foundation
Washington, DC

Charles C. Savitt
President
Island Press
Washington, DC

Susan E. Sechler
Senior Advisor
The German Marshall Fund
Washington, DC

Nancy Sidamon-Eristoff
Washington, DC

Peter R. Stein
General Partner
LTC Conservation Advisory Services
The Lyme Timber Company
Hanover, NH

Diana Wall, Ph.D.
Director and Professor
Natural Resource Ecology Laboratory
Colorado State University
Fort Collins, CO

Wren Wirth
Washington, DC